AF357338

Assessing the Impact of Climate Change on Urban Cultural Heritage

Assessing the Impact of Climate Change on Urban Cultural Heritage

Editor

Yasemin D. Aktas

MDPI • Basel • Beijing • Wuhan • Barcelona • Belgrade • Manchester • Tokyo • Cluj • Tianjin

Editor
Yasemin D. Aktas
Civil, Environmental and
Geomatic Engineering
University College London (UCL)
London
United Kingdom

Editorial Office
MDPI
St. Alban-Anlage 66
4052 Basel, Switzerland

This is a reprint of articles from the Special Issue published online in the open access journal *Atmosphere* (ISSN 2073-4433) (available at: www.mdpi.com/journal/atmosphere/special_issues/climate_heritage).

For citation purposes, cite each article independently as indicated on the article page online and as indicated below:

LastName, A.A.; LastName, B.B.; LastName, C.C. Article Title. *Journal Name* **Year**, *Volume Number*, Page Range.

ISBN 978-3-0365-1832-9 (Hbk)
ISBN 978-3-0365-1831-2 (PDF)

Contents

About the Editor

Yasemin D. Aktas

A structural/conservation engineer by training, Yasemin is currently a lecturer at the Engineering and Architectural Design course jointly developed between Civil, Environmental and Geomatic Engineering (CEGE), Bartlett School of Architecture and Bartlett Institute of Environmental Design and Engineering, teaching sustainable building design and adaptive reuse. She is also the deputy academic director of the UK Centre for Moisture in Buildings (UKCMB). Her research interests include climate induced hazards and their mitigation, urban climate modelling, the impact of built environment on the urban microclimate, interrelations between climatic and geophysical hazards, and physical vulnerability to climatic impact with specific emphasis on heritage materials and building envelopes.

Preface to "Assessing the Impact of Climate Change on Urban Cultural Heritage"

This book is a printed edition of the Special Issue titled "Assessing the Impact of Climate Change on Urban Cultural Heritage" hosted at the *Atmosphere* journal. This topic has been chosen in light of cities' ever-growing role and immense potential in the climate adaptation and mitigation discourse and the particular challenges regarding urban heritage making and conservation. It is critical to recognise the complex set of factors governing the physical, social and political future of urban heritage in cityscapes in constant transformation and in an era of planetary urbanisation.

The 10 papers (seven research papers, two reviews and one opinion piece) that comprise the issue give a broad cross-section of the issues pertinent to this important topic –accounts on practices and conceptual/methodological improvements in energy retrofit and reuse, risk mapping, urban planning, climate vulnerability assessment, and community engagement by 38 authors from seven countries are used to delineate the implications of current and likely future climates on heritage materials and systems, knowledge and practice gaps, as well as steps that need to be taken to ensure both their safeguarding and their valorisation to achieve climate resiliency.

I hope that this collection will be of interest to the relevant research community and is a useful contribution to the discourse going forward.

Yasemin D. Aktas
Editor

atmosphere

Editorial

Cities and Urban Heritage in the Face of a Changing Climate

Yasemin Didem Aktas [1,2]

1   UCL Department of Civil, Environmental and Geomatic Engineering (CEGE), London WC1E 6BT, UK;
    y.aktas@ucl.ac.uk
2   UK Centre for Moisture in Buildings (UKCMB), London WC1H 0NN, UK

Citation: Aktas, Y.D. Cities and Urban Heritage in the Face of a Changing Climate. *Atmosphere* **2021**, *12*, 1007. https://doi.org/10.3390/atmos12081007

Received: 13 July 2021
Accepted: 31 July 2021
Published: 5 August 2021

Publisher's Note: MDPI stays neutral with regard to jurisdictional claims in published maps and institutional affiliations.

Urbanisation is defined as the process where ever more people leave rural areas to live in cities. Currently more than 55% of the world population is estimated to live in cities [1] and despite the current downward trend in urbanisation rates, the UN projections are that the urban population by 2050 will be around 68% [2]. While the definition of where the rural ceases and urban starts is a matter of lively scholarly debate, by its rough meaning as "dense, built-up, 'man-made' areas" [3], cities are where risks and vulnerabilities concentrate: the large and ever-increasing urban populations require large and complex networks and infrastructures, whose partial or complete failure may quickly exacerbate risk under a given scenario of isolated and cascading disasters [4]. When coupled with other issues especially prominent in urban areas, including deep social and economic inequalities/exclusions, high energy demands, compactness and inaccessibility, both slow and fast onset hazards can have more intense and widespread consequences, making urban resilience an extremely complex and hard-to-achieve goal.

Climate-induced hazards are complex in their formation and progression. Despite our best efforts as researchers to better understand, model and forecast climatic hazards to estimate risks, cities worldwide are under the interacting and compound attack of excessive heating, air pollution, droughts, floods, storms, and so on. These threats undermine infrastructure, and endangers communities of very significant sizes. This impact is magnified in conjunction with their geographic, technical, socio-economic and political context, and through dynamic interdependencies between components of urban systems, especially when urbanisation is too rapid, poor or unplanned.

Along with the rest of the urban infrastructure, urban heritage is also under attack from various intensifying climatic stressors. The influence of changing temperature and humidity cycles, precipitation regimes and wind patterns coalesces with constant transformation of the cityscapes to redefine climate-induced hazards in urban areas. *Urban* here should be considered as a "lens", magnifying, reducing or otherwise distorting the impact of climatic variables on the built environment and beyond (see Table 1 for a brief overview for some potential ways this lens works).

Table 1. A brief summary of the main climatic variables affected by global climate change, typical urban contributors affecting the impact of a given climatic variable, the resulting hazard, and some direct impacts of these on heritage fabrics. The arrows indicate how each listed urban contributor will typically affect a given climatic variable (second column) and the trends leading to each hazard (third column).

Climatic Variables	Typical Urban Contributor	Hazard	Some Direct Impacts on Heritage Fabrics
Temperature	Increased anthropogenic heating (e.g., traffic and buildings-induced), reduced evapotranspiration ($\uparrow$) Higher thermal admittance materials, air pollution, urban morphology ($\uparrow/\downarrow$)	Urban heat island (UHI) ($\uparrow$) Urban cool island ($\downarrow$)	Impact on strength and stiffness of the fabric through varied daily/seasonal temperature fluctuations; material and integrity loss due to cracking, spalling and similar weathering; impact on the (de)sorptive characteristics of the building materials
Humidity	Increased anthropogenic moisture generation ($\uparrow$) Reduced evapotranspiration, land-use changes leading to increased surface runoff over impermeable surfaces ($\downarrow$)	Urban moisture island (UMI) ($\uparrow$) Urban dry island ($\downarrow$)	Impact on strength and stiffness of the fabric through varied daily/seasonal humidity fluctuations; corrosion, biodeterioration, and biological attack, leading to material decay and loss
Precipitation	Suitable aeresols, and high urban temperatures encouraging cloud formation ($\uparrow$) Reduced evapotranspiration ($\downarrow$)	Soil saturation, flooding ($\uparrow$)	Corrosion, biodeterioration, efflorescence, leading to material decay and loss; moisture enrichment within the fabric, hence moisture-induced decay of building materials; mould growth in-wall/indoors; rising damp
		Drought ($\downarrow$)	Differential settlement related structural problems, or foundation damage
		Wind-driven rain ($\uparrow$) Storms ($\uparrow$)	Moisture enrichment within the fabric, which may lead to moisture-induced decay of building materials and mould growth in-wall/indoors; mechanical forcing on the structural system
Wind	Urban morphology including surface roughness and geometry of street canyons in relation to prevalent wind direction ($\uparrow/\downarrow$) Thermal influences ($\uparrow/\downarrow$)	High winds; storm surges in coastal areas ($\uparrow$)	Surface erosion; additional mechanical forcing on the structural system; in case of storm surges, flooding
		Stagnation episodes ($\downarrow$)	Exacerbation of the impact of UHI, UMI and air pollution

Is It All Doom and Gloom?

No. In stark juxtaposition to all this, cities are also where an immense potential to physically, culturally, and politically mitigate the impact of climate change exists. The innovation potential of cities, if harnessed through new forms of institutional organisations and governance, can help greatly to sustainability and climate resiliency efforts [5]. While the role and responsibilities of local governments around the world on climate mitigation and adaptation can be diverse depending on the overall organisational hierarchies within countries and the relevant legislation, they nonetheless have immense powers to address many issues locally and directly, and often in liaison and collaboration with other stakeholders. This includes non-governmental and non-profit organisations, academia and the private sector. Cities are hubs for developing, experimenting with and implementing low energy solutions for climate-induced hazards and the relevant socio-economic drivers behind cities' vulnerability in the face of these. This opens up new avenues for co-creating integrated responses to challenges for climate resilience and creates a hopeful alternative to the dystopian views of cities' futures under climate change [6].

This Special Issue

The collection of the papers in this Special Issue gives a broad cross-section of diverse problems facing urban heritage under the impact of a changing climate, and the methods and tools that can be used to address these. Urban heritage studies with specific emphasis on climatic impact require a highly crossdisciplinary outlook, bringing together not only a multitude of scientific and technological disciplines but also public engagement and policy, and urban and heritage theory, among others, as evidenced here.

Sardella et al. [7], in their comprehensive, pan-European study, discuss the methodology employed for a web-based GIS tool for the visualisation and analysis of the vulnerabilities of cultural heritage under current and different future climate scenarios. To this end, the authors undertake a thorough review of global and regional climate models to identify regional climate projections at different resolutions to map the climatic hazards in central and Mediterranean Europe. The tool is available at https://www.protecht2save-wgt.eu/, accessed on 13 July 2021. Brimblecombe et al.'s study [8] is another contribution focussing on risk-mapping: highlighting the benefits of such mapping to inform the management plans by the custodians of heritage, they map the risk of extreme temperatures, flooding, earthquake, fire, and sediment disasters (debris flow, slope failure and landslides) for Tokyo. Drawing on inspirational artwork and historic photography, the authors discuss the trends in the frequency and intensity of individual hazards/climatic parameters.

Ulu and Durmus Arsan [9] aim to identify the energy performance baseline for 22 historic and contemporary heritage buildings in Basmane District in Izmir, Turkey, and explore retrofit options to improve the performance through a methodology integrating an on-site survey, building performance modelling and a retrofit impact assessment. The authors highlight the importance of case-by-case approach in developing the retrofit solutions for future use of heritage buildings, while ensuring their energy efficiency.

The paper by Aktas et al. [10] looks into the role that building stocks play in shaping urban microclimates. To this end, the authors report their findings from monitoring multiple land-use areas in Kuala Lumpur, including a heritage site at the heart of the city centre, Kampung Baru, composed of vernacular Malay homes. The paper then discusses the outdoor thermal comfort and energy use potential in diverse urban settings. The paper also touches upon the risks associated with the reconceptualization of heritage to achieve certain political and economic goals. Carroll and Aarrevaara [11] further expand on the relations between climatic impact on heritage buildings and urban planning through a questionnaire activity aimed at probing the planning professionals' perceptions and experiences of the phenomenon. In the specific case of Finland, the authors note a good level of understanding and appreciation of climatic risks by town planners; however, they conclude that they cannot always prioritise heritage structures when tackling such risks.

Fouseki et al. [12] tackles the important question of how the residents of heritage homes make decisions around energy efficiency, thermal comfort and conservation through 59 semi-structured interviews in Greece, Mexico and the UK. With a system dynamics approach, the authors identify one-directional and iterative relationships between features, perceptions of and responses to these, demonstrating that heritage conservation is a socially and culturally dynamic practice. Noticing some strong differences in different countries and in urban–rural environments, the authors highlight that the policies should account for the unique nature of each context, and that only by doing so can they support the energy efficiency and climate resiliency of heritage.

Orr and Cassar [13] focus on wind-driven rain (WDR) indices. The authors aim to develop a new index able to better express "shorter, more intense and more consistent WDR events" than the existing semi-empirical indices, and further expand on urban complexity and seasonality, among others. Importantly, they also adopt the frequency of occurrence of gutter overspill to extend it the use of their developed index to a risk/impact indicator.

The issue includes two review contributions: Basu et al. [14] provides an extensive review of decay mechanisms of stones, lime mortar and bricks that are expected to be exacerbated under the climatic trends for the UK. The paper also includes a comprehensive section on the methods which can be employed for decay monitoring of building stones, making it a complete, geologically focussed reference resource for climatic impact on urban stone-built heritage. Jahed et al. [15] offers a critical review of the UK and Turkey's energy retrofitting policy frameworks aimed at built heritage. To this end, the paper makes a chronological, multisectoral analysis of regulatory and financial schemes as well as the outcomes of voluntary programmes and competitions to identify the incentives and constraints which shape the overall energy retrofitting realm in these countries.

Last but not the least, the issue also has an opinion piece: Pender and Lemieux's exceptionally insightful paper [16] encourages a paradigm shift in our way of addressing indoor thermal comfort through "layered systems". The authors provoke the reader to re-examine design and assessment principles where heat loss by radiation is the main player, rather than air temperature, to be able to fight back against the commodification of comfort through HVAC systems, which are only counterproductive for the sustainability endeavour. Touching upon so many key topics, from professional education/training to vernacular design principles, this paper aims to innovate our understanding as to how heritage structures can contribute to achieving climate resiliency.

I hope the research community will find this Special Issue of interest.

Funding: This research received no external funding.

Conflicts of Interest: The author declares no conflict of interest.

References

1. The World Bank. Urban Population (% of total Population). Retrieved from The World Bank—Data. Available online: https://data.worldbank.org/indicator/SP.URB.TOTL.IN.ZS (accessed on 13 July 2021).
2. United Nations. World Urbanization Prospects: The 2018 Revision. United Nations, 2018. Available online: https://population.un.org/wup/ (accessed on 13 July 2021).
3. Harding, A.; Blockland, T. *Urban Theory: A Critical Introduction to Power, Cities and Urbanism in the 21st Century*; SAGE: London, UK, 2014.
4. Hunt, J.C.; Aktas, Y.D.; Mahalox, A.; Moustaoui, M.; Salamanca, F.; Georgescu, M. Climate change and growing megacities: Hazards and vulnerability. *Eng. Sustain.* **2018**, *171*, 314–326. [CrossRef]
5. Leichenko, R. Climate change and urban resilience. *Curr. Opin. Environ. Sustain.* **2011**, *3*, 164–168. [CrossRef]
6. Bulkeley, H. *Cities and Climate Change*; Routledge: London, UK, 2013.
7. Sardella, A.; Palazzi, E.; von Hardenberg, J.; del Grande, C.; de Nuntiis, P.; Sabbioni, C.; Bonazza, A. Risk Mapping for the Sustainable Protection of Cultural Heritage in Extreme Changing Environments. *Atmosphere* **2020**, *11*, 700. [CrossRef]
8. Brimblecombe, P.; Hayashi, M.; Futagami, Y. Mapping Climate Change, Natural Hazards and Tokyo's Built Heritage. *Atmosphere* **2020**, *11*, 680. [CrossRef]
9. Ulu, M.; Durmus Arsan, Z. Retrofit Strategies for Energy Efficiency of Historic Urban Fabric in Mediterranean Climate. *Atmosphere* **2020**, *11*, 742. [CrossRef]
10. Aktas, Y.D.; Wang, K.; Zhou, Y.; Othman, M.; Stocker, J.; Jackson, M.; Hood, C.; Carruthers, D.; Latif, M.T.; D'Ayala, D.; et al. Outdoor thermal comfort and building energy use potential in different land use areas in tropical cities: Case of Kuala Lumpur. *Atmosphere* **2020**, *11*, 652. [CrossRef]
11. Carroll, P.; Aarrevaara, E. The Awareness of and Input into Cultural Heritage Preservation by Urban Planners and Other Municipal Actors in Light of Climate Change. *Atmosphere* **2021**, *12*, 726. [CrossRef]
12. Fouseki, K.; Newton, D.; Murillo Camacho, K.S.; Nandi, S.; Koukou, T. Energy Efficiency, Thermal Comfort, and Heritage Conservation in Residential Historic Buildings as Dynamic and Systemic Socio-Cultural Practices. *Atmosphere* **2020**, *11*, 604. [CrossRef]
13. Orr, S.A.; Cassar, M. Exposure Indices of Extreme Wind-Driven Rain Events for Built Heritage. *Atmosphere* **2020**, *11*, 163. [CrossRef]
14. Basu, S.; Orr, S.A.; Aktas, Y.D. A Geological Perspective on Climate Change and Building Stone Deterioration in London: Implications for Urban Stone-Built Heritage Research and Management. *Atmosphere* **2020**, *11*, 788. [CrossRef]
15. Jahed, N.; Aktas, Y.D.; Rickaby, P.; Bilgin Altinoz, A.G. Policy Framework for Energy Retrofitting of Built Heritage: A Critical Comparison of UK and Turkey. *Atmosphere* **2020**, *11*, 674. [CrossRef]
16. Pender, R.; Lemieux, D.J. The Road Not Taken: Building Physics, and Returning to First Principles in Sustainable Design. *Atmosphere* **2020**, *11*, 620. [CrossRef]

Article

Energy Efficiency, Thermal Comfort, and Heritage Conservation in Residential Historic Buildings as Dynamic and Systemic Socio-Cultural Practices

Kalliopi Fouseki *, David Newton, Krisangella Sofia Murillo Camacho, Sohini Nandi and Theodora Koukou

UCL Institute for Sustainable Heritage, University College London, London WC1E 6BT, UK; d.newton.12@alumni.ucl.ac.uk (D.N.); ucftkmu@ucl.ac.uk (K.S.M.C.); sohininandi@gmail.com (S.N.); tkoukou1989@gmail.com (T.K.)
* Correspondence: kalliopi.fouseki@ucl.ac.uk

Received: 11 May 2020; Accepted: 3 June 2020; Published: 8 June 2020

Abstract: With buildings being responsible for nearly a quarter of global greenhouse gas emissions, intensive building decarbonization programs are in place worldwide, with unintended consequences for historic buildings. To this end, national and international guidance on energy efficiency for historic buildings advocate for the adoption of a 'whole house approach' that integrates heritage values in energy efficiency plans. Most guidance, though, relies on non-evidence based, pre-assumptions of residents' heritage values. And yet, unless we understand how and why residents negotiate their decisions between energy efficiency, thermal comfort, and heritage conservation, such guidance will not be applicable. Despite the urgency to decarbonize the building stock, research on how inhabitants of old buildings make such decisions is extremely limited. It is also case-study specific, often lacking the required depth. To address this gap, this paper offers the first international, in-depth study on the topic. It does so through a rigorous double-coded, thematic analysis of 59 in-depth semi-structured interviews (totaling 206,771 words) carried out in Greece, Mexico, and the UK. The thematic analysis is combined with system dynamic analysis, essential for unveiling what parameters affect inhabitants' decisions over time. Drawing on theories on the dynamics of social practices, we conclude that the process of decision-making on energy efficiency, thermal comfort improvement, and heritage conservation is a socio-cultural, dynamic practice, the change and continuation of which depends on how the following elements are connected or disconnected: materials (e.g., original features), competencies (e.g., restoration skills), resources (e.g., costs), values, space/environment (e.g., natural light), senses (e.g., thermal comfort), and time (e.g., years living in the house). The connection or disconnection of those elements will depend on (a) the nature of the context (e.g., rural, urban, conservation area); (b) the listing status; (c) age and construction materials of building; (d) local climate; and (e) ownership status.

Keywords: heritage values; energy efficiency; thermal comfort; heritage conservation; original features; system dynamics; social practices; decision-making; historic building

---

1. Introduction

It has been almost five years since the Paris Agreement on Climate Change was signed as a response to the urgent need to reverse global warming trends. The Agreement emphatically states that, if greenhouse emissions are not reduced dramatically, the global temperature will exceed 1.5 °C, with disastrous environmental, economic, and social consequences [1]. In Europe alone, buildings are responsible for approximately 40% of energy consumption and 36% of carbon dioxide emissions [2], with the majority of buildings located in cities where 73% of the European population resides [3] and with 23% of the building stock dating before 1945 [4]. Although a small number of this building stock is listed, the majority of buildings are non-listed. And yet, non-listed, traditional buildings are also imbued with heritage values that may impede the implementation of certain energy efficiency interventions [5,6], or, as we will argue, inspire alternative ways of energy efficiency. Heritage, listed and non-listed, buildings, are thus a special building stock which requires a distinct approach to energy efficiency that considers both its heritage values and the needs of its users. It is within these lines that guidance on how to improve the energy efficiency in historic buildings is currently being developed at national and international level. In the UK, for instance, the 2018 Guidance on 'How to Improve the Energy Efficiency in Historic Buildings' published by Historic England (HE) asserts that one of the unintended consequences of 'getting energy efficiency measures wrong' includes, among others, 'harm to heritage values and significance' [7] (summary). To avoid unintended consequences, the Guidance advocates for a 'whole building approach' that uses the understanding of a building 'its context, its significance, and all the factors affecting energy use as a starting point for devising an energy efficiency strategy' [7] (summary). The assumption is that older buildings are significant because they add 'distinctiveness, meaning and quality to the places people inhabit, and provide a sense of continuity and identity' [7] (p. 7). On the European level, the 'European Standards on Energy Efficiency in Historic Buildings' describe the procedure of selecting appropriate measures to improve the energy performance for a given historic building [8] (p. 13). The standards apply to listed and non-listed buildings of all types and age perceived as heritage. The term 'values' in this document encompasses 'aesthetic, historic, scientific, cultural, social or spiritual' values [8] (p. 5). Although the document is intended for both listed and non-listed buildings, in reality, most recommendations are applicable only in designated heritage. For instance, the standards recommend collaboration between owners and heritage authorities clarifying that refurbishment or repairs are viewed as 'non-heritage interventions if they do not respect heritage significance that is based on evidence' [8] (p. 8). However, as aforementioned, the significant majority of old buildings are residential and not listed. As a result, it is exceptionally rare for inhabitants to consult heritage agencies before they repair or refurbish their residences. More importantly, in both documents, 'heritage values' are based on pre-assumptions reflecting the perspectives of heritage professionals rather on evidence on users' attitudes. This is due to the lack of in-depth studies on the meanings owners and occupants associate with historic buildings (listed or non-listed).

Historic England (HE) and European standards on Energy Efficiency in Historic Buildings (EN) guidance constitute two of the most recent attempts at the national and international level to address energy efficiency interventions in the context of heritage buildings. Because it is an issue that only recently is being remarked at a policy level, it is normal for these documents to have gaps that need to be filled in so that the provided guidance and standards are meaningful and impactful in the real-life world. The three main knowledge gaps that this paper attempts to start filling are: (a) lack of evidence on what exact values (social and cultural meanings) occupants of listed, as well as non-listed buildings attached to their residences; (b) how those values change over time; and (c) to what extent do changing values affect residents' decisions or energy efficiency and thermal comfort interventions. The paper will approach this question through a cross-cultural in-depth study carried out in five different geographical regions, the first of its kind. It draws on rich, qualitative data derived from 59 semi-structured interviews totaling 206,771 words, which were conducted in: (a) 'neo-classical' stone listed buildings in Athens (Greece); (b) Victorian and Edwardian brick, non-listed buildings in Walthamstow (London); (c) 1940s Swedish-type timber structures scattered in rural England; (d) listed stone buildings in the world heritage site of Mexico City; and; (e) brick, listed and non-listed, mostly Victorian, buildings in conservation areas of Cambridge. The paper does so by using the method of system dynamics, a methodological approach that results in re-conceptualizing heritage, energy, and thermal comfort as social and cultural, dynamic and interconnected practices. By thinking of heritage, energy, and thermal comfort as socio-cultural, dynamic, interconnected practices, better guidance can be provided for the sustainable future of historic, residential buildings. By adding Mexico into the analysis, we attempt to offer a more international approach to this subject matter. Unlike the UK and Greece, where there are growing attempts to develop policies and guidance specifically on energy efficiency in historic buildings, Mexico lacks a national, comprehensive evaluation program on the environmental performance of historic buildings, and yet there are intensive efforts to contribute to the decarbonization of the built environment more generally. Mexico is member, for instance, of the Global Energy Efficiency Accelerator Program and, as such, intensive energy efficient work has taken place in Mexico City. However, this is not done through a heritage lens.

As aforementioned, there is a lack of in-depth studies regarding attitudes of inhabitants of heritage buildings towards energy efficiency. This may be explained by the fact that recruiting and interviewing residents in their premises is a time-consuming and resource-heavy process. For cross-cultural and cross-geographical studies, in particular, the involvement of local researchers is vital. Another possible reason is that the focus in studies related to energy efficiency in historic buildings has mainly been placed on the development of technical solutions (e.g., References [9–11] since, as shown above, the heritage values are perceived by professionals as a non-negotiable pre-condition upon which the guidance is shaped. Therefore, peoples' attitudes inhabiting historic building towards energy efficiency have been understudied [12]. And yet, unless users' attitudes to energy efficiency in relation to heritage values is understood, 'there are no guarantees for achieving the planned level of energy efficiency' [12] (p. 188).

Fouseki and Cassar [6] were among the first to identify the need for research that would enable understanding the dilemmas that residents of old buildings face between thermal comfort improvement, energy efficiency, and conservation of heritage features. Six years later, a growing, but still limited, number of in-depth, qualitative studies in this area has emerged (e.g., [13–17], providing a few first insights into the dynamic change of heritage values and the ways they drive or prohibit residents' choices on energy efficiency and thermal comfort. For instance, Fouseki and Bobrova [13] have shown how cultural values associated with original features decline over time as the need for thermal comfort and affordable energy become a priority, resulting in the replacement of deteriorated original features (especially original windows) by modern materials. The authors have also observed how, in recent years, this trend is reversed, especially in conservation areas or areas going through 'gentrification', where the market preference is moving towards the restoration or even replication of original building features. The limited existing studies inevitably focus on single case studies located in a confined, geographical area. More international studies are therefore needed in order to develop implementable international guidance and standards.

2. Theory

Shove et al. [18] (p. 17) argue in their book the 'Dynamics of Social Practice' that social practices emerge, persist, shift, and disappear when connections between materials, competencies, and meanings are being made or broken. Materials include things, technologies, tangible physical entities, and materials of which objects are made. Competencies connote skills and know-how techniques and meanings refer to symbolic meanings, ideas, and aspirations. Our analysis advances this premise by unveiling new critical elements affecting the interaction and continuation of two socio-cultural, systemic and dynamic practices: (1) the practice of 'valuing' and 'conserving' heritage and (2) the practice of improving the thermal comfort and energy performance of an old building. By mapping the ways in which these two practices link to each other over time, we argue that these practices emerge, persist, shift, disappear, and, occasionally, revive when connections between materials, competencies, values (meanings), space/environment, senses, time, and resources (economic, human, etc.) are being made or broken. Space and environment refer to the interior space, as well as interior and exterior environmental factors. Senses denote the sense of thermal comfort and satisfaction or lack of satisfaction with the buildings' performance. Time connotes temporal dimensions linked with the age of the building, the time already spent living or expected to inhabit the residence. Resources refer to costs needed to implement an energy efficiency, thermal comfort improvement, or heritage conservation action. The idea that social practices are systems composed of interconnections of elements which, if broken, will affect the continuation of the system is compatible with principles of the 'whole house approach' advocated by HE. By re-conceptualizing heritage, energy, and thermal comfort as social and cultural, dynamic, and interconnected practices, we open up new avenues for future interdisciplinary and synergetic research that will move beyond isolated studies of individual components of a social system, making the research outcomes more informative for developing future national and international guidance on the decarbonization of the historic built environment.

3. Methods and Materials

3.1. Selecting the Case Studies

The main criterion for selecting the case studies was familiarization of the fieldwork researchers with the local context, personal research interests of the involved researchers, funder's requirements, and access to case studies under examination. It is worth mentioning here that this papers presents the results of ongoing research the aspiration of which is to accumulate over years of knowledge across different areas in the world. As a result, we gradually build case studies, the selection of which is proposed by the involved researchers at each time. At this stage, we are interested in diversity, hence aiming for collecting knowledge from diverse cultural and geographical settings, as well as diverse building materials. The ultimate goal is to create a global atlas of qualitative data in each specific area by bringing together local researchers from the across the globe.

3.2. Collecting the Data

As aforementioned, the paper provides the first systematic and cross-cultural study of inhabitants' energy efficiency and thermal comfort mechanisms in listed and non-listed heritage buildings. It does so by drawing on a rich dataset consisted of 59 in-depth, semi-structured interviews, which totals 206,771 words. The interviews were conducted in (a) 'neo-classical' stone listed buildings in Athens (Greece); (b) Victorian and Edwardian brick, non-listed buildings in Walthamstow (London); (c) 1940s Swedish-type timber structures scattered in rural England; (d) listed stone buildings in the world heritage site of Mexico City; and (e) mostly brick Victorian buildings in conservation areas of Cambridge (Table 1).

Table 1. Number and time of interviews per location.

Location & Climate Zone	No. Interviews & No Owners in Brackets	Period of Interviews
Mexico City (World Heritage area) (subtropical)	8 (6)	Winter 2018
Walthamstow, London (oceanic)	8 (5)	Winter & Spring 2016
Athens (Neoclassical Stone Buildings) (Mediterranean)	13 (12)	Summer 2016
Timber Swedish post-war structures in rural England (oceanic)	12 (11)	Summer 2018
Cambridge (oceanic)	18 (16)	Spring & Summer 2017

Participants in the study were recruited through 'snowballing sampling', recommended by personal contacts, and through the participants themselves. Although 'snowballing sampling' may impose issues of sample representativeness, it is one of the most suitable sampling methods for studies involving 'difficult to reach' participants [19]. In our case, entering one's private household, raises issues of safety and trust, essential for in-depth interviews. Bearing this in mind, we aimed for interviewed participants representing different age groups, ethnicities, economic status, and length of years living in the area in order to capture the diversity of responses. The interview guide comprised of three parts. The first part aimed at understanding the cultural values (meanings) that residents attribute to their buildings. Questions were combined with 'photo production' with interviewees taking pictures of those aspects they valued the most [20]. This method enabled to delve into greater depth on the complex subject of 'values'. The second part intended to explore residents' energy efficiency and thermal comfort actions over the years. The final part investigated residents' future aspirations and attitudes towards energy efficiency (including renewable technologies) solutions experimented elsewhere. At this stage, 'photo elicitation' [20] was used by showing pictures of alternative energy efficiency means (e.g., solar tiles, wind turbines, different types of insulation, etc.).

3.3. Coding the Data

The interview data were coded on the qualitative analysis software NVivo. The data were first coded by each individual researcher separately and then by the lead author who discussed the coding with each of the researchers. By conducting double-coding, we minimized inevitable interpretation biases. Each interview was uploaded as a separate file (case) and assigned an identity code comprised of the location's name initial letter and the additive number: A (Athens) C (Cambridge), T (Timber structures), W (Walthamstow), and M (Mexico). Following the principles of grounded-theory according to which the data drive the theory [21], interview data were initially coded through an open coding process, identifying as many variables and themes as possible related to the key research questions [22]. Six hundred and eighty-two codes (nodes) were created, which were then clustered under wider themes through axial coding (Figure 1).

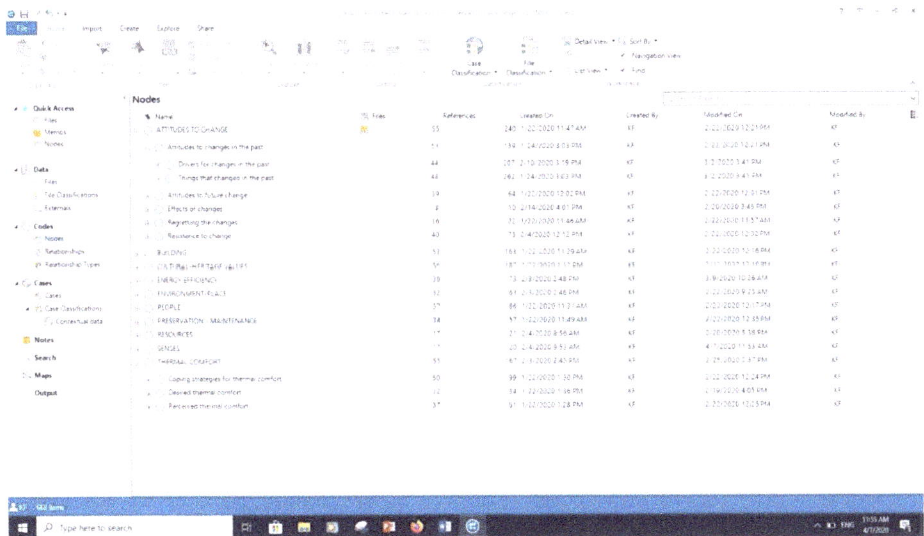

Figure 1. Snapshot of a coding tree extrapolated from Nvivo.

During the coding process, cause and effect relationships between nodes were identified using NVivo's function of relationships [23] (Figure 2).

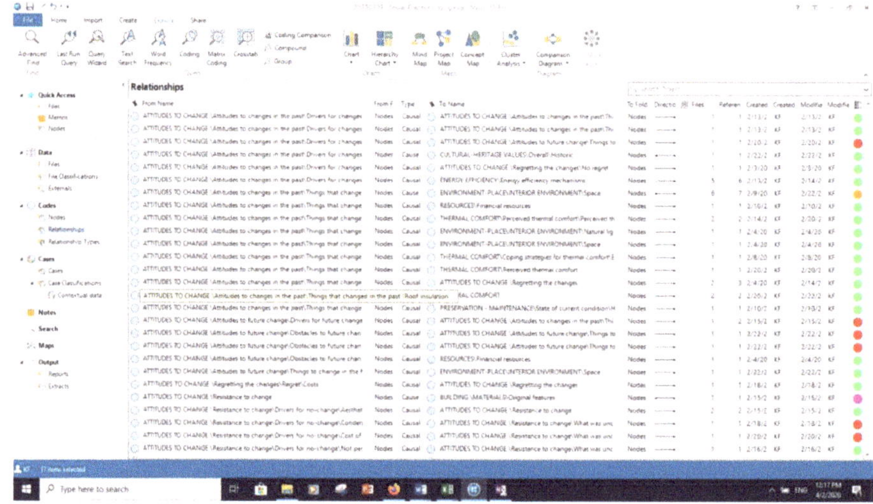

Figure 2. Snapshot of relationships identified on Nvivo.

Four types of relationships were recorded. One-directional, cause and effect, 'negative' relationships were marked in red. These relationships indicate an antithetical interrelationship between the nodes. For instance, the more the 'original features' deteriorated the less the 'perceived thermal comfort is'. One directional, cause and effect, 'reinforcing' relationships were marked in green. An example of a reinforcing relationship is the following: The 'more years a resident plans to stay in the property' (length of tenure), the more likely is to 'restore the original features'. Cause and effect, iterative relationships (loops), were marked in purple, if 'reinforcing', or in orange, if they were 'balancing'. For instance, a reinforcing loop is that the 'more the natural light', the 'bigger the space looks like', and the 'bigger the space', the 'more the natural light'. An example of a balancing loop is: the 'lower the perceived thermal comfort is', the 'more likely to insulate the roof' is, which will then improve the perceived thermal comfort. In total, 209 relationships between nodes were identified.

A list of attributes was created enabling cross-tabulations and comparisons that could further elaborate the relationships. The attributes include location, building age (19th century, early 20th century, and 1940s), construction materials (brick, stone, concrete mixed with brick, concrete mixed with stone, timber), desired thermal comfort (between 20 and 25 degrees, less than 20 degrees, more than 25 degrees), length of living in the property (1–5 years, 6–10 years, 11–30 years, more than 30 years), length of time planning to live in the property (indefinite, planning to move soon, 1–5 years), listed status (listed, non-listed, partially listed (only façade), non-listed but in a conservation/protection area), ownership status (owner, landlord, private tenant, council tenant), type of area (conservation urban area, non-conservation rural area, non-conservation urban area, world heritage area), and type of building (detached house, semi-detached house, terraced house, flat in block of apartments).

3.4. Mapping the Data through System Dynamics

Based on our theoretical assumption that a decision process on heritage conservation, thermal comfort and energy efficiency improvements is a socio-cultural, dynamic, and systemic practice, we applied system dynamics in order to unfold the dynamic interconnections of the components mobilized during this decision-making process. The method of system dynamics is commonly utilized to explore the dynamic interconnections of the components of a system [23]. The term implies a 'complex entity' consisted of interconnected elements which change over time. This 'entity' can be a social practice, a building, a city, etc. The underlying premise is that changes on any of those elements will affect the entire system [24] because a complex system comprises non-linear, multiple, interconnected loops which change over time, with some loops disappearing or re-appearing under certain conditions [25] (p. 107). The loops are cause and effect relationships [26] (p. 120) which can exponentially grow (reinforcing loops) or start declining bridging the gap between a desired and an actual goal (balancing loops) [25] (p. 133). The cause and effect relationships identified in the previous research stage were mapped on Vensim, creating a causal loop diagram [26] (p. 119). Each cause-effect relationship is indicated with + or − depending on whether the relationship is positive and reinforcing (e.g., the more ... the more) or balancing (e.g., the more ... the less). The diagram presented here (Figure 3) is the aggregate (summative) representation of the dynamic interrelationships identified during our analysis. In other words, the diagram does not illustrate all 209 relationships, as this would be too complex to communicate in one diagram, but an aggregate version which summarizes variables (for instance, all original features are depicted as one variable) and illustrates the most common thermal comfort, energy efficiency, and heritage conservation actions.

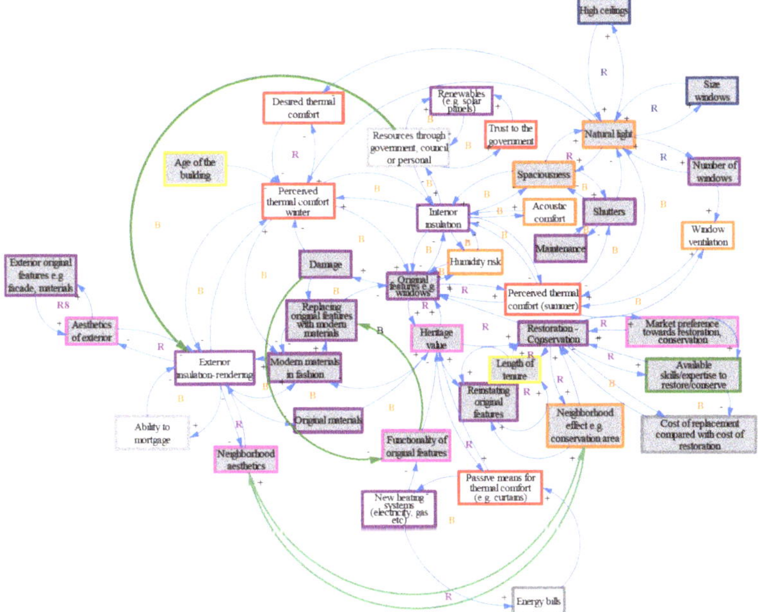

Figure 3. Color scheme: Senses = red; Materials = purple; Values = pink; Resources = grey; Time = yellow; Competencies = green; Space/environment = orange. The green arrows are purely used for communication purposes as they intersect other communicating arrows. Variables linked directly to heritage are filled in grey. R signifies reinforcing loops and is colored in purple, while B (in orange) refers to the balacing loops. The symbol + indicates the reinforcing relation (the more ... the more), while the symbol – indicates a balancing relationship (the less ... the less).

4. Results

4.1. Residents' Social and Cultural Values towards Their Historic Residences

Figure 3 depicts the aggregated, dynamic interconnections of the main factors influencing inhabitants' decisions on thermal comfort and energy performance interventions in conjunction with the heritage values they attach to their residences. A detailed presentation of each segment of the diagram follows below.

The reinforcing relationship 1 (R1), which lies at the center of the diagram, indicates that the more the 'original features' the higher 'the heritage value' attached to the building (Figure 4). A wide array of heritage values emerged, which were clustered into three groups including: (a) values attached to the interior of the house; (b) the exterior of the building; and (c) overarching values (Figure 5). As expected, the responses vary among geographical areas, including areas located in the same country. While in current guidance documents it is mainly certain external features of an old building (such as façade, windows) that are implicitly prioritized for heritage conservation, the interviewees assign cultural and social values to a diverse range of exterior and interior architectural features (Figure 5). In the case of Swedish timber houses in rural England, for instance, it is the timber structures that residents resist to change 'because it is heritage' (T1). In Cambridge conservation areas, on the other hand, we observe that heritage conservation priorities align with those imposed by the listing system (such as original windows and original fittings). In Walthamstow, Mexico, and Athens, it is 'the high ceilings … ' allowing 'lots of light' and the resulting feeling of spaciousness as new 'buildings press you down' (A4). It is worth noting here that the respondents in Mexico did not prioritize specific individual architectural features as the most important, other than the spaciousness attributed to the high ceilings.

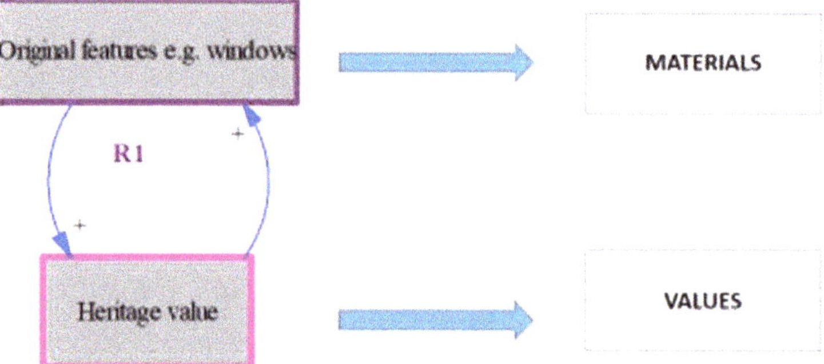

Figure 4. Reinforcing relationship (R1) between original features and heritage values.

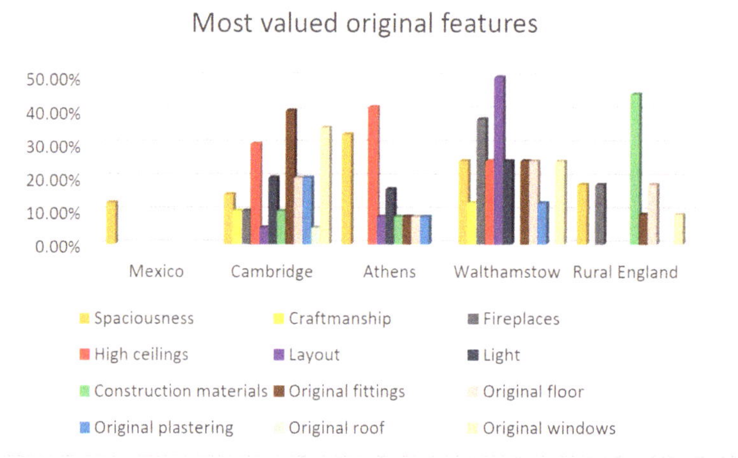

Figure 5. Most valuable heritage aspects across different regions related to the interior and exterior of the buildings.

The variation on the values attached to different heritage elements is striking (Figure 6). In Cambridge, the most predominant values are 'aesthetics' and 'originality'—echoing the values of the listing process. C12, for instance, notes how they endeavored to maintain all original features, including "the original stained glass … the original paneling and also the circle in the middle of the door. And it's really beautiful". Contrarily, in the case of timber Swedish houses in rural England, it is the 'uniqueness' of the timber structure and feelings of 'nostalgia' evoked by the presence of the fireplace: "I always liked the open fire as we all sat around the fire" (T2). In Athens, the neoclassical buildings reminiscence a lifestyle that is now lost in a densely inhabited city. In Walthamstow, interviewees noted a sense of moral responsibility and ethical guardianship towards the preservation of certain original features: "Well, like I said, I suppose it's because, I don't know, it kind of puts you back, in touch with kind of the house it was meant to be, it just kind of feels right" (W6). Interestingly, in Athens and Walthamstow, interviewees described the building as a living body the 'character' and 'soul' of which needs to be respected: "There is a comfort of space, all the rooms are easily accessible. It was built with love and it has its own soul" (A6).

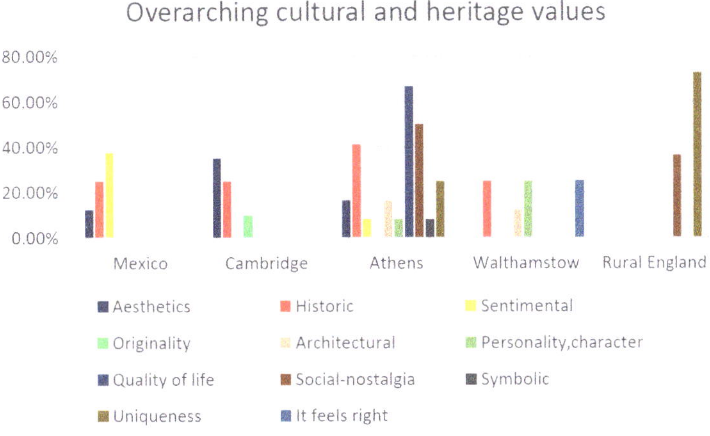

Figure 6. Comparison of heritage values among the different regions

Moreover, an interesting variation was observed between the set of values related to the exterior (what is seen from the outside) and the interior of the house (what is experienced inside), especially in Mexico, Walthamstow, and rural England, where 'aesthetic' and 'architectural values' of the façade are the most referenced values. However, 'coziness/homeliness' and social values are mostly referenced and relate to the interior of the building. This variation explains why inhabitants in these areas tend to prioritize the preservation of original features that evoke such feelings (such as fireplaces), while they are inclined to replace features, such as windows, with modern ones in order to improve thermal comfort and energy efficiency (Figure 7).

4.2. The Dynamic Interactions between Heritage Values, Heritage Conservation, and Sense of Thermal Comfort Interventions

Figure 8 demonstrates that when original features are imbued with heritage significance, while contributing to a thermally comfortable environment, especially in summer months (R3), restoration and conservation of those features is a preferred option (R2): 'In summer, it's a delight, I really tell you, it's a delight to get inside from the street and find natural air conditioning thanks to the height of ceilings and thanks to the walls that keep a delightful temperature, it's very nice' (M8).

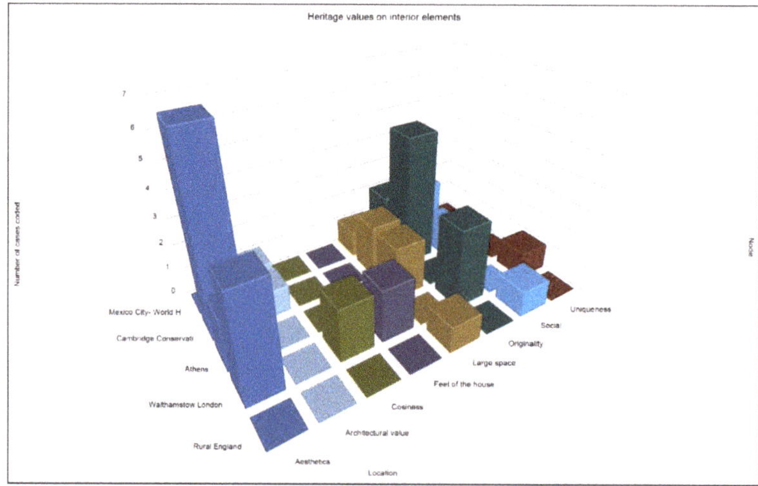

Figure 7. Heritage values attached to interior elements per number of cases. The aim of the figure is not to illustrate statistically significant variations, as this would require a much larger sample per case, but rather to illustrate what is the dominant response in each case, providing a hint on variations.

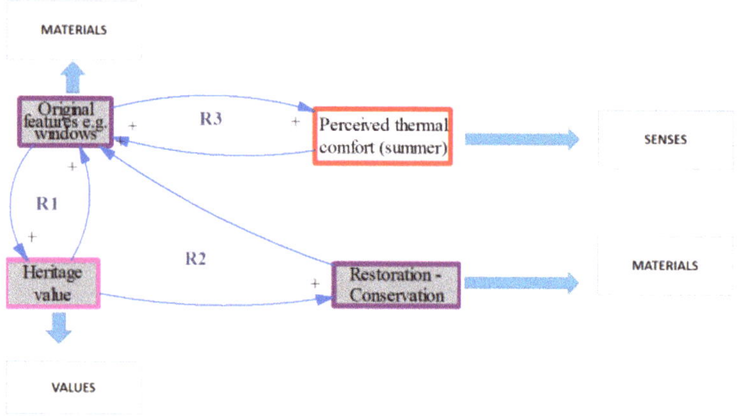

Figure 8. Reinforcing interrelationships of original features, heritage values, conservation, and perceived thermal comfort (summer).

R4 loop (Figure 9) depicts a positive interrelationship between desired and perceived thermal comfort levels, especially in winter months. In other words, the lower the desired temperature in winter, the better the perceived thermal comfort. Among the respondents who indicated desired levels of thermal comfort, 9 respondents (all located in the UK) were satisfied with a temperature of less than 20 degrees. On the contrary, 2 respondents (both from Mexico) indicated a desired temperature of 26 degrees and above, while the majority, 16 respondents, were satisfied with a temperature between 20 and 25 degrees. An additional interesting trend was noted when the desired thermal comfort levels (senses) were cross-tabulated with the age of the building (time). Lower desired thermal comfort levels in winter months were more likely to be recorded in buildings dating in the 1940s (6 out 13 cases). On the contrary, in earlier structures, the desired thermal comfort is higher (17 out of 30 cases). A cross-tabulation between desired thermal comfort and type of building (e.g., detached, semi-detached, etc.) did not reveal any significant variations.

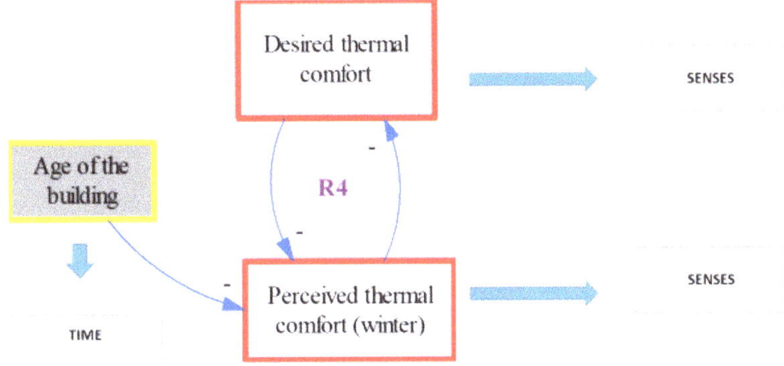

Figure 9. Reinforcing relationship between desired and perceived thermal comfort.

While the original features (materials) constitute an attractive element at the early months of inhabiting an 'old' building, over time material degradation and damage (materials) affect negatively the perceived thermal comfort (senses), especially during winter times (B1) (Figure 10). One of the actions undertaken by residents to address the issue of thermal comfort at this stage is installing interior insulation. Their decision to insulate the interior walls of their property depends on the levels of acoustic comfort and thermal comfort, especially during the summer times (R9, B3) (senses and environment), the available space (B6) (space), the number of original decorative features (such as ceiling plastering) (R12) (materials, values), and state of physical condition (B1). The risk of high humidity (R11) (space/environment) and its impact on the original features (B1) (materials) are also taken into account during the decision of installing interior insulation. Although available government funding and/or council ownership (B4) (resources) encouraged owners to opt for internal insulation, this option became less favorable as soon as the governmental financial support ended. It is worth noting here that the installation of interior insulation was not a popular solution among the respondents. It was reported only in 5 cases, with 3 deriving from semi-detached houses in Cambridge where available space is bigger and the remaining 2 from Athens and rural England, respectively, where the council, in the latter case, installed external cladding. C14 installed interior insulation, even if some of the plastering had to be 'sacrificed': "So, what we did was, we looked with the architects, we looked at which walls could be insulated internally because some of the walls... the plaster was stripped as well, and there were only a couple of sections where the plaster wasn't stripped completely to the brick, one being essentially this party wall that ran along. Most of the plaster there was kept, but in a lot of other places it wasn't particularly staying on, so we just took it all off". C5, on the other hand, seems to have regretted the decision to install internal insulation: "At one point we were putting insulation in the walls, when we did a renovation last year and we put a whole load of insulation in the walls and sadly we therefore lost a lot of room and the features … it became ridiculous the balancing of it because you know the walls were costing an awful lot of money and reducing the character and not actually contributing that much to the protection from the harsh outside". Interestingly, 3 timber house residents in rural England and 4 in Walthamstow positively approach the future prospect of interior insulation, although the ultimate decision, as described by W6, will depend of multiple factors: "When I moved in here I was interested in insulation, not just for the heating but also for the noise... I've just done some, it was really expensive, that the costs would take forever to recoup and also you did lose a fair amount of your kind of space and because like I said they're not big houses you know, um, all of those three things together I was like 'mmm, no it's not something'.

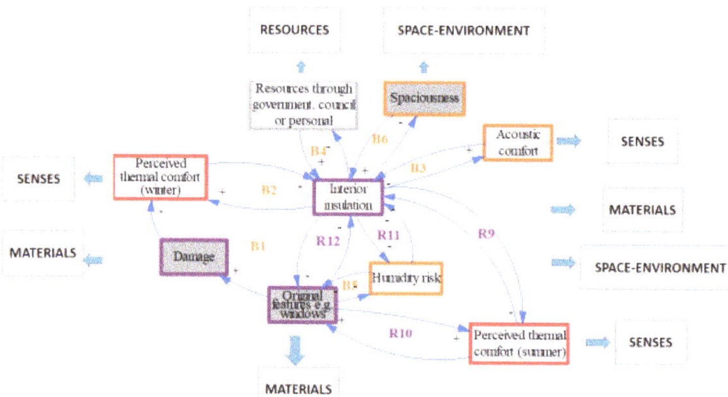

Figure 10. Loops related to the installation of interior insulation as reported by 5 respondents.

4.3. Dynamic Interactions between Interior Environment and Original Features

The two previous sections have focused on presenting the change of heritage values over time and how this aligns with certain actions on improving the interior thermal comfort. This section explores the dynamic interconnections of the interior environmental factors (such as light and levels of ventilations) in conjunction with the desire or not to preserve certain original features, such as the original windows.

More specifically, Figure 11 illustrates a reinforcing relationship between the size and number of original windows (materials) and the natural light (environment) (R14, R15) which turns into a balancing relationship once thermal and acoustic comfort are disturbed. As aforementioned, light associated with original windows and high ceilings (R13) (materials) is one of the most valuable aspects, especially among those inhabiting small-size houses (space) (R16) or flats in densely inhabited areas. Light does not only contribute to the aesthetics of the interior space but also to a warmer environment during winter, as reported in all cases, although it can also drive thermal discomfort during summer times. The appreciation of natural light associated with high ceilings and original windows influences the choice of passive thermal comfort strategies, such as the use of shutters and window ventilation during the summer months. The use of shutters, for instance, becomes critical in regulating the light in summer, as well as in maintaining the heat in winter (B7). The use of shutters though depends on affordability and space availability (B8): "But then shutters … they are expensive to get nice ones, they take up room so they take up more room within the house as well, you know it's not that big a flat so we were kind of sorting you know the maximum light, maximum space. My parents have their shutters do take off the light cause even though when they are pushed back they still do take up some light. It's quite a light room which is nice for a flat" (W7). High maintenance requirements, as with any original features, is an additional consideration (B9): "[Shutters] would be all-right, as long as they weren't on every single window, probably on the window which lets in the most light cause sometimes that can be annoying on a really, really sunny day, when you are trying to have a lie-in but the curtains are too thin … but the only thing is that if they were wooden and if they got stuck on the window still and they made this creaking noise, I hate that … " (W2).

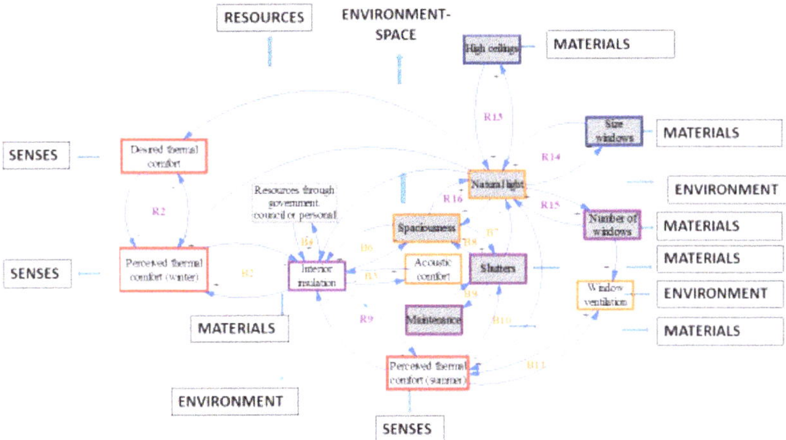

Figure 11. Interconnections between natural light, shutters, and windows for ventilation.

An interesting co-relation was observed between levels of natural light (space/environment) and desired thermal comfort levels, which, as shown above, also affects the perceived thermal comfort. In Mexico and Athens (two cities with hot summers), lower desired thermal comfort levels were recorded among those respondents who stated that natural light was adequate (Figure 12).

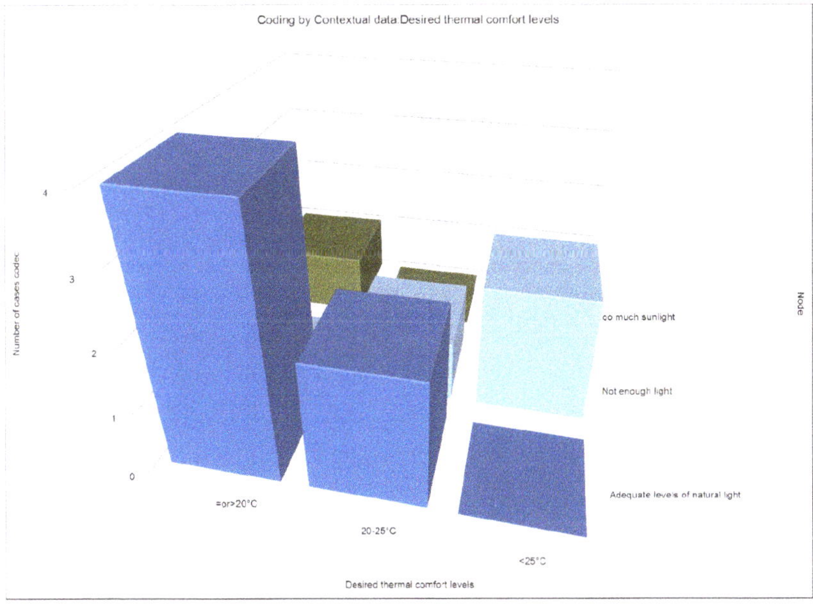

Figure 12. Cross-tabulation of desired thermal comfort levels with levels of natural light in Athens and Mexico.

Another typical example of the aforementioned dynamic interrelationship is the dynamics of original flooring. Gaps in the deteriorated original floors are perceived by interviewees as contributors to heat loss: "Yeah I like the floorboards, I mean they are a bit of state to be honest, especially down there, they obviously had wood worm and they had been badly replaced but we uncovered them and sanded them … I like having the bare floor boards. My main concern about this is that's it's really, kind of, you feel the kind of breeze, so the cold air rise in winter … So that's the only thing which having lived here for five years now it's put me off, but I like the kind of look" (W6). When original floors are not viewed as a heritage element that merits preservation, they are either being replaced by new floors or being covered by laminate flooring (especially in rented properties). In Walthamstow, Cambridge, and Athens, the original flooring is appreciated for its heritage significance in recent years. In these cases, thermal comfort interventions, such as underfloor heating or underfloor insulation, are the most preferred options often combined with the restoration of original floor through sanding (especially in Walthamstow and Cambridge).

4.4. The Dynamic Interconnections of Heritage Values Attached to the Exterior Environment with Thermal Comfort and Energy Efficiency Interventions

One of the most expected energy efficient intervention employed at the exterior of the building is the external rendering and/or the installation of exterior insulation. However, interestingly, external rendering and external insulation occurred only in 3 cases in rural England and 3 cases in Cambridge. In addition, 8 respondents, also from both areas, declared that they do not intend to install external insulation in the future. On the contrary, in Walthamstow 50% of the respondents would examine this option.

This diversity in attitudes towards external insulation can be explained by the fact that the decision for external insulation—similarly to internal insulation—depends on multiple factors. In cases where exterior original features (materials) are perceived of high aesthetic value (value), it is less likely to install external insulation and rendering (materials) (R12) (Figure 13). However, if the exterior is perceived of low aesthetics, then the chances to render and insulate externally are higher.

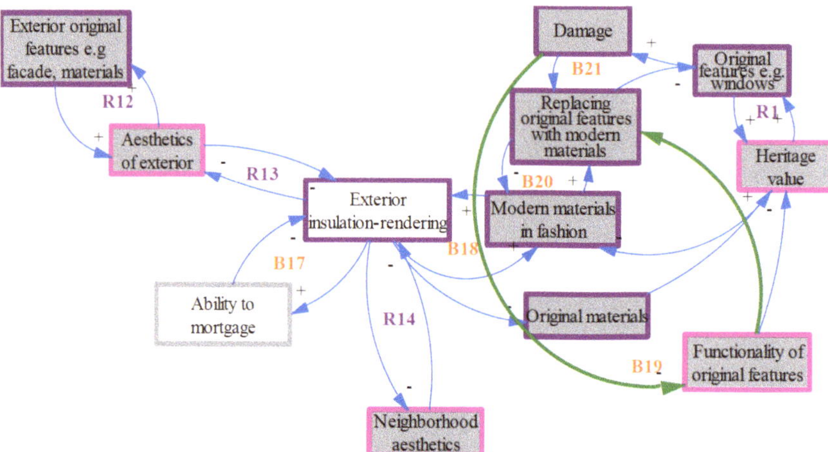

Figure 13. The dynamics of exterior insulation.

C12 summarizes the factors contributing to the choice of external rendering, including the low aesthetics of the exterior and the need to improve the thermal comfort, combined with available government financial support: "So once the Green Deal was out there, in Cambridge particularly, it seemed the obvious thing to do. As I said, because our house was already pebble dashed, and not very pretty actually on the outside and it was painted a pretty nasty color, by the previous owners, we felt that it would only actually improve the look of the house. Actually, it has improved the look of the house by quite a lot, because it was looking a bit tired, bits had fallen off and had been repaired. So for us, that was definitely a big yes to get that done, it has improved the house a lot" (C12).

Similarly, the aesthetics of the neighborhood especially in conservation areas can affect significantly the decision on external insulation (R14): "Well, the thing is I would never consider that for my house because it would mean changing the whole look of the house, to have it put on the outside. And the terrace that we live on, there's six terraces altogether, they were all built about the same time. It would look awful if one had that done" (C14).

The case of timber houses brought into light an additional variable—the ability to receive a mortgage in the UK—due to susceptibility of timber to fire. It is because of this reason that extensive cladding has been observed in previous years when the timber houses were still owned and managed by the relevant councils (B17). The impact of aesthetics is, unexpectedly, observed in the context of modern materials (plastic, cement) which, although currently viewed 'as ugly' in heritage buildings, they were in fashion in the 1970s, 1980s, and 1990s. For example, the widespread replacement of old windows with uPVC windows was a fashionable trend in the 1970s and 1980s [27]. "[We] replaced the original wooden windows with uPVC when my aunt died and left some money as they were advertised being good windows, the PVC. Originally, we had big sized windows made of wood. No dimension was changed when new windows were put" (T2). "[We changed the windows] in order to keep with modern trends. Also, the original ones just had single sheet of glass, quite thin. We actually [changed them] for aesthetic reasons but also helped in making [the house] warm" (T1). However, the popularity of modern materials has been declining in the last twenty years, with original materials becoming currently the new trend (B18). This trend depends on the physical condition of original materials. If the gradual deterioration of original materials is not managed (materials), then the perceived thermal comfort (senses/environment) is negatively affected with original features being replaced by modern materials while imitating old features (B19, B20): "I mean I ended up replacing the single glazed panes with double glazed panes on the first floor because the actual wood itself was so rotten you had to take out virtually the wooden frames and then they said to me well what sort of glass do you want put back so that prompted me to replace the fixed bit so they were done by a builder and then the double glazed part the plastic part and the insect screen was done by a guy who does that sort of thing who was recommended by next door" (C3). It is worth noting here that, in a few cases, we observed deterioration of the 'modern materials', such as the glass and plastic frames of uPVC windows, which consequently affect both the aesthetics and the thermal performance of the windows. Although no change was made at the time of the interviews and therefore no change is depicted in the diagram, it will be interesting to trace attitudes in future repairs.

4.5. Renewables and Passive Measures for Thermal Comfort Interconnected with Heritage Values

Modern materials and technologies associated with renewables (such as solar panels), reduced energy bills and financial support from local and national governments (B25) are in favor unless there is lack of trust towards the respective government (B26), as observed in Athens (Figure 14). If the existing heating and cooling systems result in high energy bills (B24), then passive approaches, such as the use of shutters in the summer or heavy curtains in the winter, are adopted. In Athens and Mexico City, for instance, the respondents make little, if any, use of air-conditioning and electric fans, relying on passive cooling measures, such as opening windows or using shutters. They view such measures as particularly effective due to the traditional construction: "During the afternoon and the night when it's really hot I switch on the A/C but in general I keep the shutters closed, with open windows so that the air can keep flowing. I drink cold water as well" (A6).

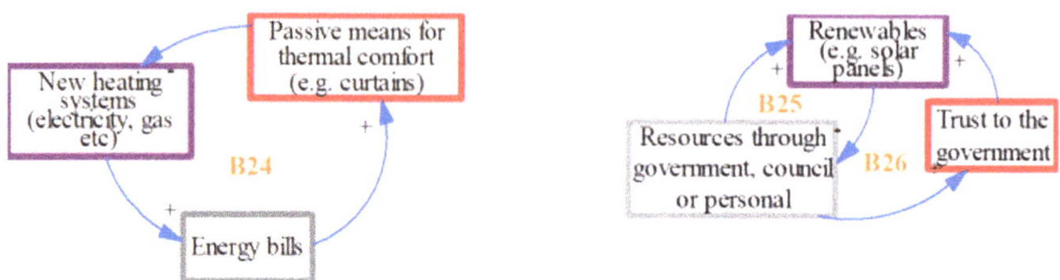

Figure 14. Interactions between passive heating and cooling means, energy bills, and renewables.

During the winter months, passive heating strategies (such as using double curtains or closing windows) are observed mainly in Mexico City and in Cambridge, where the designation status of the area requires the preservation of original windows (Figure 15).

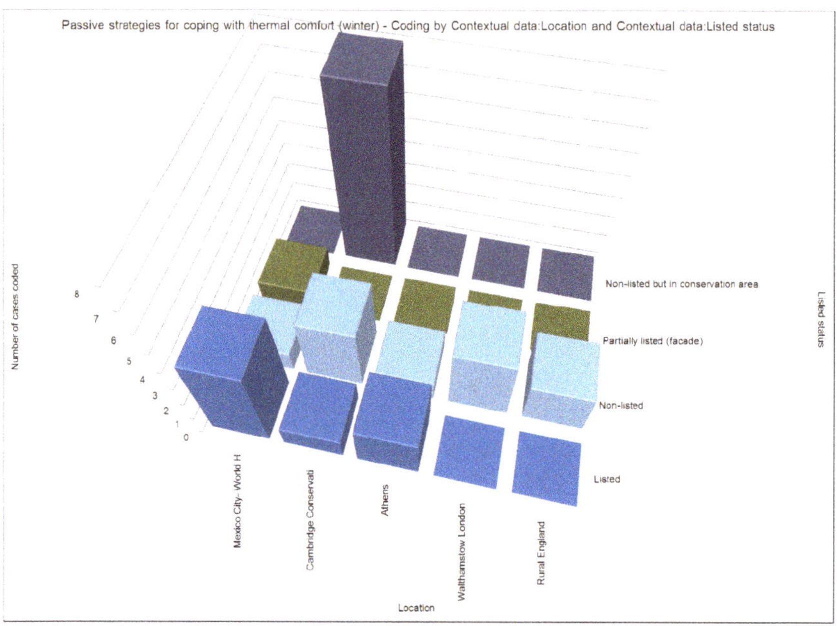

Figure 15. Cross-tabulation between passive thermal comfort strategies and listed status.

4.6. Conserving, Restoring, Replacing or Replicating Original Features? A Dilemma Driven by the Need for Thermal Comfort, Energy Efficiency and Property Values

So far, we have seen which factors from the interior and exterior building system interact with the heritage values assigned by the residents and their need to improve the thermal comfort and energy efficiency of the building. In this section, we illustrate a wider array of social factors that can influence the decision of a resident (especially of owners) to conserve, replace, or replicate certain original features, while facing the dilemma between thermal comfort, energy efficiency, and heritage conservation. B19 and B21, once again, reveal the critical role of material degradation in the decision-making process. Over time, material degradation results in the reduced functionality of original features (values) which encourages their replacement by modern features or the conversion of certain features (such as the fireplace) into decorative features. Over time, the remaining, decorative features acquire heritage significance prompting their restoration (R15) or replication and reinstatement (R17) (Figure 16).

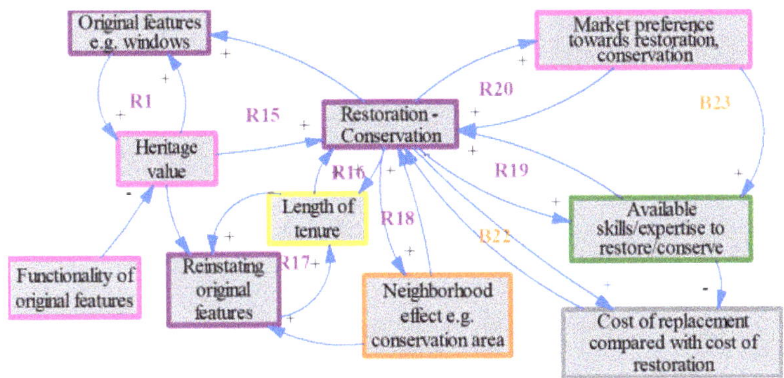

Figure 16. The dynamic effects of neighborhood on heritage conservation.

The decision to restore, reinstate or replicate depends not only on how much of the original material survives (materials) but also on how many years the occupants have been living or wish to live in the property (R16) (time), the surroundings of the neighborhood area (R18) (space/environment), the cost of restoration versus replacement (B22) (resources), the available expertise skills (competencies) (R19), and the market preference in the area (R20) and (B23) (values). The following interview excerpt summarizes the aforementioned relationships: "Yes, I really like the original sash windows … but they were in very bad condition. I mean the frames, the wood was rotten, so they then they needed to be either completely restored or taken out and restoring sash windows is a very pricey job and even if you do it properly you spend a lot of money on it, you don't get the same insulation you would get with PVC … So we went for double-glazing PVC … having double glazing has made a huge difference, I mean with the heating on, we used to be still freezing cold in that room with the winter, it's just plain cold. I: When did you take the decision to change them? We only did it last year, we couldn't afford it. We probably would have done it sooner although we weren't necessarily agreeing, because we liked the fact that it was sash windows and Jason was thinking it would be nicer to restore them, but when we saw how much it was going to cost and also finding a specialist, whereas with the PVC you could almost anyone to quote, so it was less than half price in the end. And the decision you know, it was also because those houses, especially at the time, they weren't expensive houses, they weren't £200,000 and we just felt that spending money and restoring windows if you're in a conservation area in a nice neighborhood, or even you know, like in the village or somewhere where the Real Estates move premium it makes sense, but spending money on restoring sash windows in this area it didn't really add with the price of the house … That would not have added any value to the house mmm and to be honest it would probably hurt it, because people looking for housing in this area tend to be families who want to have a warm house, not necessarily a conservation type of house" (W5).

This relationship is particularly evident in the dynamics of the use of fireplaces. Thermal comfort practices during the winter period relied in the past on the use of fireplaces and the existence of partition walls and heavy doors (materials) which prevented the cold air to diffuse across the entire house. Due to the cost of coal, the use of fireplaces was limited in the most commonly used rooms (mainly the living room): "When built, each room had a separate fireplace for coal or logs. My parents used the fire in the living room and the kitchen range. I think this was what they could afford. The bedrooms I remember as cold in winter. Frost inside the windows!!" (T6). Over time, increased smoke pollution in urban areas and the introduction of electric and gas heaters (materials) led to the abandonment of fireplaces. During periods of economic crisis though, such as the one faced recently by Athens, fireplaces are still being used (winter). When fireplaces lost their initial, functional role, as aforementioned, they either acquired a decorative role (shift) or were removed completely (stopped existing) in order to enhance the space: "The smaller [property we rent out] has all its fireplaces removed to gain space by the previous owners who lived there for 25 years. They were from Ghana so they felt similarly to us about the original features" (W3). In recent years, fireplaces or parts of those (as tiles) are being reinstated: "Unfortunately, all original features had gone in this house. All chimney breasts had been removed and we thought to put them back but the cost was too much, we don't have the original fireplaces, we did find some original tiles in one of the rooms in such a bad condition where one of the fireplaces was, so we had to put new ones on top which is a shame, you know, if we could salvage them we would have done so but we kept the front door which I think is original. We had to put a new bannister because the people had put an iron, I don't know, it looked like [...] something really weird and it was definitely not original" (W10).

To summarize the results, the reinforcing relationships, which are mainly located at the center of the diagram, relate to cultural and heritage values attached to original features which are still functional. Once functionality is compromised due to material degradation and issues of thermal comfort emerge, the reinforcing relationships evolve into balancing ones, revealing the attempts of residents to bridge the gap between perceived and desired thermal comfort. During this process, there is inevitably loss of original features, especially of those perceived as less important by the residents. At the same time, improvement actions on thermal comfort are often linked with attempts to reduce energy bills. Over time, once cultural values associated with original features are revived, there is constant growth of attempts to restore or even replicate original features, leading to new reinforcing relationships, as shown at the central, right part of the diagram (Figure 3). Listing status and/or increased property values associated with original features are contributing factors to this relationship. The continuation of restoration and replication practices will depend on available expertise (competencies), affordability (resources), neighborhood effect (environment), heritage and economic values (values), and length of time planned to be spent in the residence (time). As a result of the revival of heritage values, residents adopt alternative, passive heating and cooling strategies, such as use of shutters, double curtains, and passive ventilation.

5. Discussion

Why is this study significant and for whom? Firstly, this paper offers the first international, qualitative investigation of the unexplored dynamic nature of heritage values and heritage conservation and its relation to energy efficiency and thermal comfort, in the context of what we could call 'everyday heritage', that is, heritage being experienced in everyday life, which is not necessarily designated. The current limited research in this subject area is case-study specific, which can easily be explained by the fact that intensive and extensive resources are needed in order to carry out this type of studies. A global perspective is essential not only for informing international standards and guidance on energy efficiency in historic buildings but also for providing new insights into 'what heritage' merits preservation from peoples' perspectives [28] (p. 18). Secondly, the paper introduces a new theoretical approach for unpacking how heritage emerges, shifts, disappears, and revives. By arguing that heritage conservation is a social and cultural dynamic practice, the continuation or disruption of which depends on the interconnection between senses, materials, competencies, space/environment, resources, time, and meanings, prompts us to move beyond the narrow identification of heritage as a 'significant object' of aesthetic, architectural, and historic value [29]. Accordingly, future policy guidance should move beyond the prescriptive listing of pre-assumed values and energy efficiency measures and include guidance on how to approach, conceptualize and identify the values and the attitudes of users towards energy efficiency. For heritage professionals and policy-makers specializing on energy efficiency, understanding the dynamic and systemic nature of decision-making about heritage and energy efficiency can better guide the preliminary steps of significance assessment and choice of energy efficiency measures. The decision-making process in this area is indeed 'characterized by complex, context-specific assessments that can never be adequately captured in policy of a necessarily more general kind' [17] (p. 349).

Moreover, because 'inhabitants' assessments of the character of buildings are implicated in their use of these spaces, the relationship between increased efficiency and reduced energy consumption is far from straightforward' [17] (p. 349). This implies that the introduction of new technologies may 'reduce inhabitants' ability or desire to adapt their own behavior to it' [17] (p. 349). Although our respondents were generally receptive towards renewable technologies (under the condition that they were supported financially by local or national governments), we noticed resistance or skepticism towards the installation of solar panels in the case of Athens. Resistance to new technologies can possibly explain why, in cases of designated areas or listed buildings in Mexico, Walthamstow, and Athens, inhabitants are willing to adapt to colder or hotter temperatures by using passive heating and cooling measures. In view of the above, energy efficiency measures, as the 'whole house approach' prompts, should take place in conjunction with in-depth understandings of the social and cultural process in which inhabitants make decisions on energy efficiency, thermal comfort, and heritage conservation.

6. Conclusions

With this paper, we endeavor to offer the first systematic, international study of inhabitants' decision-making processes related to energy efficiency and thermal comfort in relation to heritage conservation in historic, residential buildings. It does so by employing a rigorous double-coding analysis of 59 in-depth semi-structured interviews (totaling 206,771 words) carried out by an international team of researchers. The analysis was combined with system dynamics, a useful tool for unveiling the complex and dynamic interrelationships of parameters contributing to social systems. Using the theoretical lenses of the Dynamics of Social Practices developed by Shove et al. [18], we further advanced the dynamics of socio-cultural practices by arguing that a socio-cultural practice emerges, shifts, evolves, disappears, or re-appears if the following elements are connected or disconnected: materials, competencies, resources, values, space/environment, senses, and time. We concluded with this theory by examining the dynamics of decisions on thermal comfort improvements, energy efficiency interventions and heritage conservation in residential historic, listed, and non-listed buildings.

We showed that a decision on these practices is a socio-cultural practice itself that changes, evolves, disappears, or re-appears as a result of the connection (or disconnection) of the following key elements: materials (e.g., original features, energy technologies), competencies (e.g., restoration/conservation skills), resources, values, space/environment, senses, and time (e.g., time spent living in the house). We demonstrated how originally the building (materials) used to operate a functional and social purpose (that of housing) with cultural values associated mainly with feelings of 'homeliness' and 'coziness' (values). Features, such as the fireplace (materials), were critical in creating such feelings. Over time, the emergence of modern energy technologies and modern materials (materials) combined with material degradation led to the loss of functionality (values) of some of those elements (such as the fireplace). The loss of functional value drove the removal of certain features, especially in densely inhabited urban areas, partially due to the need to create more living space. It also forced the replacement of deteriorated original features (such as windows, floors, doors) by modern ones (materials) which, at some point in time—especially during the 1980s—were viewed as 'aesthetically pleasing' (values). Over time, lost or deteriorated original features (materials) became heritage (values) as owners endeavored to restore them or even replicate them. The option of restoration is, though, subject to costs (resources) and available expertise, which is often lacking (competencies); but, as it becomes a trend, there is also a revival of craftsmanship skills. The situation is less dynamic in listed buildings where original features—especially of the exterior—are under protection. With original features regaining their heritage significance, a growth in the adoption of passive thermal comfort and energy efficient strategies was noted, especially in designated heritage areas and in areas where energy fuel costs are high (such as Mexico and Athens).

By arguing that heritage conservation is a social and cultural dynamic practice the continuation or disruption of which depends on the interconnection between senses, materials, competencies, space/environment, resources, time, and meanings encourages future guidance and policies to move beyond the prescriptive listing of pre-assumed values and energy efficiency measures and include guidance on how to approach, conceptualize, and identify the values and the attitudes of users towards energy efficiency. By using system dynamics, we were motivated to look at the dynamic interconnections that we would have missed if we had followed a conventional, thematic analysis. In addition, the development of a system on Vensim can allow the creation of simulation models which could be used by future decision and policy-makers in deciding which interventions will be more impactful in the long-term. Therefore, this comparative study is certainly not a final outcome but rather just the beginning of future similar studies. Given the global and diverse climatic challenges, we would like to advocate for the development of an international dataset of in-depth interviews which illuminate how residents themselves decide on thermal comfort and energy efficient strategies in relation to cultural and heritage values they attribute to their residences. Moreover, these datasets should be complemented and analyzed in conjunction with environmental and building physics data so that a 'whole house' model of how perceptions and decisions affect (or are being affected) by the actual interior and exterior environment of a building can be built. By developing such models, researchers, as well as decision and policy-makers on energy efficiency in the historic built environment, will be in a better place to assess scenarios and create guidance that is impactful and implementable. However, we acknowledge that this type of work requires extensive resources. We would thus like to conclude this paper by proposing the initiation of a global partnership scheme that brings together universities, local and national stakeholders, and users in order to create a dynamic global dataset that can feed local, national, and international policies and guidance on energy efficiency in historic buildings.

Author Contributions: K.F. carried out the fieldwork in Walthamstow and supervised the fieldwork in the remaining case studies. She also carried out the overarching, integrated thematic analysis on Nvivo, the development of the aggregate system dynamics model on Vensim and the writing of the paper. D.N., K.S.M.C., S.N., and T.K. carried out the fieldwork in Cambridge, Mexico, rural England, and Athens, respectively. They also undertook the first layer of coding analysis in each of the aforementioned cases under the supervision of K.F. In addition, they reviewed this paper and made comments where necessary. All authors have read and agreed to the published version of the manuscript.

Funding: This research has been supported by various funders. The fieldwork in Walthamstow (London) was supported by a Beacon Bursary awarded by the UCL Public Engagement. The fieldwork in Mexico has received funding from SENER-CONACyT Mexico and the EPSRC Doctoral Training Centre in Science and Engineering in Arts, Heritage and Archaeology (EP/L016036/1).

Acknowledgments: The authors would like to thank UCL Culture for supporting with a Beacon Bursary the research in Walthamstow (London), as well as SENER-CONACyT Mexico and EPSRC Doctoral Training Centre in Science and Engineering in Arts, Heritage and Archaeology (SEAHA), for supporting the research fieldwork in Mexico. More importantly, the authors would like to thank the anonymous participants in the study who volunteered their time in explaining in detail to us how they make decisions on energy efficiency and thermal comfort.

Conflicts of Interest: The authors declare no conflict of interest.

References

1. United Nations. Goal 13: Take Urgent Action to Combat Climate Change and Its Impacts. Available online: https://www.un.org/sustainabledevelopment/climate-change/ (accessed on 23 April 2020).
2. Rodriguez-Maribona, I.; Gunnar, G. Energy Efficiency in European Historic Urban Districts—A Practical Guidance. 2016. Available online: http://www.buildup.eu/sites/default/files/link-files/effesus_booklet_final-version.pdf (accessed on 23 April 2020).
3. United Nations Department of Economic and Social Affair. World Urbanization Prospects, 2014 Revision. Available online: https://www.un.org/en/development/desa/publications/2014-revision-world-urbanization-prospects.html (accessed on 23 April 2020).
4. Dol, K.; Haffner, M. *Housing Statistics in the European Union 2010*; Ministry of the Interior and Kingdom Relations: The Hague, The Netherlands, 2010. Available online: https://www.researchgate.net/publication/334030779_housing-statistics-in-the-european-union-2010 (accessed on 24 April 2020).

5.	Eriksson, P.; Hermann, C.; Hrabovszky-Horváth, S.; Rodwell, D. EFFESUS Methodology for Assessing the Impacts of Energy-Related Retrofit Measures on Heritage Significance. *Hist. Environ.* **2014**, *5*, 132–149. [CrossRef]

6.	Fouseki, K.; Cassar, M. Energy Efficiency in Heritage Buildings-Future Challenges and Research Needs. *Hist. Environ.* **2014**, *5*, 95–100. [CrossRef]

7.	Historic England. Energy Efficiency and Historic Buildings: How to Improve Energy Efficiency. 2018. Available online: https://historicengland.org.uk/images-books/publications/eehb-how-to-improve-energy-efficiency/heag094-how-to-improve-energy-efficiency/ (accessed on 23 April 2020).

8.	TC346. EN-16883. *Conservation of Cultural Heritage-Guidelines for Improving the Energy Performance of Historic Buildings*; Comité Europeen de Normalisation: Brussels, Belgium, 2017.

9.	Cornaro, C.; Puggioni, V.A.; Strollo, R.M. Dynamic simulation and on-site measurements for energy retrofit of complex historic buildings: Villa Mondragone case study. *J. Build. Eng.* **2016**, *6*, 17–28. [CrossRef]

10.	Rohdin, P.; Milic, V.; Wahlqvist, M.; Moshfegh, B. On the use of change-point models to describe the energy performance of historic buildings. In Proceedings of the 3rd International Conference on Energy Efficiency in Historic Buildings (EEHB2018), Visby, Sweden, 26–27 September 2018; pp. 182–190.

11.	Castele, D.S.; Webb, A.L. Insulating the Walls of Historic Buildings. *Apt Bull. J. Preserv. Technol.* **2019**, *50*, 37–44.

12.	Berg, F.; Flyen, A.C.; Godbolt, Å.L.; Broström, T. User-driven energy efficiency in historic buildings: A review. *J. Cult. Herit.* **2017**, *28*, 188–195. [CrossRef]

13.	Fouseki, K.; Bobrova, Y. Understanding the change of heritage values over time and its impact on energy efficiency. In Proceedings of the 3rd International Conference on Energy Efficiency in Historic Buildings (EEHB2018), Visby, Sweden, 26–27 September 2018; pp. 11–21.

14.	Koukou, T.; Fouseki, K. Heritage values and thermal comfort in Neoclassical residential buildings of Athens, Greece: Tension or co-existence? In Proceedings of the 3rd International Conference on Energy Efficiency in Historic Buildings (EEHB2018), Visby, Sweden, 26–27 September 2018; pp. 463–471.

15.	Newton, D.; Fouseki, K. Heritage values as a driver or obstacle for energy efficiency in Victorian and Edwardian buildings. In Proceedings of the 3rd International Conference on Energy Efficiency in Historic Buildings (EEHB2018), Visby, Sweden, 26–27 September 2018; pp. 530–538.

16.	Adams, C.; Douglas-Jones, R.; Green, A.; Lewis, Q.; Yarrow, T. Building with History: Exploring the relationship between heritage and energy in institutionally managed buildings. *Hist. Environ. Policy Pract.* **2014**, *5*, 167–181. [CrossRef]

17.	Yarrow, T. Negotiating heritage and energy conservation: An ethnography of domestic renovation. *Hist. Environ. Policy Pract.* **2016**, *7*, 340–351. [CrossRef]

18.	Shove, E.; Pantzar, M.; Watson, M. *The Dynamics of Social Practice: Everyday Life and How It Changes*; Sage: Thousand Oaks, CA, USA, 2012; ISBN 978-0857020437.

19.	Beauchemin, C.; González-Ferrer, A. Sampling international migrants with origin-based snowballing method: New evidence on biases and limitations. *Demogr. Res.* **2011**, *25*, 103–134. [CrossRef]

20.	Banks, M.; Zeitlyn, D. *Visual Methods in Social Research*; Sage: Thousand Oaks, CA, USA, 2015; ISBN 978-1446269756.

21.	Glaser, B.G.; Strauss, A.L. *The Discovery of Grounded Theory*; Aldine: Chicago, IL, USA, 1967; ISBN 0-202-30260-1.

22.	Corbin, J.; Strauss, A. *Basics of Qualitative Research: Techniques and Procedures for Developing Grounded Theory*; Sage: Thousand Oaks, CA, USA, 2014; ISBN 9781412997461.

23.	Eker, S.; Zimmermann, N. Using textual date in System Dynamics Model Conceptualization. *Systems* **2016**, *4*, 28. [CrossRef]

24.	Sterman, J. *Business Dynamics: Systems Thinking and Modelling for a Complex World*; McGraw-Hill Education: New York, NY, USA, 2000; ISBN 978-0071179898.

25.	Forrester, J.W. Lessons from system dynamics modelling. *Syst. Dyn. Rev.* **1987**, *3*, 136–149. [CrossRef]

26.	Randers, J. *Elements of the System Dynamics Method*; Wright Allen Press: Cambridge, MA, USA, 1980; ISBN 978-0915299393.

27.	Holmes, M.J. Material substitution in the UK window industry, 1983–1987. *Resour. Policy* **1990**, *16*, 128–142. [CrossRef]

28. Yarrow, T. How conservation matters: Ethnographic explorations of historic building renovation. *J. Mater. Cult.* **2019**, *24*, 3–21. [CrossRef]
29. Smith, L. *Uses of Heritage*; Routledge: London, UK, 2006; ISBN 978-0415318310.

atmosphere

Article

Risk Mapping for the Sustainable Protection of Cultural Heritage in Extreme Changing Environments

Alessandro Sardella [1], Elisa Palazzi [2], Jost von Hardenberg [2,3], Carlo Del Grande [4], Paola De Nuntiis [1], Cristina Sabbioni [1] and Alessandra Bonazza [1,*]

[1] Institute of Atmospheric Sciences and Climate, National Research Council of Italy (ISAC-CNR), Via Gobetti 101, 40129 Bologna, Italy; a.sardella@isac.cnr.it (A.S.); p.denuntiis@isac.cnr.it (P.D.N.); c.sabbioni@isac.cnr.it (C.S.)

[2] Institute of Atmospheric Sciences and Climate, National Research Council of Italy (ISAC-CNR), Corso Fiume, 4 10133 Torino, Italy; e.palazzi@isac.cnr.it (E.P.); jost.hardenberg@polito.it (J.v.H.)

[3] DIATI, Politecnico di Torino, Corso Duca degli Abruzzi 24, 10129 Torino, Italy

[4] Studio Associato Ambiente Terra, Via Monte Calderaro 2700B, Castel San Pietro Terme, 40024 Bologna, Italy; carlo.delgrande@ambienteterra.it

*   Correspondence: a.bonazza@isac.cnr.it; Tel.: +39-05-1639-9576

Received: 30 May 2020; Accepted: 26 June 2020; Published: 1 July 2020

Abstract: Cultural heritage is widely recognized to be at risk due to the impact of climate change and associated hazards, such as events of heavy rain, flooding, and drought. User-driven solutions are urgently required for sustainable management and protection of monumental complexes and related collections exposed to changes of extreme climate. With this purpose, maps of risk-prone areas in Europe and in the Mediterranean Basin have been produced by an accurate selection and analysis of climate variables (daily minimum and maximum temperature—Tn and Tx, daily cumulated precipitation—RR) and climate-extreme indices (R20mm, R95pTOT, Rx5 day, CCD, Tx90p) defined by Expert Team on Climate Change Detection Indices (ETCCDI). Maps are available to users via an interactive Web GIS (Geographic Information System) tool, which provides evaluations based on historical observations (high-resolution gridded data set of daily climate over Europe—E-OBS, 25 km) and climate projections (regional climate models—RCM, ~12 km) for the near and far future, under Representative Concentration Pathways (RCP) 4.5 and 8.5 scenarios. The tool aims to support public authorities and private organizations in the decision making process to safeguard at-risk cultural heritage. In this paper, maps of risk-prone areas of heavy rain in Central Europe (by using R20mm index) are presented and discussed as example of the outputs achievable by using the Web GIS tool. The results show that major future variations are always foreseen for the 30-year period 2071–2100 under the pessimistic scenario (RCP 8.5). In general, the coastal area of the Adriatic Sea, the Northern Italy, and the Alps are foreseen to experience the highest variations in Central Europe.

Keywords: extreme events; climate projection; Central Europe; ProteCHt2save; climate risk indices; heritage climatology; cultural heritage safeguarding; preparedness

---

1. Introduction

It is widely recognized that climate change is creating continuous and new challenges for the protection and conservation of cultural heritage. Monumental complexes, archaeological sites, and historic buildings with related collections are at risk as a consequence of the impacts of slow and extreme climate changes, particularly in urban areas, where the effect of multiple pressures is amplified. Research on the quantification of climate change impacts on heritage assets and on the development of future scenarios for setting up protection strategies with a long-term perspective has undoubtedly strengthened during the last 15 years, focusing in particular on slow damage processes, such as surface recession and thermal stress on marble, biological accumulation on architectural surfaces, and metal corrosion [1–6]. In this respect, the EU FP6 Project Noah's Ark (2004–2007) produced a vulnerability atlas with maps by applying the global and regional Hadley climate models (grid resolutions of 295 × 278 km and 50 × 50 km, respectively), along with guidelines for cultural heritage protection towards climate change, coupling climatology with conservation science expertise and acquiring unique knowledge in delivering future projections of damage of outdoor cultural heritage induced mainly by slow climate changes [1]. The scientific approach developed within Project Noah's Ark constituted the basis for the research enhancement carried out in the FP7 Project Climate for Culture (2009–2014) [7,8]. Within this project, hazard and damage projections were forecasted to assess the impact of the slow ongoing climate change rather than extreme events on historic building envelopes, as well as on artwork preserved indoors. In addition, projections of sea level rise—a potential threat to many coastal regions and to their cultural heritage—up to the year 2100 were calculated using a simulation with a global climate model and data from a regionally coupled atmosphere–ocean model regional model (REMO) run on the horizontal grid of 12.5 km, EUR-11) [9,10]. Recent H2020-funded projects (HEritage Resilience Against CLimate Change on Site—HERACLES, Safeguarding cultural heritage through Technical and Organisational Resources Management—STORM) have focused on the development of information and communications technology (ICT) systems and solutions to strengthen the resilience of cultural heritage against climate change effects and natural hazards, and have started to focus on the impacts of extreme events [11].

The research done within these projects and related publications allowed us to highlight the existing gaps in the knowledge that need to be overcome in this sector, suggesting that solutions should be based on a user-driven approach by meeting the requirements and needs of the different targets of stakeholders involved in the protection and management of cultural heritage at risk from climate change. Among these gaps, the need for further development of damage functions and for identification of extreme climate indices specifically devised to quantify the impact on heritage assets, as well as the generation of future projections with high spatial resolution, should be considered of paramount importance. In addition, it is clear that there is an urgent requirement to invest resources and efforts in the production of tools and solutions to enhance the preparedness of cultural heritage to face events linked to hydrometeorological and climatic extremes (such as storms, floods, drought periods, and heat waves) [12].

Within this framework, the ongoing Interreg Central Europe ProteCHt2save Project (Risk assessment and sustainable protection of cultural heritage in changing environment, 2017–2020) aims to improve the capacities of the public and private sectors to mitigate the impacts of climate change and natural hazards on cultural heritage sites, including monumental complexes, historic buildings, and related collections in urban and coastal areas in Central European countries. The project focuses primarily on the development of feasible and tailored solutions to build cultural heritage resilience to extreme events linked to climate change by supporting regional and local authorities with preparedness measures and evacuation plans for emergencies.

This overall objective is achieved by performing the following activities:

(1) Identification of risk-prone areas in Central Europe where cultural heritage is exposed to extreme weather and climate events (heavy rain, flood, drought);
(2) Determination of elements for the vulnerability assessment of cultural heritage, specifically monumental complexes and related collections in historic centers;
(3) Set up of evacuation plans and preparedness measures for cultural heritage safeguarding.

One of the major outputs is a Web GIS risk mapping tool used for the identification of risk-prone areas and vulnerable cultural heritage areas exposed to extreme events linked to climate change, particularly heavy rains, flood, and fire due to drought periods.

In ProteCHt2save, seven pilot heritage sites were selected to test the measures and strategies for protection of at-risk cultural heritage developed in the project. The pilot actions carried out were linked to climate change and the associated variability was linked to hydro-meteorological and climate extremes:

(1)   Flood events in large basins (sites in the Czech Republic and Austria);
(2)   Fire due to drought periods (site in Croatia);
(3)   Extreme events of heavy rain (sites in Italy, Hungary, Croatia, Poland, and Slovenia).

The first pilot action targeted the testing of preparedness strategies for monumental complexes in historic city centers affected by flooding and heavy rain. The second stage of pilot actions tested evacuation plans as measures of emergency phases in museums in historic buildings facing sea flooding, fire due to drought, and heavy rain.

In this contribution, we illustrate the methodological approach followed for the map production integrated in the Web GIS tool, with particular reference to the climate modelling framework used and the climate extreme indices selected.

In addition, potential areas of increased risk of heavy rain regarding cultural heritage in Central Europe in the near and far future under two emission scenarios are discussed as examples of the potential use and outputs of the ProteCHt2save Web GIS tool.

It should be highlighted that while within the ProteCHt2save Project the major area of interest is Central Europe, this tool has been developed to allow end-users to assess the changes of climate extremes and their impacts on cultural heritage at the European and Mediterranean levels.

The provided maps turned out to be significant tools in support of policy and decision makers for the development of measures and strategies of preparedness, with short- and long-term perspectives aiming to protect cultural heritage, with possible applications to landscape protection, urban territorial planning, and emergency management in extreme changing environments.

2. Methodology for Heritage Climatology Mapping

2.1. Index Selection for Extreme Event Analysis

As a first step, we analyzed the changes in climate extremes, such as dry spells or intense precipitation, using indices to evaluate statistics of extreme events for temperature and precipitation and to compare them with observed extremes. In particular, we used standard indices defined by the Commission for Climatology/World Climate Research Programme/Technical Commission for Oceanography and Marine Meteorology (CCI/WCRP/JCOMM) Expert Team on Climate Change Detection Indices (ETCCDI), whose definition can be found on the Climdex Project web site [13]. The index computation was performed in R using the standard climdex.pcic.ncdf library (https://github. com/pacificclimate/climdex.pcic.ncdf), which includes a bootstrap procedure for the percentile-based indices according to Zhang et al. [14].

For the mapping of heritage climatology in ProteCHt2save, we selected 5 extreme climate indices among the 27 standardized indices mentioned above. As shown in Table 1, they are related to the following extreme events: heavy rain, flooding, drought, and extreme heating. These indices were selected to evaluate statistics of extreme events for temperature and precipitation and to compare them with observed extremes. In addition to the indices, daily precipitation and maximum and minimum temperature were also taken into consideration (Table 2).

Table 1. Relevant extreme climate indices selected for heavy rain, flooding, drought, and extreme heating.

Index	Definition End Description	Related Extreme Event	Unit
R20mm	Very heavy precipitation days Number of days in a year with precipitation greater than or equal to 20 mm/day.	Heavy rain	days

Table 1. *Cont.*

Index	Definition End Description	Related Extreme Event	Unit
R95pTOT	Precipitation due to extremely wet days The total precipitation in a year cumulated over all days when daily precipitation is larger than the 95th percentile of daily precipitation on wet days. A wet day is defined as having daily precipitation ≥ 1 mm/day. A threshold based on the 95th percentile selects only 5% of the most extreme wet days over a 30-year-long reference period.	Heavy rain	mm
Rx5day	Highest 5-day precipitation amount Yearly maximum of cumulated precipitation over consecutive 5-day periods.	Flooding	mm
CDD	Maximum number of consecutive dry days Maximum length of a dry spell in a year, which is the maximum number in a year of consecutive dry days with daily precipitation smaller than 1 mm/day.	Drought	days
Tx90p	Percentage of extremely warm days Percentage of days in a year when daily maximum temperature is greater than the 90th percentile. A threshold based on the 90th percentile selects only 10% of the warmest days over a 30-year-long reference period.	Extreme heating	days

Table 2. Selected climate variables.

Code	Climate Variables	Description	Unit
Tn	Tmin	daily minimum temperature	°C
Tx	Tmax	daily maximum temperature	°C
RR	Precipitation	daily cumulated precipitation	mm

2.2. Climate Modelling

In this study, numerical climate model simulations were analyzed to study the possible future evolution of the climate system. In particular, an ensemble of global climate models (GCMs) driving an ensemble of regional climate models (RCMs) was used to provide regional projections for the European continent. Multi-model ensembles of regional climate projections were based on the WCRP Coordinated Regional Downscaling Experiment (CORDEX), considering the EURO-CORDEX initiative in particular, which provides regional climate projections for Europe at two different spatial resolutions, namely the "standard" resolution of 0.44 degrees (EUR-44, ~50 km) and a finer resolution of 0.11 degrees (EUR-11, ~12 km). Within the EURO-CORDEX experiment, seven RCMs were employed to dynamically downscale the Climate Model Intercomparison Project phase 5 (CMIP5) GCM projections using the CMIP5 Representative Concentration Pathways (RCPs) emission scenarios. When RCMs are driven by a large-scale global model, in addition to the uncertainties inherent in the specific RCM at hand, additional uncertainty is inherited from the driving GCM. In order to estimate this type of uncertainty, a common approach is to consider an ensemble of simulations performed with a given RCM driven by different GCMs. The spread among the RCM outputs provides an estimate of the effects of GCM diversity on the RCM simulations.

Within ProteCHt2save, the Euro-CORDEX simulations at 0.11° resolution were selected among those available [15].

In this study, 12 different combinations of 6 global models driving 5 regional models were taken into account to elaborate the maps related to the future projections (see Table 3).

Table 3. Combinations of numerical models applied in ProteCHt2save.

GCM	RCM	Institute
CNRM-CM5	CCLM4-8-17	CLM Community with contributions by BTU, DWD, ETHZ, UCD, WEGC (CLMcom)
CNRM-CM5	RCA4	Rossby Center, Swedish Meteorological and Hydrological Institute, Norrkoping Sweden (SMHI)
EC-EARTH	CCLM4-8-17	CLM Community with contributions by BTU, DWD, ETHZ, UCD, WEGC (CLMcom)
EC-EARTH	HIRHAM5	Danish Meteorological Institute, Copenhagen, Denmark (DMI)
EC-EARTH	RACMO22E	Royal Netherlands Meteorological Institute, Ministry of Infrastructure and the Environment (KNMI)
EC-EARTH	RCA4	Rossby Center, Swedish Meteorological and Hydrological Institute, Norrkoping Sweden (SMHI)
HadGEM2-ES	RACMO22E	Royal Netherlands Meteorological Institute, Ministry of Infrastructure and the Environment (KNMI)
HadGEM2-ES	RCA4	Rossby Center, Swedish Meteorological and Hydrological Institute, Norrkoping Sweden (SMHI)
CM5A-MR	RCA4	Rossby Center, Swedish Meteorological and Hydrological Institute, Norrkoping Sweden (SMHI)
MPI-ESM-LR	CCLM4-8-17	CLM Community with contributions by BTU, DWD, ETHZ, UCD, WEGC (CLMcom)
MPI-ESM-LR	REMO2009	Climate Service Center (CSC), Hamburg, Germany (MPI-CSC)
NorESM1-M	HIRHAM5	Danish Meteorological Institute, Copenhagen, Denmark (DMI)

Two future emission scenarios, described in detail in the latest Intergovernmental Panel on Climate Change (IPCC) assessment report (AR5) [16], were chosen:

- RCP 4.5 is a stabilization scenario in which anthropogenic radiative forcing is stabilized at 4.5 W/m^2 after year 2100, without overshooting the long-run radiative forcing target level [17];
- RCP 8.5 is a high pathway scenario characterized by increasing greenhouse gas emissions over time, for which anthropogenic radiative forcing reaches 8.5 W/m^2 at year 2100 and continues to rise for some time. This is also known as the "business as usual" scenario [18].

2.3. Future Projections and Historical Changes

RCM historical and projection simulations were analyzed to calculate anomalies; that is, changes of future climatologies with respect to past conditions. The historical model period taken into account was 1976–2005. Long-term climatologies around the mid-21st century (e.g., 2021–2050) and end of century (e.g., 2071–2100) were considered. In addition, historical observations for the 30-year-periods of 1987–2016 and 1951–1980 were analyzed using E-OBS, a state-of-the-art observational dataset based on the interpolation of in situ station data available for the European domain, which is a robust and widely used dataset that is regularly updated, with a spatial resolution of 25 × 25 km [19]. E-OBS provides long-term daily precipitation and near-surface air temperature climatology data (from 1950 to present); its spatial coverage includes all land areas in Europe and in the Mediterranean region. It is supported by a clear documentation of the methods used to derive it (interpolation techniques, underlying stations, etc.), and the underlying orography (elevation data) and individual station data are available as well.

3. Results and Discussion

Hazard maps referring to heavy rain, flooding, drought, and extreme heat were elaborated covering the European and Mediterranean areas on the basis of data from the selected combination of models, specifically:

- maps of past and future changes in precipitation and (minimum and maximum) temperature;
- maps of past and future changes related to climate extreme indices.

For each of the 5 climate extreme indices and for the 3 climate variables, the past changes were calculated as the difference between the period 1987–2016 and the period 1951–1980, using E-OBS (spatial resolution 25 × 25 km), while future changes were calculated as the difference between the period 2021–2050 and the period 1976–2005 (near future projection) and as the difference between the period 2071–2100 and the period 1976–2005 (far future projection), under both RCP 4.5 and 8.5 scenarios (spatial resolution 12 × 12 km).

A total of 8 maps with historical observations were elaborated, while 4 maps were produced for each combination of models, leading to 36 maps for each of the climate extreme indices and variables considered. As final result, a total of 384 maps related to future climate simulations were produced by individual model projections.

Being aware that each individual GCM/RCM model has its own uncertainties, we kept the entire ensemble and considered all members and their statistics, in particular calculating the minimum, mean, and maximum values of the model ensemble, creating a further 96 maps (Figure 1) [20,21].

This process was also performed in order to provide a useful tool for non-specialized users in climate modelling and to meet the requirements of public authorities, territorial agencies, and policy and decision makers involved in the management of at-risk cultural heritage and urban territorial planning.

All maps reported the position of the ProteCHt2save pilot sites: Troja (Czech Republic), Krems (Austria), Bielsko (Poland), Pecs (Hungary), Kocevje (Slovenia), Ferrara (Italy), and Kastela (Croatia). The maps could include further heritage sites located in Europe and in the Mediterranean Basin, and can be interactively visualized and downloaded in the ProteCHt2save Web GIS tool for the risk mapping described in Section 3.1.

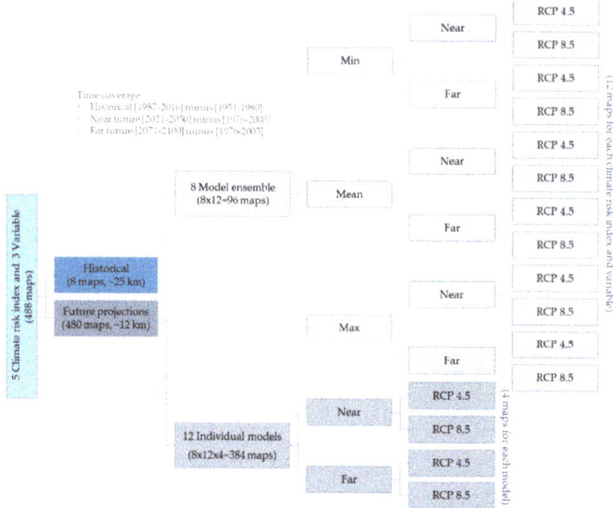

Figure 1. Diagram showing the structure of the climate maps creation.

In this paper, the maps realized at the Central European scale for heavy rain using the R20mm climate extreme index are illustrated and discussed as an example of the products available within the Web GIS tool and its usefulness and potential applicability for at-risk cultural heritage management.

Figure 2 shows the changes of the R20mm index in the past between the 30 year periods 1987–2016 and 1951–1980 by using the E-OBS observational dataset. The map shows that the major changes occur on the Croatian coast of the Adriatic Sea; in the mountain areas at the borders between Italy with Austria and France, with the maximum increase observed >10 days (red); and in Northern Italy, reaching values no greater than 10 days (blue).

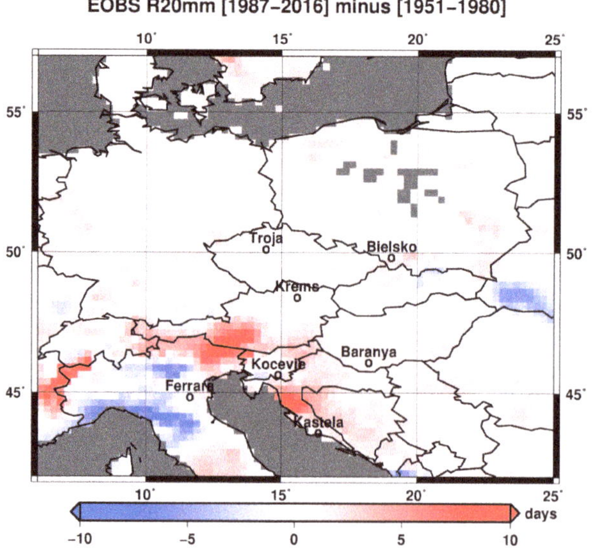

Figure 2. Map related to historical change between the two 30-year periods (1987–2016 and 1951–1980), considering the R20mm climate risk index (dataset E-OBS, spatial resolution 25 × 25 km).

A complete set of maps derived from the model ensemble statistics (minimum, mean, and maximum of the ensemble in the left, middle, and right columns, respectively) for the future changes in the R20mm climate extreme index is shown in Figure 3. Firstly, the analysis highlights that the model spread is huge, with both negative (as shown by the ensemble minimum output) and positive (as shown by the ensemble maximum output) projected changes. In general, the coastal area of the Adriatic Sea, Northern Italy, and the Alps will continue to experience the highest variations. It should also be noted that major variations are always foreseen in the far future (2071–2100) under the pessimistic scenario (RCP 8.5). Observing the ensemble mean projection for the far future in the pessimistic scenario, a general increase is foreseen for almost all of Central Europe, with the exception of the East Coast of the Adriatic Sea and Northern Italy. It should be pointed out that for an exhaustive assessment of the potential threats, an evaluation of the results from the multi-model ensemble statistics taking into consideration the mean value and its spread should be performed, and in the case of significant differences among the individual models, a deeper analysis that considers all single models beyond the ensemble statistics is recommended, in addition to an evaluation of the data and their variations in the specific area of interest [21].

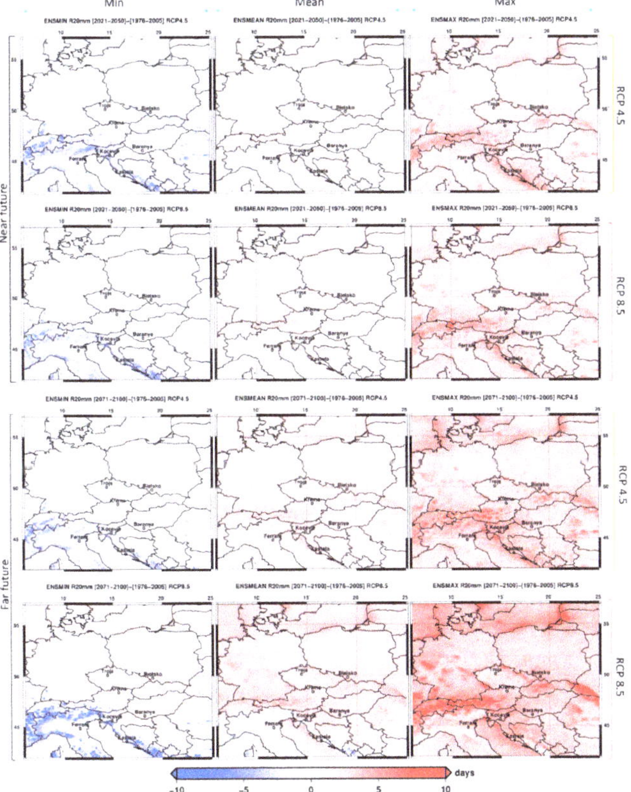

Figure 3. This figure shows a complete set of climatic projection simulations related to the model ensemble statistics of the R20mm climate extreme index. Minimum, mean, and maximum variations under RCP 4.5 and 8.5 for the near future (2021–2050) and far future (2070–2100) are reported (spatial resolution = 12 × 12 km).

With respect to the previous applications, the use of 12 combinations of GCM/RCM models achieving a resolution of 0.11° (EUR-11, ~12 km) undoubtedly constitutes an advancement in the research on the evaluation of climate change impacts on cultural heritage.

3.1. Significance of Changes for Cultural Heritage

Research conducted up to now on the climate change impact on cultural heritage has clearly demonstrated that water-derived parameters are driving factors, being involved in several deterioration processes, such as structural damage and erosion in the case of extreme rain events; surface recession on carbonate stones due to chemical dissolution; decohesion and fracturing in the case of salt crystallization; and freeze–thaw cycling and biological accumulation [22,23]. Events of very heavy precipitations days (here studied using the R20mm ETCCDI index) are widely recognized to constitute a major threat to building structures, particularly vernacular architecture, and to be the cause of pluvial flooding [1]. The type and magnitude of damage created on a building by climate change depends on the building material and its state of conservation, on the surrounding urban or rural landscape, and on the geometry of the building, which is essential in determining the degree of exposure of architectural surfaces to (and shelter from) climate parameters, such as rain, wind, and solar radiation [24,25]. Building materials are differentially susceptible to damage processes imposed by climate change, which include stone, mortar, and brick. Chemical composition and porosity, with the associated texture and surface roughness, are recognized as crucial geological controls on weathering [26]. In the specific case of events of heavy rain, erosion and loss of materials are major effects for stones, while swelling and shrinkage due to moisture gradients can easily occur on wood, with associated biodeterioration [27]. Physical features of the heritage site under pressure are of paramount importance in determining the vulnerability to climate change; therefore, an exhaustive assessment of the risk cannot disregard an analysis of the existing criticalities at the building level, where managerial issues also have to be taken into consideration.

The work done in this perspective for the ProteCHt2save pilot sites is available in the Web GIS tool, where physical and managerial criticalities are highlighted for each case study. The setup of the methodology for vulnerability ranking of different categories of cultural heritage, such as monumental complexes, archaeological, and natural sites, was one of the major activities in ProteCHt2save and is the focus of the recently funded Interreg Central Europe Project STRENCH (STRENgthening resilience of Cultural Heritage at risk in a changing environment through proactive transnational cooperation) This project will capitalize on and further implement the Web GIS tool developed in ProteCH2save, in addition to the outputs produced in H2020 HERACLES and in the Interreg Central Projects BhENEFIT (Built heritage, Energy and Environmental Friendly Integrated Tools), RUINS (Sustainable reuse, preservation and modern management of historical ruins), and HiCAPS (Historical Castle ParkS). This approach will allow us to identify risk-prone areas of cultural heritage exposed to extreme climate events, as a function of exposure, vulnerability, and hazard for different heritage categories.

3.2. Web GIS Tool

A Web GIS Tool for risk mapping was designed and implemented in order to:

(1) create an online platform that visualizes in an interactive way all the climate risk maps produced, as described in the previous section;
(2) support policy and decision makers in the identification of risk areas and vulnerabilities for cultural heritage in Europe and in the Mediterranean Basin, exposed to extreme events linked to climate change.

The GIS platform was designed in order to provide user-friendly graphical interfaces published on a website (Web GIS) to meet and satisfy the needs of a large number of users.

The need to visualize and obtain geocoded cartographic data online led us to create a tool that can publish and make information available on the web. Therefore, a Web GIS tool was designed and open source applications and cartographic bases were chosen for its implementation.

In particular, the system architecture comprises:

- Leaflet (https://leafletjs.com/index.html);
- OpenStreetMap (https://www.openstreetmap.org/);
- Mapbox (free up to 25,000 monthly active users, https://www.mapbox.com/).

The website of the "ProteCHt2save Web GIS Tool for Risk Mapping" [28], which we developed, is composed of 6 pages, including explanations on the utilized climate models, climate extreme indices, and variables; and illustrations cards for the pilot sites with major information on the features of the monuments and historic buildings taken into consideration, their state of conservation, past disasters that occurred in the area, and measures undertaken for protection (Figure 4). A special focus on the methodology followed to determine the criticalities and rank the vulnerability of each site is also given and a section is completely dedicated to the map creation. Maps related to historical observations and climate projections can be produced by selecting all combinations of individual models and ensemble statistics described in the present contribution. The location of the ProteCHt2save pilot sites can be visualized in addition to the distribution of the UNESCO World Heritage Sites in Central Europe, thanks to the collaboration with the Project Joint Programming Initiatives on Cultural Heritage—Protection of European Cultural Heritage from Geo-hazards (JPI-CH-PROTHEGO) (2015–2018). The tool is accessible at https://www.protecht2save-wgt.eu/.

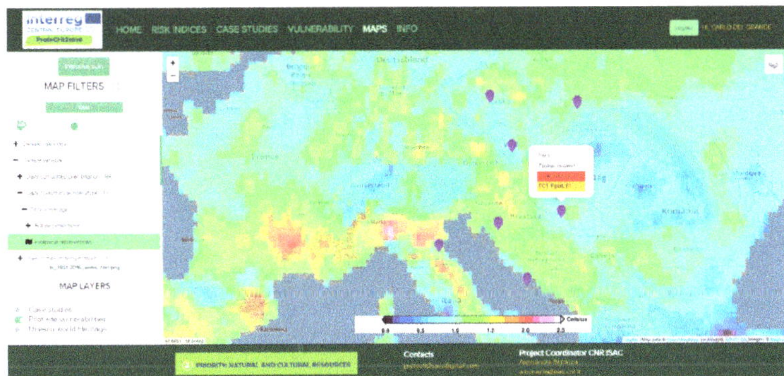

Figure 4. Screenshot from the Web GIS tool website showing a hazard map obtained with the historical observations of daily maximum temperature. The vulnerability rate for the case study in Hungary (Pecs) is also illustrated.

4. Conclusions

In this contribution, a methodological approach for the assessment of the potential impacts of climate extreme events (heavy rain, flooding, and drought) on cultural heritage is illustrated. Maps of risk-prone areas in Europe and in the Mediterranean Basin have been elaborated by an accurate selection and analysis of climate variables (Tn, Tx, RR) and climate extreme (standard) indices (R20mm, R95pTOT, Rx5day, CCD, Tx90p).

Maps are available to users via an interactive Web GIS tool, which provides evaluations based on historical observations (E-OBS) and climate projections (near and far future); it is possible to select a series of combinations of individual models and ensembles statistics with a spatial resolution of 12 × 12 km, under RCP 4.5 and RCP 8.5 scenarios.

This tool was conceived and designed in order to contribute to still-existing gaps in knowledge on climate change impacts on cultural heritage by increasing the spatial resolution of the projections as much as possible and by providing insights into the formulation of risk expression by considering the exposure determined by the areal distribution of heritage assets and by laying the foundations for the development of a methodology of vulnerability ranking. A user-driven approach is at the base of all the procedures and the system is open to further implementation and data integration in order to support policy and decision makers in the management of at-risk cultural heritage, in addition to further applications, such as urban territorial planning and landscape protection.

Author Contributions: Conceptualization, A.B. and A.S.; methodology, A.S., E.P., J.v.H., and C.D.G.; software, C.D.G.; formal analysis, A.S., E.P., and J.v.H.; investigation, A.S., C.D.G., P.D.N., and A.B.; resources, A.B., E.P., and J.v.H.; data curation, A.S., E.P., and J.v.H.; writing—original draft preparation, A.S. and A.B.; writing—review and editing, A.B., A.S., E.P., J.v.H., C.D.G., P.D.N., and C.S.; visualization, A.S. and P.D.N.; supervision, A.B.; project administration, A.B.; funding acquisition, A.B. All authors have read and agreed to the published version of the manuscript.

Funding: This research was funded by Interreg Central Europe Programme, grant number CE1127. Within the "Risk Assessment and Sustainable Protection of Cultural Heritage in Changing Environments, ProteCHt2save" project.

Acknowledgments: Authors wish to thank the team members from Partners Institutions of ProteCHt2save Consortium for their fruitful collaboration and support in the project implementation activities.

Conflicts of Interest: The authors declare no conflict of interest.

References

1. Sabbioni, C.; Brimblecombe, P.; Cassar, M. *Atlas of Climate Change Impact on European Cultural Heritage*; Anthem Press: London, UK, 2010; p. 160. ISBN 978927909800-0.

2. Bonazza, A.; Messina, P.; Sabbioni, C.; Grossi, C.M.; Brimblecombe, P. Mapping the impact of climate change on surface recession of carbonate buildings in Europe. *Sci. Total Environ.* **2009**, *407*, 2039–2050. [CrossRef] [PubMed]

3. Bonazza, A.; Sabbioni, C.; Messina, P.; Guaraldi, C.; De Nuntiis, P. Climate change impact: Mapping thermal stress on Carrara marble in Europe. *Sci. Total Environ.* **2009**, *407*, 4506–4512. [CrossRef]

4. Gomez-Bolea, A.; Llop, E.; Arino, X.; Saiz-Jimenez, C.; Bonazza, A.; Messina, P.; Sabbioni, C. Mapping the impact of climate change on biomass accumulation on stone. *J. Cult. Herit.* **2012**, *13*, 254–258. [CrossRef]

5. Ciantelli, C.; Palazzi, E.; Von Hardenberg, J.; Vaccaro, C.; Tittarelli, F.; Bonazza, A. How can climate change affect the UNESCO cultural heritage sites in Panama? *Geosciences* **2018**, *8*, 296. [CrossRef]

6. Sesana, E.; Gagno, A.S.; Bonazza, A.; Hughes, J.J. An integrated approach for assessing the vulnerability of World Heritage Sites to climate change impacts. *J. Cult. Herit.* **2020**, *41*, 221–224. [CrossRef]

7. Hujibregts, Z.; Kramer, R.P.; Martens, M.H.J.; van Schijndel, A.W.M.; Schellen, H.L. A proposed method to assess the damage risk of future climate change to museum objects in historic buildings. *Build. Environ.* **2012**, *55*, 43–56. [CrossRef]

8. Kramer, R.; van Schijndel, J.; Schellen, H. Inverse modeling of simplified hygrothermal building models to predict and characterize indoor climates. *Build. Environ.* **2013**, *68*, 87–99. [CrossRef]

9. Leissner, J.; Kilian, R.; Kotova, L.; Jacob, D.; Mikolajewicz, U.; Broström, T.; Ashley-Smith, J.; Schellen, H.; Martens, M.; van Schijndel, J.; et al. Climate for Culture: Assessing the impact of climate change on the future indoor climate in historic buildings using simulations. *Herit. Sci.* **2015**, *3*, 38–52. [CrossRef]

10. Jacob, D.; Elizalde, A.; Haensler, A.; Hagemann, S.; Kumar, P.; Podzun, R.; Rechid, D.; Remedio, A.R.; Saeed, F.; Sieck, K.; et al. Assessing the transferability of the regional climate model REMO to different coordinated regional climate downscaling experiment (CORDEX) regions. *Atmosphere* **2012**, *3*, 181–199. [CrossRef]

11. Padeletti, G.; HERACLES Consortium Staff. Heritage Resilience Against Climate Events on Site—HERACLES Project: Mission and Vision. In *Transdisciplinary Multispectral Modeling and Cooperation for the Preservation of Cultural Heritage. TMM_CH 2018. Communications in Computer and Information Science*; Moropoulou, A., Korres, M., Georgopoulos, A., Spyrakos, C., Mouzakis, C., Eds.; Springer: Cham, Switzerland, 2020; Volume 961, pp. 360–375, ISBN 978-3-030-12956-9. [CrossRef]

12. Bonazza, A.; Maxwell, I.; Drdácký, M.; Vintzileou, E.; Hanus, C.; Ciantelli, C.; De Nuntiis, P.; Oikonomopoulou, E.; Nikolopoulou, V.; Pospíšil, S.; et al. *Safeguarding Cultural Heritage from Natural and Man-Made Disasters a Comparative Analysis of Risk Management in the EU*; European Union: Brussels, Belgium, 2018; p. 207, ISBN 978-92-79-73945-3. [CrossRef]

13. Climdex. Available online: www.climdex.org (accessed on 28 May 2020).

14. Zhang, X.; Hegerl, G.; Zwiers, F.W.; Kenyon, J. Avoiding Inhomogeneity in Percentile-Based Indices of Temperature Extremes. *J. Clim.* **2005**, *18*, 1641–1651. [CrossRef]

15. Euro-CORDEX Simulations. Available online: https://euro-cordex.net/imperia/md/content/csc/cordex/20180130-eurocordex-simulations.pdf (accessed on 28 May 2020).

16. IPCC. *Climate Change 2013: The Physical Science Basis. Contribution of Working Group I to the Fifth Assessment Report of the Intergovernmental Panel on Climate Change*; Stocker, T.F., Qin, D., Plattner, G.-K., Tignor, M., Allen, S.K., Boschung, J., Nauels, A., Xia, Y., Bex, V., Midgley, P.M., Eds.; Cambridge University Press: Cambridge, UK; New York, NY, USA, 2013; 1535p. [CrossRef]

17. Thomson, A.M.; Calvin, K.V.; Smith, S.J.; Kyle, G.P.; Volke, A.; Patel, P.; Delgado-Arias, S.; Bond-Lamberty, B.; Marshall, A.W.; Clarke, L.E.; et al. RCP4.5: A pathway for stabilization of radiative forcing by 2100. *Clim. Chang.* **2011**, *109*, 77–94. [CrossRef]

18. Riahi, K.; Gruebler, A.; Nakicenovic, N. Scenarios of long-term socio-economic and environmental development under climate stabilization. *Technol. Forecast. Soc. Chang.* **2007**, *74*, 887–935. [CrossRef]

19. European Climate Assessment & Dataset. Available online: http://www.ecad.eu/download/ensembles/ensembles.php (accessed on 30 May 2020).

20. Rangwala, I.; Palazzi, E.; Miller, J.R. Projected Climate Change in the Himalayas during the Twenty-First Century. In *Himalayan Weather and Climate and Their Impact on the Environment*; Dimri, A., Bookhagen, B., Stoffel, M., Yasunari, T., Eds.; Springer: Cham, Switzerland, 2020; pp. 51–71, ISBN 978-3-030-29683-4. [CrossRef]

21. Palazzi, E.; von Hardenberg, J.; Terzago, S.; Provenzale, A. Precipitation in the Karakoram-Himalaya: A CMIP5 view. *Clim. Dyn.* **2015**, *45*, 21–45. [CrossRef]

22. Grossi, C.M.; Brimblecombe, P.; Harris, I. Predicting long-term freeze–thaw risks on Europe built heritage and archaeological sites in a changing climate. *Sci. Total Environ.* **2007**, *377*, 273–281. [CrossRef] [PubMed]

23. Grossi, C.M.; Brimblecombe, P.; Menendez, B.; Benavente, D.; Harris, I.; Deque, M. Climatology of salt damage on stone buildings. *Sci. Total Environ.* **2011**, *409*, 2577–2585. [CrossRef] [PubMed]

24. Tang, W.; Davidson, C.I. Erosion of limestone building surfaces caused by wind-driven rain: 2. Numerical modelling. *Atmos. Environ.* **2004**, *38*, 5601–5609. [CrossRef]

25. Tang, W.; Davidson, C.I.; Finger, S.; Vance, K. Erosion of limestone building surfaces caused by wind-driven rain: 1. Field measurements. *Atmos. Environ.* **2004**, *38*, 5589–5599. [CrossRef]

26. Grossi, C.M.; Bonazza, A.; Brimblecombe, P.; Harris, I.; Sabbioni, C. Predicting twenty-first century recession of architectural limestone in European cities. *Environ. Geol.* **2008**, *56*, 455–461. [CrossRef]

27. Haugen, A.; Bertolin, C.; Leijonhufvud, G.; Olstad, T.; Broström, T. A Methodology for Long-Term Monitoring of Climate Change Impacts on Historic Buildings. *Geosciences* **2018**, *8*, 370. [CrossRef]

28. ProteCHt2save Web GIS Tool. Available online: https://www.protecht2save-wgt.eu/ (accessed on 28 May 2020).

 *atmosphere*

Article

Mapping Climate Change, Natural Hazards and Tokyo's Built Heritage

Peter Brimblecombe [1,2,3,*], Mikiko Hayashi [4] and Yoko Futagami [5]

1   School of Energy and Environment, City University of Hong Kong, Tat Chee Avenue, Kowloon, Hong Kong
2   Department of Marine Environment and Engineering, National Sun Yat-Sen University, Kaohsiung 80424, Taiwan
3   Aerosol Science Research Center, National Sun Yat-Sen University, Kaohsiung 80424, Taiwan
4   Center for Conservation Science, National Institutes for Cultural Heritage Tokyo National Research Institute for Cultural Properties, Tokyo 110-8713, Japan; hayashi03@tobunken.go.jp
5   Department of Art Research, Archives and Information Systems, Tokyo National Research Institute for Cultural Properties, Tokyo 110-8713, Japan; futa@tobunken.go.jp
*   Correspondence: p.brimblecombe@uea.ac.uk

Received: 26 May 2020; Accepted: 24 June 2020; Published: 28 June 2020

Abstract: Although climate change is well recognised as an important issue in Japan, there has been little interest from scientists or the public on the potential threat it poses to heritage. The present study maps the impact of emerging pressures on museums and historic buildings in the Tokyo Area. We examine a context to the threat in terms of fluctuating levels of visitors as a response to environmental issues, from SARS and COVID-19, through to earthquakes. GIS mapping allows a range of natural and human-induced hazards to be expressed as the spatial spread of risk. Temperature is increasing and Tokyo has a heat island which makes the city hotter than its surroundings. This adds to the effects of climate change. Temperature increases and a decline in relative humidity alter the potential for mould growth and change insect life cycles. The region is vulnerable to sea level rise, but flooding is also a likely outcome of increasingly intense falls of rain, especially during typhoons. Reclamation has raised the risk of liquefaction during earthquakes that are relatively frequent in Japan. Earthquakes cause structural damage and fires after the rupture of gas pipelines and collapse of electricity pylons. Fires from lightning strikes might also increase in a future Tokyo. These are especially relevant, as many Japanese heritage sites use wood for building materials. In parallel, more natural landscapes of the region are also affected by a changing climate. The shifting seasons already mean the earlier arrival of the cherry blossom and a later arrival of autumn colours and a lack of winter snow. The mapping exercise should highlight the spatial distribution of risk and the way it is likely to change, so it can contribute to longer term heritage management plans.

Keywords: earthquakes; fire; floods; historic sites; landslides; museums; insects; sea level rise; typhoons; visitors

---

1. Introduction

Our heritage is under threat. The need to protect Japanese tangible heritage from disasters has been addressed by government funding increases: JPY 2905 million in 2019, to JPY 3907 million in 2020. While natural hazards are well recognised issues in Japan and climate science is strong, there has been relatively little interest there from scientists or the public on the potential threat a changing climate poses on heritage, especially in the way they alter the frequency of meteorologically driven hazards. The present study maps the impact of external pressures on museums, historic buildings such as temples and shrines in the Tokyo Area. This is one of the most populous metropolitan areas in the world, which includes several prefectures of the Kantō Region of Japan, as well as Yamanashi

Prefecture. The Tokyo Metropolis is elongated from east to west, stretching from mountains which stand in the west to Tokyo Bay to islands scattered over the Pacific Ocean; although in this study islands are excluded, yet it still covers an area of some 1790 square kilometres. The *Japan Meteorological Agency* (JMA) places Tokyo in the EJP climate zone (Pacific side of eastern Japan), with hot and humid summers and cold winters. In winter, the wind from the Siberian continent causes heavy snow on the Sea of Japan side, but when the air crosses the mountains to Tokyo and reaches the Pacific side to Tokyo, it becomes dry air. The winter is the driest season, thus the season is sunny and fairly mild, but the city experiences hot, humid and rainy summers.

Japan is an island country in East Asia (Figure 1a) and Tokyo, excluding islands, is located between 35°30′05″ to 35°53′54″ North latitude and 138°56′35″ to 139°55′07″ East Longitude (Figure 2a). The topography gradually decreases in elevation from the western mountains to the alluvial lowlands, ending at Tokyo Bay (Figure 1c). The terrain can be regarded as defined by the Tamagawa River catchment, and the geographical characteristics are completely different between east and west, with the boundary near Ome City, the estuary settlement in the Kantō Mountains. There are accurate 5 m-mesh data as a quantitative representation of terrain within the Digital Elevation Model (DEM) provided by the Geospatial Information Authority of Japan [1].

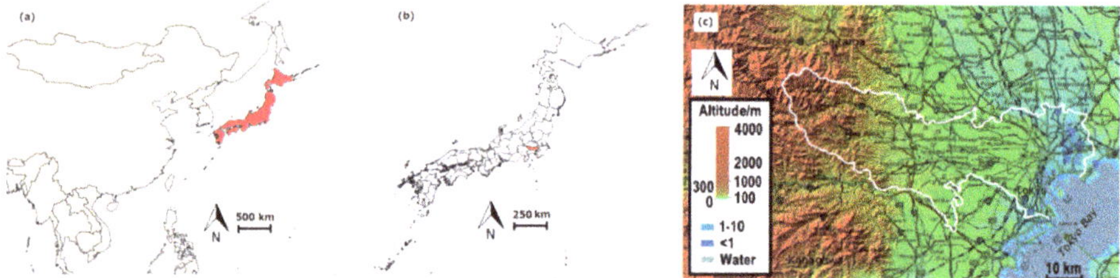

Figure 1. (**a**) Japan's position in East Asia. (**b**) Tokyo's position in Japan. Source: National Land Numerical Information (https://nlftp.mlit.go.jp/ [2]) (**c**) Topography of the Tokyo region discussed in this paper, with a white line denoting the boundary of the Tokyo Area.

Tokyo, as a major city with a lengthy history, has a great wealth of heritage. There are numerous historic buildings and sites shrines, temples and great buildings, perhaps some favourites being: (i) Asakusa Shrine, (ii) Tōgū Palace, (iii) Tokyo Station, (iv) Sensō-ji Niten-mon, (v) Nezu Shrine, (vi) Enyuu-ji Temple, (vii) Former Iwasaki House, (viii) Eitai Bridge, (ix) Shofuku-ji Temple Jizōdō. Add to this many historic buildings; these marked as 83 points in pink (Figure 2a,b) and all of them are designated buildings and structures as National Treasure or Important Cultural Properties. Tokyo has hundreds of museums and galleries, some of the favourites being: (i) Tokyo National Museum, (ii) Edo-Tokyo Museum, (iii) Ghibli Museum, (iv) Nezu Museum, (v) Hara Museum of Contemporary Art, (vi) Mori Art Museum, (vii) Tokyo Metropolitan Art Museum, (viii) National Museum of Western Art, and (ix) the National Museum of Emerging Science and Innovation. Museums are marked as 232 points in green (Figure 2a,c). Maps of heritage and the risk imposed by climate change are often seen as important tools for the strategic management of heritage, e.g., [3,4], with some at a national or regional level [5–8]. Among a few examples of Japanese studies in this field, [9,10] integrated a database of the nationally designated cultural properties and a map of the active faults with GIS to estimate the seismic risk of each property. Difficulties in such mapping involve the problem of scaling because data are typically collected at national or regional levels rather than at city scale. It is also problematic to tune the data, often collected for other purposes, to heritage; e.g., seismic risk is defined for buildings in general, not heritage, or that meteorological information is collected for many purposes, so the risks imposed on heritage requires considering the notion of heritage climate [3,11].

Special threats to the region are flooding and the increased pressures that climate change may present in terms of both river and sea floods [12,13] and the failure of sea defences [14]. As Tokyo is a coastal city, sea level rise makes low lying areas additionally vulnerable to increased flooding, with Estaban et al. [15], arguing that "The combined effect of an increase in typhoon intensity and sea level rise could pose significant challenges to coastal defences around Tokyo Bar around the turn of the twenty-first century." There are also risks from earthquakes, which are regularly mapped in the city as this affects a range of issues in addition to the threat to built heritage, e.g., real estate prices [16]. The Bureau of City Planning, Tokyo Metropolitan Government maps earthquake threat in their regional risk measurement survey on earthquakes, but it is also the subject of damage forecasts. Although air pollution and acid rain represent threats, they are not treated in this paper. Tokyo has worked hard to improve its air quality [17] and the study of acid rain has a long history in Japan [18], so despite concerns about the high level of threat especially from Chinese emissions [19] these have been much reduced in the current decade [20].

Figure 2. Location of (**a**) Historic buildings (pink) and museums (green) in Tokyo, (**b**) major historic buildings and (**c**) museums in central Tokyo. Note: Source: National Land Numerical Information (https://nlftp.mlit.go.jp/ [2]).

2. Materials and Methods

This study uses data on climate and natural hazards relevant to looking at long term pressures on built heritage in the Tokyo area. In addition to a range of research articles, we make use of government and municipal reports to assess the magnitude of the threat from a changing environment. A range of climate projections available from the Japanese Meteorological Agency [21] are especially useful in estimating likely future change. Geographic information system (GIS) allows the threat to be mapped.

We used the software QGIS, originally *Quantum GIS*, which allows users to analyse and edit spatial information in multiple raster formats and as vector data; which can be stored as either point, line, or polygon features. We particularly adopted 3.6 Noosa developed by the QGIS community. The software provides viewing, editing and analysis capability to overlap a range of natural hazards in the area and their changes over time. Vector data of points and areas, such as the location of built heritage (historical sources and museums), maps of Tokyo, the area of inundation by flooding and area of potential sediment disasters, were obtained from the National Land Numerical Information Download Service [2]. Areas of inundation are calculated for each river under designated storm impacts. Where vector data of point and area data were not available, we have resorted to simple counts from manually overlapped data.

3. Results

Pressures on our heritage are much discussed, and the impact of external local and global events is clear from Figure 3, where the numbers of foreign visitors to Japan are plotted. Pressures from epidemics, financial collapse and geophysical events cause fluctuating visitor numbers. The most recent crisis of COVID-19 has had a special impact on the heritage sector, where it has greatly affected visitor numbers and the attendant loss of income, though in some cases the most fragile sites experienced a welcome respite from heavy flows, allowing natural sites to recover a little [22]. There will also be problems with the subsequent cleaning of the viral contamination from properties and interior surfaces.

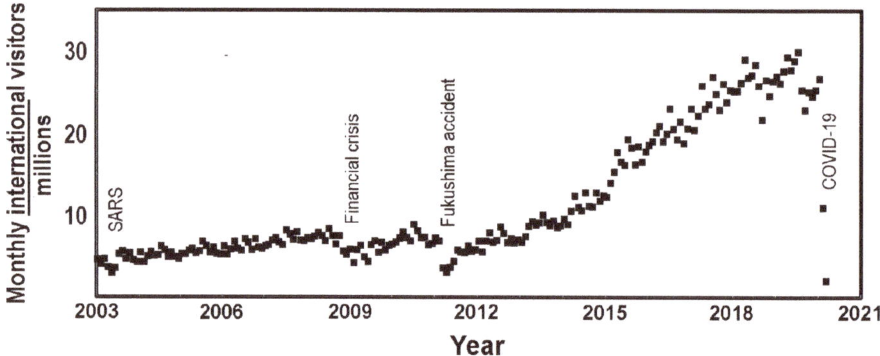

Figure 3. International visitors to Japan from Japan National Tourism Organization.

3.1. Temperature and Urban Heat Island

Temperature is the best understood parameter under a changing climate; likely to be affected through the addition of radiatively active gases to the atmosphere. Globally, the temperature is likely to have increased by about 0.8 °C through the 20th century. The Intergovernmental Panel on Climate Change (IPCC) *Fifth Assessment Report* suggests that surface temperature will rise a further 0.3 to 1.7 °C during the current century. In Japan from 1931 to 2017, the average temperature in Tokyo increased at a rate of 3.2 °C/century [23]. Projecting the regional climate of Eastern Japan between the periods 1980–1999 and 2076–2095 suggests that the annual mean temperature in Tokyo, where there is a strong heat island, will increase by more than 3 °C in almost a century.

In recent years, high temperatures in summer have been noted in large cities such as Tokyo, and public interest has been engaged by the added contribution the urban heat island makes to the increase in the number of heat stroke patients. The annual mean temperature in central Tokyo has increased about 3.2 °C/century, which is 4.4 times faster than the global mean temperature (0.73 °C/century) and 2.1 times faster than the Japanese mean temperature of 1.21 °C/century (JMA, 2019). These increases are partly due to the development of the urban heat island. The current trends for this effect, studied by Lee et al. [24], suggest that the summer heat island in Tokyo is increasing at 0.85 °C/century, while Manila, Seoul and Mumbai were slower, at 0.15, 0.2 and 0.36 °C/century, respectively. The heat island clearly substantially adds to the effect of greenhouse warming. The speed of temperature change in Tokyo has drawn popular comment [25] and the urban heat island is generally seen to arise from the decrease in green and water areas, the increase in ground covered with asphalt and concrete, the increase in heat (exhaust heat) generated from cars and buildings and the poor ventilation by wind in the street canyons [26]. Rising temperatures in summer have been a particular problem, because the heat island has helped reduced the comfort of urban life and affected human health. Urban influences are recognised, not only in the production of the heat island, but in other climate effects such as: urban rainfall, wind circulation, number of fog days, relative humidity, etc. [27,28]. These have impact not only human health, but also materials and the management of built heritage. Although Tokyo has a strong heat island, there are potentials to mitigate this, through the urban greening of buildings or increasing the area of open water in central Tokyo [29,30].

Temperature increases and a decline in relative humidity alter the potential for mould growth and change insect life cycles. The total number of hours with above 30 °C are shown in Figure 4. This will affect comfort within naturally ventilated buildings and put more pressure on the cooling systems in museum environments. It can also affect mould growth when accompanied by humid conditions, and will favour insect infestations. In Japan, the overwintering survival of insects has increased along with earlier appearances in the spring, an increase in the number of generations each year, lengthening of the reproductive season, etc. However, insects can also be susceptible to heat stress when temperatures increase to high values from about 28 to 32 °C [31]. The northward movement of insects through the Japanese archipelago has been widely described. The distribution range of many insects have increased, including those of dragonflies [32] and the Great Mormon butterfly, *Papilio memnon* [33]. The southern green stink bug, *Nezara viridula* has moved away from the coasts to warmer regions [34]. Looking at parallels in other cities, such as London, it is possible to see a notable increase in insects such as the webbing clothes moth, *Tineola bisselliella* [35], which may be partly due to climate change, but also related to food availability, habitat and urban activities [36]. The changes in insect populations can be modelled as a function of the future climate [37,38]. Such changes create a characteristic problem with wood, as a material, which is especially important in Japanese buildings. The extensive use of wood makes the heritage particularly vulnerable to insect infestation [39]. In Tokyo, the urban heat island may cause an increase in insect numbers and variety. Additionally, there is often a greater availability of food in cities, and insects can be transported on objects or display materials. Stag beetles are popular as pets in Japan; notably, *Dorcus* spp. is an invasive insect from East and Southeast Asia, which can be released or escape [40].

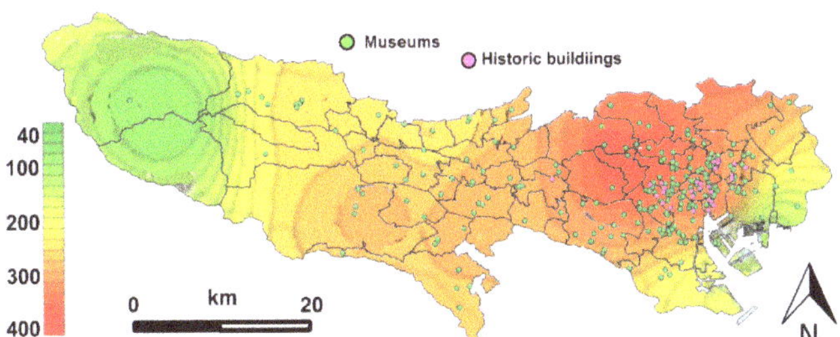

Figure 4. The total number of hours with temperatures above 30 °C averaged 2003–2007 over the Tokyo area. Note: Base map extracted from RealEstateJapan and JMA.

3.2. *Storms, Typhoons and Floods*

Historic floods in Tokyo are well described, for example in the great floods of 1742, the Zen monk Enjū was engaged in a pilgrimage and trapped by the intense rain, so left an account of damage to religious buildings [41]. This historic event led the Sumida River to flood, eroding the piles of the Ryōgoku Bridge, as well as damaging Eitai and Shin'ō Bridges (Figure 5a), while overflowing levees caused extensive flooding in the Kasai district [42]. The Heavy Rain Event during July 2018 [43], although not affecting

Tokyo so strongly, caused tremendous damage elsewhere. There were 221 fatalities, 6296 buildings were completely destroyed, 8929 houses were inundated above the floor level, and two key buildings (National Treasures), 35 important cultural property buildings, and 24 registered cultural property buildings were affected [44].

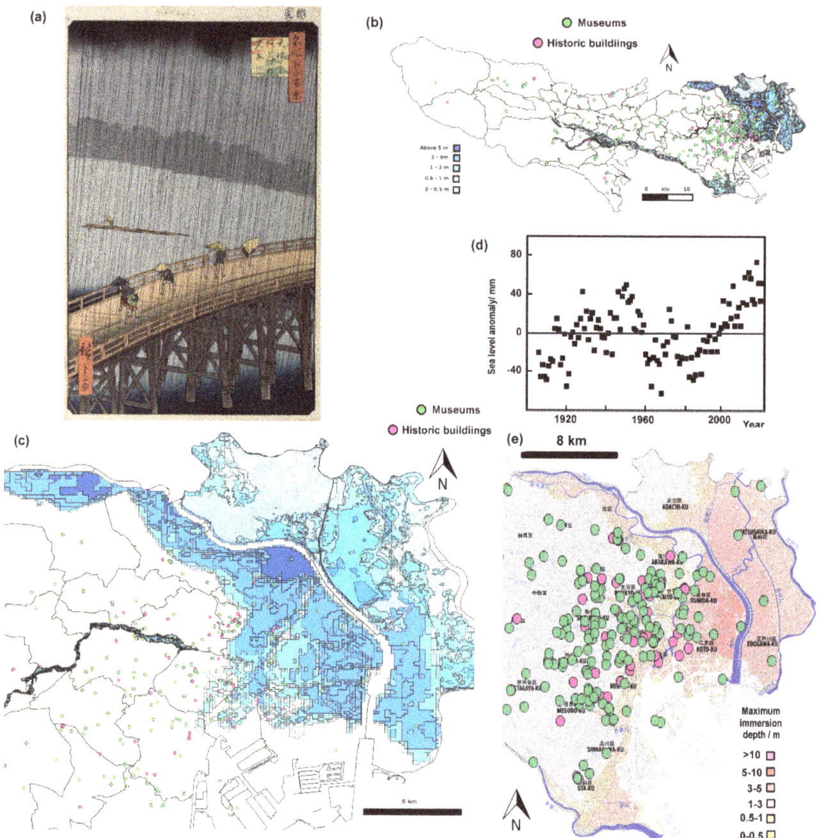

Figure 5. (**a**) *Sudden Shower Over Shin-Ohashi Bridge and Atake* by Hiroshige https://upload.wikimedia.org/ wikipedia/commons/c/cc/Hiroshige_Atake_sous_une_averse_soudaine.jpg [45] (**b**) Area of inundation under flood conditions of Table 1. Source: (https://nlftp.mlit.go.jp/ksj/index.html [2]) (**c**) Enlargement of the central area of the inundation. (**d**) Annual mean sea level anomalies for Japan; source: Japan Meteorological Agency). (**e**) Immersion depth for the central parts of Tokyo; sources: JPC 2018, The Nikkei Shimbun, 30 March 2018, TBS News, 30 March 2018.

Changes in rainfall patterns over Japan mean that rainfall is likely to increase more than 10% over the 21st century [46]. Over the period 1976–2012, the number of hours of heavy rain (>50 mm) have increased across all regions of Japan, which is likely to continue into the future [47]. However, the picture for Tokyo is not particularly clear, although it was noted some decades ago that the urban heat island enhances convective storm activity over the city [48]. There was a period with many hours of heavy precipitation during the 1940s, though the 1990s seem to lie within the range of variation for the 20th century [49]. Nevertheless, the general view remains that there will be an increase in falls of heavy rain over the coming century [39]. In addition, there has been considerable public concern over increased frequency of stormwater flooding and its impact on residential housing [50]. Oddly, despite increases in total precipitation amount, the number of dry days with daily precipitation less than

1mm is expected to increase in almost every region of Japan. In Tokyo, a declining water table or soil moisture has the potential to expose sensitive buried archaeological sites or weaken the foundations of buildings.

The GIS mapping shows the potential risk to historic sites and museums. It reveals a high flood risk along the river and the presence of a large area of eastern Tokyo where land has elevations below mean tide level (Figure 5b,c). Historic buildings such as the Sensō-ji Temple complex and many museums will be flooded during infrequent but large storms and, in the future, vulnerability will probably increase more than previously thought under a changing climate. Figure 5b,c show the extent of the inundation risk to museums and historic buildings. The risk is calculated adopting the conditions listed in Table 2 for each important river in the catchment. The eastern part of Tokyo can be seriously affected, with about 15% of museums at risk from high water. Some 13 museums are likely to be affected when water levels rise 2.0–5.0 m (Table 2). It is thus important to take measures to protect against flooding, especially that storage locations are required to be on upper floors, with water—and fire—proof doors. The collections in art museums are especially sensitive to damper environments. There are four art museums under threat from inundation, but three occupy galleries present on the 4th floor or above, so risk is much reduced. The Museum of Contemporary Art Tokyo and the Tokyo Metropolitan Archives are sensitive to flooding. There are 12 historic buildings in Tokyo at risk of inundation, such as Asakusa Shrine and Sensō-ji Niten-mon (Table 1), so mitigation measures are especially relevant. Longer periods of inundation can cause more damage to museums and historic buildings through a range of secondary hazards, such as mould growth and insect attack. Saraswat et al. [51] present an overview of stormwater runoff management in Tokyo to guide optimal measures and management policies within the city's governance. An increase in the future sea level (Figure 5d) would also make Tokyo vulnerable and may cause sea defences in Tokyo Bay to fail by the end of the 21st century, with increased typhoon intensity adding further threat [13].

Modelling suggests that the frequency of typhoon landfalls will decrease (Figure 6a) and the mean value of the typhoon central atmospheric pressure will not change significantly. An important point is that the arrival probability of stronger typhoons will increase (bottom right Figure 6b) under future climate scenarios [52]. This means that flooding is a likely outcome of such increasingly intense falls of rain. Wind speeds in Tokyo in the future (2075–2099) are compared with those of the recent past (1979–2003) in Figure 6b, suggesting that the probabilities of the occurrence of higher annual maximum wind speeds will increase (i.e., exceeding 30 m s^{-1}), while medians of the annual maximum wind speeds decrease [53]. Figure 6c shows the relative typhoon wind risk in Japan, as the number of buildings likely to be damaged each year under the current and projected future climates. The decrease in frequency is associated with a decline in relevant typhoon events in Japan and although there is an increase in typhoon intensity, it is not enough to compensate for this decrease [53]. Dangers of wind damage in typhoons across East Asia have recently been examined along with factors that affect the intensity of damage and its extent, along with the potential to mitigate future impact [54].

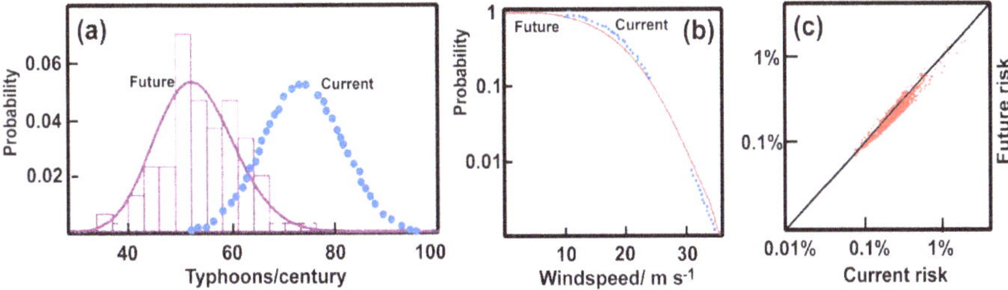

Figure 6. (a) Probable number of typhoons in Tokyo in 100 years from the stochastic typhoon model. Dotted and solid lines indicate present and future projections, respectively; see also [13]. (b) Exceedance probabilities for annual maximum wind speeds in Tokyo under current and projected climate. (c) Percentage wing risk for residential buildings in Japan under current and future scenarios [53].

Table 1. Design storms (i.e., a hypothetical discrete rainstorm) characterized by a specific duration in each river. Note full details in MLIT 2020.

Target River	Design Storm	Rainfall	Cities Affected
Shibakawa River, New Shibakawa River	Accounts for current development of river and drainage channels, regulating ponds, etc. Rise of Shibakawa River after heavy rain that occurs every 100 years. Inundation from overflow and collapse of levees.	September 1958 flood, *Kanogawa Typhoon* or *Typhoon Ida*: 2 days rain 411 mm	Adachi, Katsushika
Kandagawa River	Inundation potential from the Kandagawa River due to the Tokai heavy rain accounting for channel maintenance. Depth from both river water and inside the levee.	September 2000 *Tokai heavy rainfall*: rain 589 mm, hourly maximum 114 mm	Chiyoda, Chuo, Shinjuku, Bunkyo, Taito, Shibuya, Nakano, Suginami, Toshima, Musashino, Mitaka
Tonegawa River	Account for current state of river channel and flood control. Simulates expectation from heavy rain once in 200 years.	Tonegawa River basin and upper Yattajima: 3 days rain 318 mm	Adachi, Katsushika, Edogawa
Edogawa River	Simulates Edogawa River overflow due to heavy rain that occurs about once every 200 years.	Upper Yattajima: 3 days rain 318 mm	Adachi, Katsushika, Edogawa
Arakawa River	Accounts for current state of the Arakawa River river channel and flood control. Simulates Arakawa River overflows due to heavy rain every 200 years.	Arakawa basin; 3 days rain 548 mm	Chiyoda, Chuo, Minato, Taito, Sumida, Koto, Kita, Arakawa, Itabashi, Adachi, Katsushika, Edogawa
Nakagawa River, Ayasegawa River	Accounts for current state of channel and flood control. Simulates Nakagawa and Ayasegawa River overflows due to 100-year heavy rain, but only inundation from Edogawa River.	Nakagawa and Ayase basins: 48 h rain 355 mm	Adachi, Katsushika
Tamagawa River, Oogurigawa River	Accounts for current state of channel and flood control. Simulates Tamagawa River overflow due to a 200-year heavy rain.	Tamagawa River basin and the upstream area of Ishihara site: 2 days rain 457 mm	Ota, Setagaya, Hachioji, Tachikawa, Ome, Fuchu, Akishima, Chofu, Hino, Kunitachi, Fukuo, Komae, Tama, Inagi, Hamura, Akiruno
Asakawa River	Accounts for development of river channels. Simulates Asakawa River overflows due to a 200-year heavy rain.	Tamagawa River basin and the upstream area of Ishihara site: 2 days rain 457 mm	Hachioji, Hino, Tama

Table 2. Elevation of museums and sites in flood risk areas.

Elevation/m	Museums	Historic Buildings
0–0.5	4	3
0.5–1.0	10	1
1.0–2.0	8	2
2.0–5.0	13	5
>5.0	0	1

3.3. Earthquakes and Fires

Japan is frequently affected by devastating earthquakes, because of the active seismicity caused by the subduction of the Philippine Sea Plate beneath the continental Eurasian Plate and Okinawa Plate, as well as dense distribution of inland active faults. There were major earthquakes in Tokyo in 1703 (Genroku earthquake), 1855 (Ansei Edo Earthquake) and 1894 (Meiji Tokyo Earthquake) [55], but it is the Great Kantō Earthquake of 1923 that has been especially influential (Figure 7a). It was so powerful that the 100-ton Great Buddha statue in Kamakura moved almost 60 cm. In the metropolis, the earthquakes often led to fires if they struck at lunchtime, when many people were cooking meals. Today, fires are more often caused by the rupture of gas pipelines and collapse of electricity pylons. However, at the same time, energizing fire is typical on the occasion of earthquake as it happened in the Great Hanshin-Awaji Earthquake in 1995. Extensive coastal reclamation has raised the risk of soil liquefaction during earthquakes, which are so frequent in Japan. Additionally, soil moisture and the water table are likely to affect the supporting structures of older buildings.

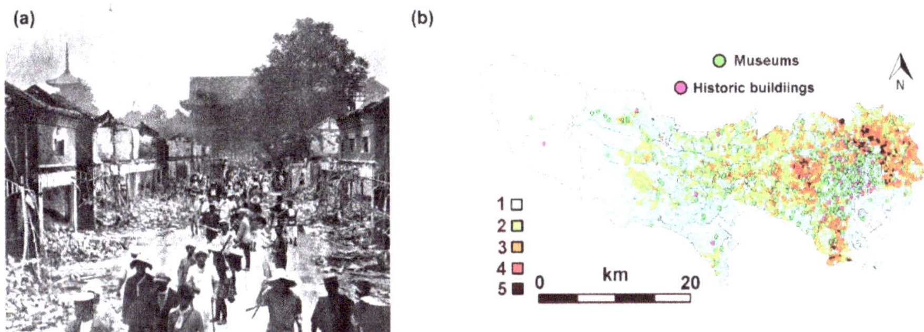

Figure 7. (**a**) Photograph of devastation in the area around Sensō-ji Temple, Asakusa after the Great Kantō earthquake of 1923, with smoke from fires in the background. Source: Public Domain File: Theosakamainichi-earthquakepictorialedition-1923-page9-crop.jpg [56] (**b**) The map of combined earthquake risks for Tokyo. Source: TMG 2018 and National Land Numerical Information (https://nlftp.mlit.go.jp/ [2]).

Figure 7b maps the potential risk to buildings from earthquakes [57]. Although it is not tuned to heritage related structures, these sites are marked on the combined risk map of Tokyo. This combination brings together the risk of building collapse with the risk of fire. Older buildings tend to be found in the historic downtown of Tokyo. Those along the Arakawa and Sumidagawa rivers offer lower earthquake resistance, as the structures typically have wooden or light-gauge steel frames. Machiya, Arakawa-ku is one of the worst affected areas in Tokyo, a district of densely packed wooden houses and narrow alleys, with little access for fire trucks. High risk areas are thus characterised by tightly packed buildings. Additionally, shaking can be amplified by ground characteristics, and valley and alluvial lowlands may lead to a higher risk. Fire risk is often enhanced in residential areas because of the wide use of open-flame appliances, and high density adds to the risk of fire spreading as these communities have fewer open spaces, such as parks and wide roads, which might act as fire breaks. Overall, such conditions are typical of historic areas where there is likely to be a high concentration of close-set older wooden structures (as mapped for central Tokyo in Figure 8a), with low resistance to fire [57]. Thus, the vulnerability of built heritage is not surprising, because it is most frequently associated with the older historic districts. Tokyo has been subjected to many fires. The most notable one is the Great Fire of Meireki, which occurred in 1657. It is rumoured that the fire was started when a cursed *kimono* was set on fire on a very windy day [58]. The fire spread quickly through the city, because of strong winds from the northwest, which illustrates the role of climate conditions in large scale fires. The conflagration that followed the Great Kantō Earthquake destroyed neighbourhoods and some significant sites, such as the Metropolitan Police Department building (Figure 8b). Such characteristics of central Tokyo contributed to the enormous loss of historic buildings and residences from fires after incendiary bomb raids during World War II.

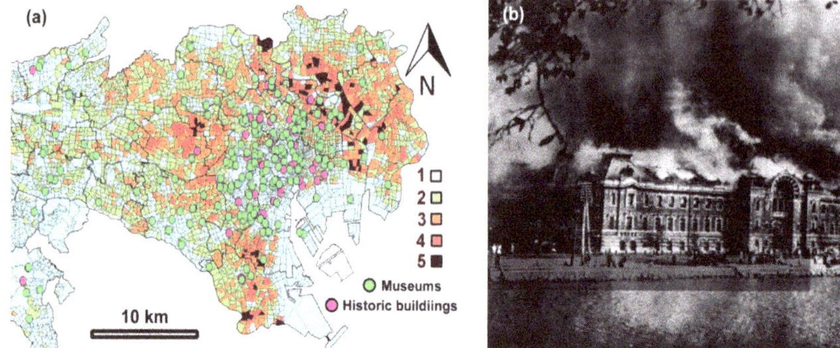

Figure 8. (**a**) Fire risk for Tokyo. Source: TMG 2018 and National Land Numerical Information (https://nlftp.mlit.go.jp/). (**b**) The Metropolitan Police Department building burning after the Great Kantō Earthquake. Source: Public Domain—File: Metropolitan Police Office after Kanto Earthquake.jpg [59].

There is the potential for increased risk of fire both after earthquakes and that induced by lightning strikes. Shindo et al. [60] show that the number of lightning flashes per year over Japan and has increased over the period 1992 to 2008. Additionally, Tokyo is likely to see enhanced convective storm activity in future, so an increased frequency of lightning strikes and fires might be expected. The potential for climate change to lead to more lightning strikes can be countered by improved protection against lightning and incorporating fire-suppression devices in buildings. The fire at Notre-Dame de Paris on the evening of 15 April 2019 initiated many concerns in Japan, yet soon after, there was a devastating fire at Okinawa's Shuri Castle in 31 October 2019. It was followed almost immediately by a fire on small thatched-roof huts at a car park in Ogi-machi, one of the components of a World Heritage property named Historic Villages of Shirakawa-go and Gokayama in Gifu Prefecture. Wooden structures and thatched roofs are especially vulnerable, so have been of great concern for many hundreds of years, so Japan developed a range of approaches to limit the extent of damage from fire [39]. These fires in 2019 encouraged expanded budgets for heritage protection in Japan, but also more modest activities; e.g., the Tokyo Fire Department offers fire prevention guidance to Zōjō-ji temple, a complex which has survived many fires since the early 17th century, even though many of its components, such as the mausoleums of the Tokugawa Shoguns, were burnt down during raids of World War II.

3.4. Debris Flow, Slope Failure and Landslides

A range of sediment disasters are induced by heavy rain, but also triggered by earthquakes. These can cause catastrophic damage to built heritage, and such sediment flows include:

1. Debris flow—soil and rock on a hillside or in a riverbed are washed after heavy or continuous rain, which can reach 20–40 km hr^{-1}
2. Slope failure—abrupt slope collapse under the influence of a rain or an earthquake. This occurs so suddenly that people fail to escape when it occurs near a residential area and can lead to many fatalities.
3. Landslide—massive quantities of soil move slowly downslope under the influence of groundwater and gravity. The large soil mass means serious damage can occur, and once started, it is extremely difficult to stop.

Sediment hazards are more probable in the western parts of Tokyo, because of the mountainous geography and shown in Figure 9 which shows the red and yellow risk areas.

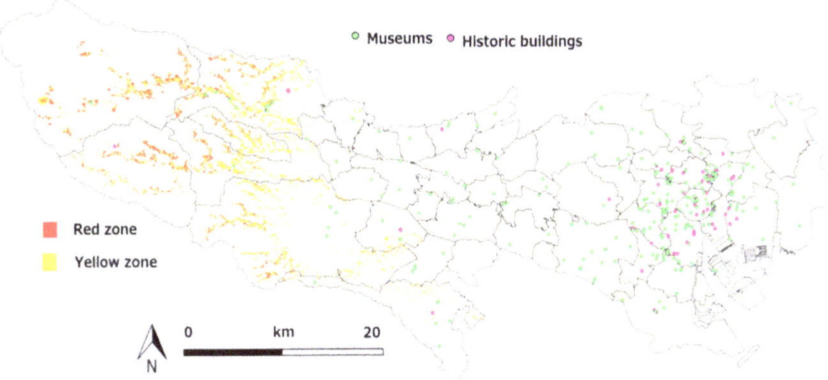

Figure 9. Sediment disaster map of the Tokyo area. *Red Zone* where slopes can collapse and damage buildings and threaten lives. *Yellow Zone* where steep slopes can collapse, risking injury or life. Source: National Land Numerical Information (https://nlftp.mlit.go.jp/ [2]).

Red Zone: Where steep slopes are likely to collapse, buildings are damaged; areas where there is a risk of significant damage to the lives or physical health of residents. In Japan, a permit is required for specific development activities and structural regulation for buildings. *Yellow Zone*: Where steep slopes are likely to collapse, risking injury or life, and when the danger is apparent, a warning and evacuation system is to be in place.

Red and yellow zones are more predominant in western Tokyo, but there are many popular places in the eastern and central districts such as the NHK Museum of Broadcasting and the Itabashi Art Museum, which even though not in red or yellow zones, there may be some risk as the sites are close to potential sediment threats.

There are nine museums in the yellow zone: seven are in the western area and two in the eastern part of Tokyo. Additionally, there are two historic buildings in yellow zone: one is in the western area and the other in the eastern part of Tokyo. The west has fewer historic buildings and museums than the east, but still there are many small red and yellow zones liable to sediment disaster in central Tokyo. This means that the risk is not only to the western part, but also the eastern, where there are relatively fewer sites likely to be threatened. In the west, which is more mountainous, the yellow zone is occupied by: Okutama Water and Green Friendship Hall, Gyokudo Art Museum, Ome Kimono Museum, Yoshikawa Eiji Museum, Ome Municipal Museum of Provincial History, Akiruno City Itsukaichi Museum and the Hamurashi Kyodo Museum. Protection can best take the form of slope maintenance, well-practiced in Hong Kong [61], which has steep terrain and receives heavy falls of rain, especially during typhoons.

3.5. Visitor Experience Under a Changing Climate

As shown in Figure 3, many external parameters affect visitor choices. Typically, wet weather can discourage travelling round the city, but on the other hand it can mean a greater tendency to go indoors to shelter [62]. Many individuals appear to actively adjust their plans throughout the day in response to rain. However, for others, attendance depends upon prior weather forecasts of rain. The duration of a visit is also likely to increase during rainy periods [63]. However, in the hotter weather of the future, visitors may want to escape from the oppressive outdoor heat, especially to air-conditioned museum interiors. In historic dwellings where mechanical ventilation is not appropriate, the heat of the interiors may be such that visitors will be driven outside and away from the poor air movement among crowds of visitors. The change in visitor behaviour is obviously not simply a matter of the changing climate and must also consider changes to visitor types and different patterns of behaviour and expectations.

Museums and historic buildings are increasingly broadening the range of visitors they attract, so it is no longer the well-informed tourist, with guidebook in hand. Plans, for example at the UK's National Trust, aim to welcome a wider variety of visitor types they categorise as: (i) active thinkers, (ii) live life to the full, (iii) spontaneous characters, (iv) young experience seekers and, among groups with children, (v) the family group, (vi) kids first families, and (vii) home and family [64]. Encouraging these new groups of visitors will inevitably change the way the heritage sites are experienced and used. We can imagine that the gentle tours through historic rooms may be less important than more active pursuits and greater use of the grounds and surroundings.

In recent years, the sunnier urban environments have led to more oxidizing pollutants and meant that the surfaces of buildings have taken on a somewhat warmer tone [65]. In addition to chemical drivers, shifting colours may be a product of biological growth. As the climate changes, viewing the beauty of a snowy Tokyo, reminiscent of the paintings of Hiroshige, is likely to be more difficult (Figure 10). The changing seasons have influenced the viewing of cherry blossoms (*sakura*) and the autumn leaves, the dates of these activities have had to shift as the warm seasons have grown longer [66].

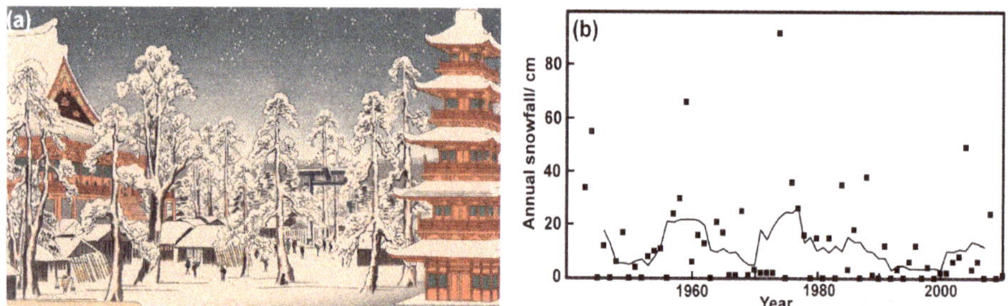

Figure 10. (**a**) Hiroshige's *Snow Scene at Kinryuzan Buddist Temple, Asakusa District* (**b**) Annual Tokyo snowfall. The line shows the 7-year running mean. Note: Image is part of *Toto Yukimi Hakkei* (8 Snow Scenes of the Eastern Metropolis): http://mercury.lcs.mit.edu/~ljnc/prints/8snow.html [67]. Data from WMO Station ID:47662 of JMA.

4. Discussion and Conclusions

This study presents maps of some natural pressures on cultural heritage in Tokyo. Accardo et al. [4] claim maps of heritage risk to be useful, because they provide a "deeper and more concise knowledge of risk intensity than may be obtained by comparing the potential danger level of any single cultural property, based on the three static/structural, human impact, and environmental factors, with its state of conservation (vulnerability index) recorded". It is easy to accept such thoughts, but the translation of maps into management policies is not trivial. Maps of potential risk are strategic, so they seem more useful to large organisations or government agencies. Such entities need to consider the broader picture and prioritise and compare risks over an extensive range of sites. Ultimately, they need to allocate limited resources among sites with competing, but necessary demands; and budgets are always tight. It is also possible that the maps can be of some value to a single museum or site, as they may hint at risks of which they are unaware, but often they will know many of the risks that confront them. In Italy, Istituto Superiore per la Conservazione ed il Restauro (High Institute for Conservation and Restoration) has developed a risk map named Carta del Rischio del Patrimonio Culturale (Risk map of Cultural Heritage) since the 1990s. It integrates the national inventory of cultural heritage and storage building of artistic objects such as museums, and hazard maps of natural/human-induced disasters, to utilise it for establishing restoration plans of cultural heritage by Soprintendenze, national authorities of cultural heritage in the regions of Italy.

Risk maps can reveal the changes in climate and pressures from natural disasters. They can be open and provide information to stakeholders and the public. The risks may become evident to more individuals and encourage risk reduction. While it is possible to create overlapping maps with GIS, the availability of data is often limited with only a few of the desirable maps are available. For instance, Tokyo tsunami data have yet to be made available in a useful format, although the municipality webpage gives tsunami hazard maps. In a publicly available GIS form, they can more readily be used to identify and visualise built heritage hazards and sites at risk in an effective way.

Risk management and assessment in general is relatively advanced in Japan, so there is a hazard map portal site (https://disaportal.gsi.go.jp/index.html [68]). In contrast, application of risk maps to cultural heritage protection from natural/human-induced disasters is not so frequent. National-scale risk maps are useful for policy making such as allocation of resources, but smaller-scale maps dealing with multiple disasters should be necessary for site managers and museum curators who take care of one or small numbers of site or museums. Up to now, disaster preparedness planning has used such material for establishing evacuation routes for visitors and objects, but less for planning mitigation strategies. Risk maps can also be useful in designing new museums by assessing risks in advance. A general problem in creating risk maps is that most of the publicly available data focusses on general built structures rather than built heritage, so it would be best to develop techniques that might tune it for that use. Historic buildings, especially, are more sensitive to hazards. In this study, museums and historic buildings designated as national treasures and important cultural properties have been considered, but it is also characteristic of large cities like Tokyo that there are many privately owned cultural properties, and they should also be stored safely. There are other areas worthy of future expansion, e.g., much of our knowledge of insects is related to those which are agricultural pests, rather than those that place heritage at risk. A better understanding of the life cycles of wood-boring insects in Japan would be useful. An understanding of natural hazards is well developed in Japan, and the science of climate change is nicely represented by the Japan Meteorological Agency, but there is only limited effort to tune this to a future heritage climatology. Similarly, a knowledge of air pollution, both indoors and out, has largely been concerned with the impact on human health. The example of Tokyo can be applied to other large cities, which often have rivers and similar characteristics, so some risks are shared. Cultural heritage is affected by climate change, as this alters the nature of many hazards. It is evident that more work is necessary to improve the health of our heritage.

Author Contributions: The following statements should be used "Conceptualization, P.B. and M.H. and Y.F.; methodology, M.H. and Y.F.; data analysis P.B.; investigation and writing P.B. and M.H. All authors have read and agreed to the published version of the manuscript.

Funding: This research received no external funding.

Acknowledgments: We would like to thank Helen Lloyd of the UK National Trust for her usual thoughtful advice.

Conflicts of Interest: The authors declare no conflict of interest.

References

1. Geospatial Information Authority of Japan (GSI): Digital Elevation Model (DEM) Download Service. Available online: https://fgd.gsi.go.jp/download/menu.php (accessed on 23 May 2020).
2. Ministry of Land, Infrastructure, Transport and Tourism (MLIT). National Land Numerical Information Download Service. Available online: http://nlftp.mlit.go.jp/ (accessed on 23 May 2020).
3. Sabbioni, C.; Brimblecombe, P.; Cassar, M. *The Atlas of Climate Change Impact on European Cultural Heritage: Scientific Analysis and Management Strategies*; Anthem Publishing: London, UK, 2010.
4. Bonazza, A.; Maxwell, I.; Drdácký, M.; Vintzileou, E.; Hanus, C. *Safeguarding Cultural Heritage from Natural and Man-Made Disasters: A Comparative Analysis of Risk Management in the EU*; ICOMOS Open Archive: Paris, France, 2018.
5. Accardo, G.; Giani, E.; Giovagnoli, A. The risk map of Italian cultural heritage. *J. Arch. Conserv.* **2003**, *9*, 41–57. [CrossRef]
6. De la Fuente, D.; Vega, J.M.; Viejo, F.; Díaz, I.; Morcillo, M. Mapping air pollution effects on atmospheric degradation of cultural heritage. *J. Cult. Herit.* **2013**, *14*, 138–145. [CrossRef]
7. Wang, J.J. Flood risk maps to cultural heritage: Measures and process. *J. Cult. Herit.* **2015**, *16*, 210–220. [CrossRef]

8. Wu, P.S.; Hsieh, C.M.; Hsu, M.F. Using heritage risk maps as an approach to estimating the threat to materials of traditional buildings in Tainan (Taiwan). *J. Cult. Herit.* **2014**, *15*, 441–447. [CrossRef]

9. Futagami, Y.; Kumamoto, T. Seismic assessment of the risks that intraplate earthquakes pose for the National Treasures of Japan—Constructing a GIS database of National Treasures that is integrated with an active fault database. *Archeol. Nat. Sci.* **2002**, *44*, 45–75.

10. Futagami, Y.; Morii, M.; Kumamoto, T. Construction and Integration of GIS Databases for Risk Assessment of Nationally Designated Cultural Properties due to Earthquakes and Typhoons in Japan. In *Digital Documentation, Interpretation & Presentation of Cultural Heritage, In Proceedings of the 22nd CIPA Symposium 11–15 October 2009, Kyoto, Japan*; CRC Press: Boca Raton, FL, USA, 2009.

11. Brimblecombe, P. Heritage climatology. In *Climate Change and Cultural Heritage*; Lefevre, R.-A., Sabbioni, C., Eds.; Edipuglia: Bari, Italy, 2010; pp. 54–57.

12. Yanai, H.; Kumano, N.; Tamura, M.; Kuwahara, Y. Basic Research for the Estimation of Flood Damage Costs using Global Coastal Dike Information-Tokyo Bay and Ise Bay. *JSCER* **2019**, *75*, 323–330. [CrossRef]

13. Yasuhara, K.; Komine, H.; Yokoki, H.; Suzuki, T.; Mimura, N.; Tamura, M.; Chen, G. Effects of climate change on coastal disasters: New methodologies and recent results. *Sustain. Sci.* **2011**, *6*, 219–232. [CrossRef]

14. Hoshino, S.; Esteban, M.; Mikami, T.; Takabatake, T.; Shibayama, T. Effect of sea level rise and increase in typhoon intensity on coastal structures in Tokyo Bay. In *Coastal Structures 2011*; World Scientific: Singapore, 2013; Volume 2, pp. 141–152.

15. Esteban, M.; Jonkman, S.N.; Hoshino, S.; Ruiz-Fuentes, M.J.; Mikami, T.; Takagi, H.; Shibayama, T.; van Ledden, M. Chapter. 33: Adaptation to Sea Level Rise in Tokyo Bay: Opportunities for a Storm Surge Barrier? In *Handbook of Coastal Disaster Mitigation for Engineers and Planners*; Esteban, M., Takagi, H., Shibayama, T., Eds.; Butterworth-Heinemann: Oxford, UK, 2015; pp. 723–747.

16. Nakagawa, M.; Saito, M.; Yamaga, H. Earthquake Risks and Land Prices: Evidence from the Tokyo Metropolitan Area. *Jpn. Econ. Rev.* **2009**, *60*, 208–222. [CrossRef]

17. de Oliveira, J.A. Why an air pollution achiever lags on climate policy? The case of local policy implementation in Mie. *Jpn. Environ. Plan. A.* **2011**, *43*, 1894–1909. [CrossRef]

18. Ishikawa, Y.; Hara, H. Historical change in precipitation pH at Kobe, Japan: 1935–1961. *Atmos. Environ.* **1997**, *31*, 2367–2369. [CrossRef]

19. Nagase, Y.; Silva, E.C.D. Acid rain in China and Japan: A game-theoretic analysis. *Reg. Sci. Urban. Econ.* **2007**, *37*, 100–120. [CrossRef]

20. Cheng, H.; Zhang, W.; Wang, Q.; Wan, W.; Du, X. *Breakthroughs, China's Path to Clean Air 2013–2017*; Clean Air Asia: Beijing, China, 2018.

21. Japan Meteorological Agency (JMA). Climate Change Monitoring Report 2018 (in Japanese). 2019. Available online: https://www.data.jma.go.jp/cpdinfo/monitor/2018/pdf/ccmr2018_all.pdf (accessed on 23 May 2020).

22. Historic England. Number 10: The Impact of Covid-19 on the Heritage Sector, Heritage Online Debate. 2020. Available online: https://historicengland.org.uk/whats-new/debate/ (accessed on 23 May 2020).

23. Japan Meteorological Agency (JMA). Climate Change Monitoring Report 2017 (in Japanese). 2018. Available online: https://www.data.jma.go.jp/cpdinfo/monitor/2017/pdf/ccmr2017_all.pdf (accessed on 23 May 2020).

24. Lee, K.; Kim, Y.; Sung, H.C.; Ryu, J.; Jeon, S.W. Trend Analysis of Urban Heat Island Intensity According to Urban Area Change in Asian Mega Cities. *Sustainability* **2020**, *12*, 112. [CrossRef]

25. RealEstateJapan. *Urban Heat Island Effect: Why It's So Hot in Tokyo & What's Being Done about It*; GPlusMedia Lnc.: Tokyo, Japan, 2020.

26. Ministry of the Environment, Government of Japan. General Rules of Heat Island Measures (in Japanese). Available online: https://www.env.go.jp/air/life/heat_island/taikou/taikou_h250508.pdf (accessed on 23 May 2020).

27. Mikami, T. Recent progress in urban heat island studies: Focusing on the case studies in Tokyo. *Euro. J. Environ. Earth Sci.* **2006**, *1*, 79–88. [CrossRef]

28. Matsumoto, J.; Fujibe, F.; Takahashi, H. Urban climate in the Tokyo metropolitan area in Japan. *J. Environ. Sci.* **2017**, *59*, 54–62. [CrossRef] [PubMed]

29. Ichinose, T.; Shimodozono, K.; Hanaki, K. Impact of anthropogenic heat on urban climate in Tokyo. *Atmos. Environ.* **1999**, *33*, 3897–3909. [CrossRef]

30. Mochida, A.; Murakami, S.; Ojima, T.; Kim, S.; Ooka, R.; Sugiyama, H. CFD analysis of mesoscale climate in the Greater Tokyo area. *J. Wind Eng. Ind. Aerodyn.* **1997**, *67*, 459–477. [CrossRef]

31. Kiritani, K. Different effects of climate change on the population dynamics of insects. *Appl. Entomol. Zool.* **2013**, *48*, 97–104. [CrossRef]

32. Doi, H. Delayed phenological timing of dragonfly emergence in Japan over five decades. *Biol. Lett.* **2008**, *4*, 388–391. [CrossRef]

33. Yoshio, M.; Ishii, M. Geographical variation of pupal diapause in the Great Mormon butterfly, Papilio memnon L. (Lepidoptera: Papilionidae), in western Japan. *Appl. Entomol. Zool.* **1998**, *33*, 281–288.

34. Tougou, D.; Musolin, D.L.; Fujisaki, K. Some like it hot! Rapid climate change promotes changes in distribution ranges of Nezara viridula and Nezara antennata in Japan. *Entomol. Exp. Appl.* **2009**, *130*, 249–258. [CrossRef]

35. Pinniger, D.B. Ten years on-from vodka beetles to risk zones. In *Integrated Pest Management for Collections*; Winsor, P., Pinniger, D., Bacon, L., Child, B., Harris, K., Lauder, D., Phippard, J., Xavier-Rowe, A., Eds.; English Heritage: Swindon, UK, 2011; pp. 1–9.

36. Brimblecombe, P.; Brimblecombe, C.T. Trends in insect catch at historic properties. *J. Cult. Herit.* **2015**, *16*, 127–133. [CrossRef]

37. Brimblecombe, P. Policy relevance of small changes in climate with large impacts on heritage. In *Cultural Heritage Facing Climate Change*; Lefevre, R.-A., Sabbioni, C., Eds.; Edipuglia: Bari, Italy, 2018; pp. 25–32.

38. Brimblecombe, P.; Lankester, P. Long-term changes in climate and insect damage in historic houses. *Stud. Conserv.* **2013**, *58*, 13–22. [CrossRef]

39. Brimblecombe, P.; Hayashi, M. Pressures from long term environmental change at the shrines and temples of Nikkō. *Herit. Sci.* **2018**, *6*, 27. [CrossRef]

40. NIES (>2010) Invasive Species of Japan, National Institute for Environmental Studies, Tsukuba-City, Japan. Available online: https://www.nies.go.jp/biodiversity/invasive/DB/detail/60500e.html (accessed on 24 May 2020).

41. Marceau, L.E. One Flood, Two "Saviours": Takebe Ayatari's Changing Discourse on the Kanpō Floods of 1742. In *Crisis and Disaster in Japan and New Zealand*; Palgrave Macmillan: Singapore, 2019; pp. 13–21.

42. Fiévé, N.; Waley, P. *Japanese Capitals in Historical Perspective: Place, Power and Memory in Kyoto, Edo and Tokyo*; Routledge: Abingdon, UK, 2013.

43. Tsuguti, H.; Seino, N.; Kawase, H.; Imada, Y.; Nakaegawa, T.; Takayabu, I. Meteorological overview and mesoscale characteristics of the Heavy Rain Event of July 2018 in Japan. *Landslides* **2019**, *16*, 363–371. [CrossRef]

44. Ministry of Education, Culture, Sports, Science and Technology (MECSST). Damage Information Due to Heavy Rain in July 2018 (22nd Report) (in Japanese). Available online: https://www.mext.go.jp/component/a_menu/other/detail/__icsFiles/afieldfile/2019/01/09/1407285_22_1.pdf (accessed on 23 May 2020).

45. Sudden Shower over Shin-Ōhashi Bridge and Atake. Available online: https://upload.wikimedia.org/wikipedia/commons/c/cc/Hiroshige_Atake_sous_une_averse_soudaine.jpg (accessed on 23 May 2020).

46. Kimoto, M.; Yasutomi, N.; Yokoyama, C.; Emori, S. Projected changes in precipitation characteristics around Japan under the global warming. *Sola* **2005**, *1*, 85–88. [CrossRef]

47. Japan Meteorological Agency (JMA). Climate Change Monitoring Report 2011 (in Japanese). 2012. Available online: https://www.data.jma.go.jp/cpdinfo/monitor/2011/pdf/ccmr2011_all.pdf (accessed on 23 May 2020).

48. Yonetani, T. Increase in number of days with heavy precipitation in Tokyo urban area. *J. Appl. Meteorol.* **1982**, *21*, 1466–1471. [CrossRef]

49. Kanae, S.; Oki, T.; Kashida, A. Changes in hourly heavy precipitation at Tokyo from 1890 to 1999. *J. Meteorol. Soc. Jpn. Ser. II* **2004**, *82*, 241–247. [CrossRef]

50. Japan Property Central (JPC). Tokyo's Latest Flood Map Puts a Third of City in Risk Zone, Japan Property Central. 2018. Available online: https://japanpropertycentral.com/2018/04/tokyos-latest-flood-map-puts-a-third-of-city-in-risk-zone/ (accessed on 23 May 2020).

51. Saraswat, C.; Kumar, P.; Mishra, B.K. Assessment of stormwater runoff management practices and governance under climate change and urbanization: An analysis of Bangkok, Hanoi and Tokyo. *Environ. Sci. Policy* **2016**, *64*, 101–117. [CrossRef]

52. Yasuda, T.; Mase, H.; Mori, N. Projection of future typhoons landing on Japan based on a stochastic typhoon model utilizing AGCM projections. *Hydrol. Res. Lett.* **2010**, *4*, 65–69. [CrossRef]

53. Nishijima, K.; Maruyama, T.; Graf, M. A preliminary impact assessment of typhoon wind risk of residential buildings in Japan under future climate change. *Hydrol. Res. Lett.* **2012**, *6*, 23–28. [CrossRef]

54. Yang, Q.; Gao, R.; Bai, F.; Li, T.; Tamura, Y. Damage to buildings and structur54es due to recent devastating wind hazards in East Asia. *Nat. Hazards* **2018**, *92*, 1321–1353. [CrossRef]

55. Furumura, T.; Takeuchi, H. Large Earthquakes Occurring beneath Tokyo Metropolitan Area and Strong Ground Motions (in Japanese). *J. Geogr. (Chigaku Zasshi)* **2007**, *116*, 431–450. [CrossRef]
56. Photograph of Bevastation in the Area around Sensō-ji Temple, Asakusa after the Great Kantō Earthquake of 1923. Available online: https://commons.wikimedia.org/wiki/File:Theosakamainichi-earthquakepictorialedition-1923-page9-crop.jpg (accessed on 23 May 2020).
57. Disaster Management Section Tokyo Metropolitan Government (TMG), Your Community's Earthquake Risk 2018. Available online: https://www.toshiseibi.metro.tokyo.lg.jp/bosai/chousa_6/download/earthquake_risk.pdf (accessed on 23 May 2020).
58. Cabinet Office Japan (CO). Disaster Management Japan. *1657 Edo Great Fire of the Meiji Era.* 2003. Available online: http://www.bousai.go.jp/kohou/oshirase/h15/pdf/2-7.pdf (accessed on 23 May 2020).
59. The Metropolitan Police Department Building Burning after the Great Kantō Earthquake. Available online: https://it.m.wikipedia.org/wiki/File:Metropolitan_Police_Office_after_Kanto_Earthquake.jpg (accessed on 23 May 2020).
60. Shindo, T.; Motoyama, H.; Sakai, A.; Honma, N.; Takami, J.; Shimizu, M.; Tamura, K.; Shinjo, K.; Ishikawa, F.; Ueno, Y.; et al. Lightning occurrence characteristics in Japan for 17 years: Observation results with lightning location systems of electric power utilities from 1992 to 2008. *IEEJ Trans. Electr. Electr. Eng.* **2012**, *7*, 251–257. [CrossRef]
61. Geotechnical Engineering Office, Guide to Slope Maintenance, Geoguide 5. Civil Engineering and Development Department HKSAR. 2018. Available online: https://www.cedd.gov.hk/filemanager/eng/content_113/eg5_20181120.pdf (accessed on 23 May 2020).
62. Grossi, C.M.; Brimblecombe, P.; Lloyd, H. The effects of weather on visits to historic properties. *Views* **2010**, *47*, 69–71.
63. Cuffe, H.E. Rain and museum attendance: Are daily data fine enough? *J. Cult. Econ.* **2018**, *42*, 213–241. [CrossRef]
64. *Audience Insight Pack*; National Trust: Swindon, UK, 2015; pp. 1–11.
65. Grossi, C.M.; Brimblecombe, P.; Esbert, R.M.; Alonso, F.J. Color changes in architectural limestones from pollution and cleaning. *Color. Res. Appl.* **2007**, *32*, 320–331. [CrossRef]
66. Brimblecombe, P.; Hayashi, M. Perception of the relationship between climate change in Japan and traditional wooden heritage in Japan. In *Public Archaeology and Climate Change*; Dawson, T., Nimura, C., Lopez-Romero, E., Daire, M.-Y., Eds.; Barnsley: Oxbow, UK, 2017; pp. 288–302.
67. Snow Scene at Kinryuzan Buddist Temple, Asakusa District. Available online: http://mercury.lcs.mit.edu/~{}jnc/prints/8snow.html (accessed on 23 May 2020).
68. Portalsite of Hazardmap. Available online: https://disaportal.gsi.go.jp/index.html (accessed on 23 May 2020).

atmosphere

Article

The Awareness of and Input into Cultural Heritage Preservation by Urban Planners and Other Municipal Actors in Light of Climate Change

Paul Carroll * and Eeva Aarrevaara

Faculty of Technology, LAB University of Applied Sciences, Mukkulankatu 19, 15210 Lahti, Finland; eeva.aarrevaara@lab.fi
* Correspondence: paul.carroll@lab.fi; Tel.: +358-414645092

Abstract: Future climate conditions need to be considered in planning for urban areas. As well as considering how new structures would best endure in the future, it is important to take into account factors that contribute to the degradation of cultural heritage buildings in the urban setting. Climate change can cause an increase in structural degradation. In this paper, a review of both what these factors are and how they are addressed by urban planners is presented. A series of inquiries into the topic was carried out on town planning personnel and those involved in cultural heritage preservation in several towns and cities in Finland and in a small number of other European countries. The target group members were asked about observed climate change impacts on cultural heritage, about present steps being taken to protect urban cultural heritage, and also their views were obtained on how climate change impacts will be emphasised in the future in this regard. The results of the inquiry demonstrate that climate change is still considered only in a limited way in urban planning, and more interaction between different bodies, both planning and heritage authorities, as well as current research on climate change impacts, is needed in the field.

Keywords: cultural heritage; urban planning; climate change

Citation: Carroll, P.; Aarrevaara, E. The Awareness of and Input into Cultural Heritage Preservation by Urban Planners and Other Municipal Actors in Light of Climate Change. *Atmosphere* **2021**, *12*, 726. https://doi.org/10.3390/atmos12060726

Academic Editor: Yasemin D. Aktas

Received: 15 March 2021
Accepted: 28 May 2021
Published: 6 June 2021

Publisher's Note: MDPI stays neutral with regard to jurisdictional claims in published maps and institutional affiliations.

1. Introduction

Urban planning involves catering for the needs of urban dwellers and those using the urban area in the future. It needs to allow for growth in population, for increasing traffic and for any other future conditions that are likely to arise. Town and city planners need knowledge from a range of fields, including future foresight skills, in addition to the required knowledge of urban sustainability, engineering and architecture. Urban planning needs to also consider the existing environment and heritage and the impacts of the possible changes due to urbanisation and other phenomena like climate change. In the case of cultural heritage buildings and areas, it is usually a matter of this being defined in advance by authorities such as museum and antiquities departments, thereby communicating to planners the necessity to incorporate their preservation as part of the urban plans. Others involved in cultural heritage preservation need local knowledge regarding the relative value of different sites, comparative knowledge of threats or damage already incurred to similar structures elsewhere, among other skills. The community's cultural values are also found to be of importance for the future preservation opportunities, which involve not only authorities, but community members and stakeholders with the assistance of participatory planning [1].

Cultural heritage in the broader sense consists of archaeological remains, cultural sites and environments together with the infrastructure they involve, including building exteriors and interiors. There is also an ethnographical distinction between such material cultural heritage and that of the immaterial type (culture, customs, ceremonies, storytelling, music, etc.). This paper deals with tangible, immovable cultural heritage [2].

National legislation gives instructions for implementing the national planning system, usually incorporating different levels of more general or detailed plans in urban areas. It has been noticed that several European countries launched an official planning system linked with the preservation of cultural heritage after the second World war [3]. From the perspective of urban planning, it is essential to comprehend that preservation needs to be considered from the viewpoint of larger areas, and not only separate buildings. The concept of preservation is dynamic and changes over time. The UNESCO convention [2] recognised monuments, groups of buildings, and sites as different forms of cultural heritage. However, Janssen et al. [3] argue that there is still a gap remaining between the integration of planning and the preservation of historical built environment. In this article, we explore this argument in the context of authorities working in both fields and the special focus is on the climate change impact in cultural heritage and its consideration in urban planning. It is not a coincidence that more attention was paid to cultural heritage in the post-war period, because increasing urbanisation and planning ideologies like modernism were also starting to threaten traditional human-scale environments, together with the impact of growing traffic networks. The importance of preservation of cultural heritage is also connected with the tradition of preservation and repairing existing buildings and neighbourhoods as a symbol of living urbanism and urban ecology [4].

Threats caused by global warming have been recognised primarily in the biggest cities, several of which are also very vulnerable due to their coastal situation [5]. World Heritage cities, especially, have been researched and described due to their possible vulnerability risks all over the world. A special index has been given by a World Bank research study to these cities, evaluating risks from low to extreme, with a division into five different categories. These risks depend on the situation of the city and also on the socio-economic status of the residents [6]. Climate change has increasingly become a factor to consider in planning the towns and cities of the future [7]. The growing interest of this field is demonstrated, for example, by Jiang et al. [8], in a review focusing on the amount of published research dealing with climate change in urban areas from 1990 to 2016. The number of publications started to increase strongly after the early 2000s, especially after 2007. The fourth IPCC Climate Assessment Conference was arranged in 2008, and together with the United Nations Environmental Programme (UNEP), the theme of "low-carbon society" was launched. Cultural heritage has an interesting position in the climate change discussion, while the preservation of historic buildings and infrastructure also benefits the communities by saving energy and reducing carbon footprint in the building sector [9].

In the case of cultural heritage, aspects of weathering and pollution damage from acid rain and flooding, for example, have long been recognised as threats, leading to preservation action [5]. The threats of increasing temperature combined with precipitation, flooding, humidity and wind have clearly been identified to cause more damage to cultural heritage [10]. The challenge of resisting the impacts of climate change on buildings that possess cultural value is often connected with changes in social structures, such as changes in the use or maintenance of a building, which together can increase the negative impact on the condition of the object. Cultural changes can also affect the use of buildings and their maintenance. UNESCO [11] (pp. 64–65) and Haugen [12] distinguishes between direct and indirect impacts of climate change on people and the environment, speculating that indirect impacts, such as policy changes related to higher energy efficiency in protected buildings, may be of even higher relevance than direct ones.

The field of climate change impacts on cultural heritage has attracted growing interest in scientific and more general articles since the early 2000s [13,14]. Additionally, Sabbioni et al. [13], for example, previously presented some steps for recognising the threats that cultural heritage is facing, especially considering UNESCO World Heritage sites, which were brought into the wider discussion within the conventions in 2007. In the literature review of Fatorić & Seekamp [1], the authors selected and reviewed 124 articles dealing with climate change and cultural heritage or cultural resources (a concept commonly used in the USA). The first paper that examined the role of cultural heritage in coastal areas and the need for sustainable strategies was published in the Journal of Cultural Heritage in 2003 [1,15]. In a study in the Netherlands, cultural heritage and environmental experts, as well as climate change experts, were interviewed about their observations with respect to culture preservation, its management, and its connection with climate change. Progressive thinking was identified in this regard, but the changing contents of management were also mentioned [16].

The extent to which those responsible for decision making and background preparations regarding cultural heritage buildings and sites, such as urban planners, are aware of the potential impact of climate change on these buildings and areas is of particular interest to the authors of this paper. In addition, from the perspective of planning, to best be able to address the threats in the future, it is important to be able to target the correct decision makers and this perceived gap of integration between planning and preservation [3]. It is the purpose here to discuss and analyse the factors involved in this aspect of the protection process. Furthermore, it is the intention to determine whether and to what extent urban planners specifically take these climate change factors into account when making short- and long-term plans for their towns and cities. Of interest to this study is how empowered urban planners feel they are in addressing these issues. One goal is also to review the level of common understanding of, and degree of preparation for climate change in the cities included in the inquiry.

The viewpoint of managing cultural heritage has been presented in previous research; for example, Philips [17] reported interviews with organisations and authorities managing cultural heritage sites in the United Kingdom, in which the main conclusion was the urgent need for updated management plans and resources for heritage sites. Sesana et al. [10] interviewed European experts in three different countries (the UK, Italy and Norway) concerning their views about the adaptation of cultural heritage to climate change and the challenges and barriers they identified. The interview results also contained suggestions for improving adaptation measures, which mostly concentrated on the management and repair of cultural heritage, but also included, e.g., cooperation with communities, stronger regulation, and development of good plans.

However, Hall et al. [18] argued that engagement with the impact of climate change on cultural heritage in planning is a challenge, due to the long-lasting time scales typical of planning processes.

The reason to explore the urban planning authorities is to widen this perspective to an area which, in principle, should consider cultural heritage as a significant element of existing environments, which are the target of urban planning actions. The input of this particular group of civil servants is important, because their statements play a major role in influencing decision makers to divert public funds towards what is regarded as necessary for preservation.

2. Literature Review

Climate change is defined by the United Nations as being part of the larger challenge of sustainable development [19], and therefore addressing the sustainability of cultural heritage sites also addresses climate change impacts on them. Leus & Velhelst [20] investigated and reported on the sustainability of urban heritage sites and emphasised that sustainability can be subjective, meaning different things to different people. This is also relevant to urban planners, who are required to make choices on matters of urgency with respect to conservation, which can be based on the perceived value of certain sites or buildings compared to others in situations where public funding is limited (private funding or heritage trusts may be an option).

UNESCO and other relevant organisations, like the International Council on Monuments and Sites (ICOMOS), have already been aware of the threatening impacts of climate change on cultural heritage for several years. At a conference devoted to the topic, ICOMOS made a resolution including the recommendation that "climate change adaptation strategies for cultural heritage should be mainstreamed into the existing methodologies for preservation and conservation of sites, buildings, settlements, landscapes, movable objects and the living traditions and that appropriate standards and protocols should be developed for the purpose" [21]. They have also been monitoring the current changes and anticipating the future conditions dealing with heritage sites. The Intergovernmental Panel on Climate Change (IPCC) has introduced scenarios up until 2100 to demonstrate the different possible outcomes of increases in temperature. The Convention concerning the Protection of the World Cultural and Natural Heritage, which was adopted by UNESCO in 1972, already formed the basis for heritage protection in general [11] (pp. 12–15). Every country has a national legislation which adapts intergovernmental decisions considering the national and cultural situation and needs [22]. The IPCC did not specifically mention cultural heritage in association with climate change impacts until 2014 in their extensive Synthesis Report, and did so only in a limited way, referring to protecting built assets in general [23]. Although several international agreements have been ratified by European countries, there are still many administrative steps required to realise the goals of supporting and protecting cultural heritage in local-level planning. National legislation is usually drafted in coherence with the relevant international treaties, as well as with the guidance or under the control of state authorities. In local policies, the position and appreciation of cultural heritage, as well as public opinion, can have more variation, which influences the attitude and opportunities to prioritise the importance of the cultural environment in urban planning processes.

The climate impact can also be divided into direct physical impact on the cultural environment and social impact on the community and its way of living, especially when indigenous built environments are considered [11] (p. 64).

Howard [24] points out that it is often important to recognise how closely the location of cultural heritae is connected with the landscape, due to several reasons effecting the choice of suitable places. For example, hill tops and steep slopes provided suitable locations for fortresses and religious places, as well as rivers and watercourses supplying connections for trade and early industry. The understanding of the geomorphological and geological context of cultural heritage sites assists in figuring out also the present challenges and future threats by climate change.

Looking at the situation in Norway, Haugen [12] concluded that cultural heritage owners and local authorities need information and training to be able to limit the negative effects of climate change, adding that paying attention to ordinary maintenance and intensifying repairs during "normal times" may give a buffer effect for extreme conditions. She explained that knowledge of local conditions and of the risk of damage from exposure to cultural heritage buildings and sites can help in designing their protection, but that changes in the type of exposure, as is the case due to climate change, lead to unpredictable problems.

Jabareen [25] stresses how city planning has significant potential to influence adaptation and mitigation measures for dealing with climate change effects; he then analysed the contents of 20 city plans from large cities in both developed and developing countries to determine the extent to which this opportunity has been availed of. Jabareen's conclusions were that, although some cities, such as New York, London and Paris, went beyond their national governments' recommendations in terms of planning for climate change, especially with respect to adaptation, for the most part, the broader opportunities that planning could offer in this regard were not availed of. He listed problems like pandemics, street violence, poverty and economic instability as all receiving more attention than climate change.

Hasse [26], while specifically describing the development of an urban water management plan to adapt in advance to the effects of extra precipitation resulting from climate change, stressed that it took a major flood event, causing extensive damage, before it becomes a public/political issue in the part of Germany in question, and only then did it attract the interest of municipal decision makers. Hasse goes further to conclude that existing planning documents need to be developed to include the pro-active approach of "Water Sensitive Urban Design" to anticipate such needs, mentioning e.g., municipal risk monitoring as a necessary addition to aid planning.

According to Carmichael et al. [27], climate change adaptation planning rarely makes cultural assets a primary concern, referring to the views of Adger et al. [28] that it is a mistake not to do so, as things which give meaning to people's lives, such as cultural heritage sites, need to be an essential part of adaptive planning.

The adaptation plans are considered crucial in historic environments for their future preservation. According to UNESCO [29], the existing action plans do not contain heritage as a definite priority, but concentrate on adaptation of climate change and mechanisms to respond to disasters, relevant examples of proper integration are rare. The lack of city vulnerability assessment methodologies which could integrate the impact of flooding and extreme precipitation with cultural heritage is also referred to [30].

Some essential concepts are used to describe the vulnerability of cultural heritage which is under the risk of some natural hazard. Risk is defined as the likelihood that historical building will be damaged by a hazard. Risk is affected by exposure and vulnerability, where exposure describes the physical location of heritage. For example, risks are usually detected near waterways, like by the sea, on a lakeshore or along a river. Historic buildings are often likely to suffer hazards due to their age, material, or method of construction. Cultural heritage has also been noticed to recover more slowly from natural hazards than the overall built environment [9].

The Republic of Ireland recently drew up a Climate Change Sectorial Plan for Built and Archaeological Heritage [31], where when referring specifically to the threats to cultural heritage, those listed as priorities for adaptation planning include flooding (inland and coastal), storm damage, coastal erosion, soil movement (landslip or erosion), changing burial preservation conditions, pests and mould, wildfires and maladaptation. This further identifies the need for a set of recommendations that goes beyond the scope of the National Monuments Act and the Planning and Development Act, because of the fact that much of the heritage in question remains in private ownership. The castle in Figure 1 is a good example of a major urban cultural heritage site in Ireland that is addressed in this plan.

Figure 1. King John's Castle in Limerick, Ireland, built from 1200–1212, is an example of a heritage building in need for monitoring and protection from the elements. Photograph by William Murphy.

The organisation Historic Scotland has updated its climate change action plan, with the earlier version covering the period 2012–2017 [32], while the more recent programme deals with the period 2020–2025 [33]. The latter programme largely discusses the means for protecting cultural heritage, including carbon management and improving energy efficiency in historic buildings. The Northern Ireland Climate Change Adaptation Programme [34] covers the period 2019–2024, and provides similar viewpoints with respect to climate change adaptation and cultural heritage.

A national cultural environment strategy was published in Finland covering the period 2014–2020, and is currently undergoing a review process. The strategy was evaluated under a separate process and one finding was that the strategy was not adapted at regional and local levels in a sufficient way. Additionally, the strategy did not contain any discussion on the impacts of and adaptation to climate change [35,36].

General awareness on the topic has been increasing, and there are many ongoing research studies addressing it. Significant interest has targeted the impacts and phenomena that climate change is causing to cultural heritage including buildings, structures and sites and what kind of management plans these areas need to mitigate climate change. On the national level, countries like the Republic of Ireland, Northern Ireland and Scotland have launched well-established mitigation and adaptation strategies that consider cultural heritage, while Finland has not recognised the impact of climate change on cultural heritage in its latest strategy paper. However, it is discussed in local level urban and town planning and in decision making, but in a more limited way.

3. Framework for an Explorative Study

The authors proposed in their earlier work a table (Table 1) dealing with evaluation of the urgency of different climate-change-driven situations in the built environment [37] (pp. 5–6). The table is based on several research studies dealing with the most common phenomena identified in climate change impacts on cultural heritage [38–40]. This table is included here in order to compare what the authors previously proposed as the relative urgency in responding to different climate change impact factors on the one hand, to those that the interviewees in the present study most frequently referred to.

Table 1. Table demonstrating causes, results, and proposed level of urgency for acting in case studies in Southern Finland, in which the urgency rating * is classified as follows: 1—A mild or minor perceivable long-term effect (100 years or more), 3—A major perceivable long-term effect (50–100 years), 5—A mild or minor perceivable short- to mid-term effect (1–50 years), and 10—A major short- to mid-term effect (adapted from Carroll & Aarrevaara 2018 [37]).

Climate Change Category	Measure or Scale	Result/Effect	Materials/Structures Affected	Proposed Urgency Rating *
Warmer climate	Rise in degrees C/year	Freeze—thaw damage	Stone Brick	3
		Rust	Metal	5
		New fauna—pests	Wood Brick	5
Longer growing season	Days/year	New/increased flora, algae, moss, root damage	Wood Brick Stone	5
Increased precipitation: rain or snow	mm/year	Humidity	Wood Brick structures	10
		Increased loads (snow)	Wood Brick Roof/roof structures (typically wood)	5–10
		Soil and material degradation	Foundation base floor	5
		Flooding (from any increased precipitation effect)	Wood Brick structures	10

Table 1. *Cont.*

Climate Change Category	Measure or Scale	Result/Effect	Materials/Structures Affected	Proposed Urgency Rating *
Severe rain incidents	mm/h	Erosion	Wood Brick stone	5
Extreme winds	m/s	Damage to structures through falling trees or wind causing damage to the roof	Metal roofs Wood & brick structures	5–10

The previous literature research indicates that in the sector of research and maintenance institutions concentrating on areas of world heritage and national heritage, the understanding of the threats caused by climate change is comprehensive [29,41].

The table above formed one set of background data compiled previously by the authors, and partly acted as a motivation for the present explorative study, which was prepared to gather information on the current situation among urban planners in cities and towns, not representing capital regions, in Finland and other European countries. The scale of urgency proposed here makes the assumption that major effects in the short or mid-term are what most need to be addressed, and therefore it can be assumed that this would have more relevance to heritage protection in present planning. Although the lowest score of 1 for urgency was defined, it was not seen as being applicable to any of the materials listed due to the prevailing accelerating rate of climate change. Table 1 was therefore prepared to serve as one aid to decision making for urban planners or other heritage protection bodies in prioritising the types of cultural heritage structures that most urgently need protection, based on their composition.

The explorative study was prepared in the format of a Webropol inquiry containing 13 different questions. The inquiry was sent in early autumn 2019 by email to a 50 urban planners or heritage authorities in Finland, Ireland, the United Kingdom (also to several recipients in Sweden, Italy and Germany, but with no answers received from these three countries) with the request to share with their colleagues and contact persons in the local city's urban planning staff. Reminders were sent two times during autumn 2019. Although the number of respondents did not turn out to be very high—there were seven responses received from Finland and three from other European countries— the inquiry still served as an exploration of a new area the authors were interested in getting in touch with.

The Finnish respondents amounted to seven in total, and the size of the cities/towns of the responses varied from less than 20,000 to more than 140,000 inhabitants.

Respondents from other countries amounted to three in total, and these were from the three small to large cities Limerick, Belfast and Glasgow, with the respondents in each case being very much involved in planning and/or cultural heritage issues. The contents of the questionnaire are given in Table 2, after which the results received are summarised and, later in this paper, discussed.

Table 2. List of the inquiry questions.

1. Background information.	Name of the respondent, contact information, organisation, position in the organisation
2. What town or municipality do your answers refer to and when was it established /how old is it?	Open question
3. Does your town/area have nationally valuable built cultural heritage sites (buildings or areas)? Please list them briefly.	Open question
4. The general attitude to cultural heritage in your town or area.	Answers by Likert scale

Table 2. *Cont.*

5. If there is public debate in your town or area about the preservation of buildings or sites, does the issue of climate change arise?	Open question
6. Have there been any studies carried out to determine the possible effects of climate change in your town/area, e.g., flooding risks? Please mention any that you are aware of.	Open question
7. List buildings or sites (from 1 to 3 different examples) in your town or area where the impact of climate change is clearly visible. Please mention in each case when the building or site was constructed and from what main materials.	Open question
8–10. For the first/second/ third building or site example mentioned above tick the observed factors caused by climate change.	Open question
11. In your opinion how should the effects of climate change be considered in areas that have valuable cultural heritage?	Open question
12. Can town planning be used to influence the preservation of buildings in terms of reducing the impact of climate change? If so, please mention in what way.	Open question
13. Please add give any additional comments on this theme that you might have.	Open question

4. Results

After asking for the background details of the respondents and their cities, the respondents were asked about the nationally valuable built cultural heritage sites and buildings (Question 3). In Finland, the number of cultural heritage sites depended on the history and the area of the city. Some cities have merged with the surrounding municipalities. In Ireland and the UK, the exact number of nationally valuable areas was not specified; therefore, it was simply designated as being above a certain amount. Table 3, below, gives, for the sake of comparison, numbers of relevant heritage sites in the different towns and cities involved in the study, although it should be noted that for Ireland and Britain the numbers come from protected structures or listed buildings, which is not exactly the same thing as a Finnish heritage site. Additionally, when the responses covered the number of "listed buildings", they did not specify different value categories for them.

Table 3. The respondents were asked about the existence of nationally valuable areas and buildings (environments) in their city area, and the results can be presented as follows. The population of each city is provided in the table.

Name of the City/Town/Year of Foundation	Inhabitants (2019)	Amount of Nationally Valuable Heritage Sites
Porvoo/1602	50,380	17
Jyväskylä/1837	142,400	15
Heinola/1839	18,667	7
Hyvinkää/1960	46,470	4
Lappeenranta/1649	72,634	12
Lahti/1905	119,823	14
Vihti/1867 *	29,158	6
Limerick/1199 (city status)	94,192	>2000 **
Belfast/1888 (city status)	311,512	>2000 ***
Glasgow/1170	620,000	>1800

(*) Independent municipality. (**) Limerick city and county. (***) All Northern Ireland.

Statements about their own perceptions of the local common attitude to cultural heritage were requested (Question 4) by giving the response options: 1. Negative 2. Fairly negative 3. Neither negative nor positive 4. Fairly positive 5. Positive. In Finland the common attitude was considered by those who responded to be relatively positive on the whole, although 2/7 answers chose N/A, 3/7 considered it fairly positive and 2/7 considered the attitude positive. Although this is only the view of those who completed the survey of what the attitudes of the general public might be, the researchers regard these answers as quite reliable based on the familiarity with such issues obtained from their work as urban planners or involvement in cultural heritage tasks. None of the respondents considered the common attitude regarding cultural heritage as negative. In the other three cities in Britain and Ireland, the answers to this were all also fairly positive. Question 5 dealt with the connection between preservation of cultural heritage and climate change: If there is public debate in your town or area about the preservation of buildings or sites, does the issue of climate change arise? This connection was not identified by the respondents.

Previous studies dealing with climate change impacts were inquired about as follows (Question 6): Have there been any studies carried out to determine the possible effects of climate change in your town/area, e.g., flooding risks? Please mention any that you are aware of. In Finland all the respondents mentioned some kind of stormwater programme prepared for the city or the town in question. The reports can be divided into flooding risk reports and stormwater management programmes. Flooding reports are prepared at the national level, where the Finnish Environmental Institute has published a map of the significant flooding risk areas 2018–2024 covering 22 different areas where flooding can be caused by sea level or water course level rises [42]. Figure 2 is from this publication.

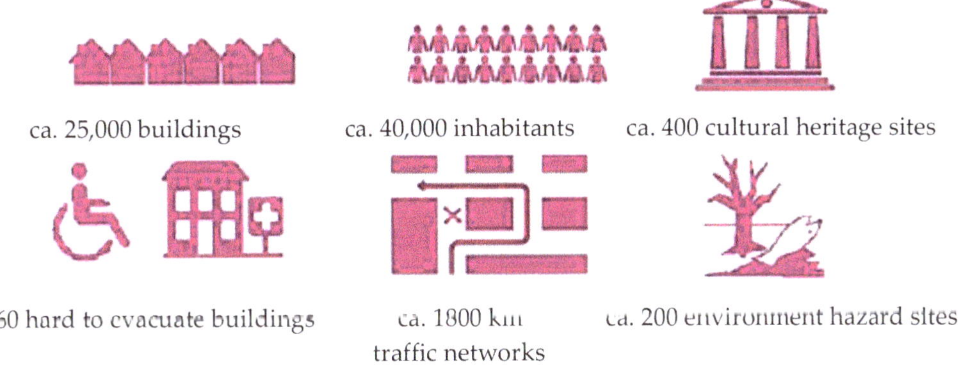

Figure 2. Indicators for significant flood risk areas in Finland (Based on FEI [42]).

The largest municipalities stated that in the early 2010s, stormwater programmes were prepared covering the whole city area, and these are already being updated. Several cities have also established the practice of adding a separate stormwater plan to all town plans. The responses to this question also expressed these new practices in Finland. It was noteworthy that no other actions, reports or plans were reported as being available or in preparation, considering the impact of climate change in urban areas.

Some Finnish respondents had noticed the increase of flooding, both in lakeshore and seashore situations. One respondent mentioned that in their city, situated on the shore of a large inland lake, the lowest altitude at which to build had previously been determined according to the highest flooding level once in 50 years. Recently, this requirement was changed to once in 100 years, due to the increase in flooding tendencies. Three other urban planners also mentioned the significant increase of flooding, and also surface water flow into the basements of town houses.

The following question (7) covered observations of buildings and sites where the impact of climate change was clearly visible. Detailed information about the impacting factors was asked for according to the following list (also based on Carroll & Aarrevaara [36]). The respondents were able to present a maximum of three different examples, but the question was mostly responded to with only one such example being referred to (Questions 8–10). The results are summarised in Figure 3.

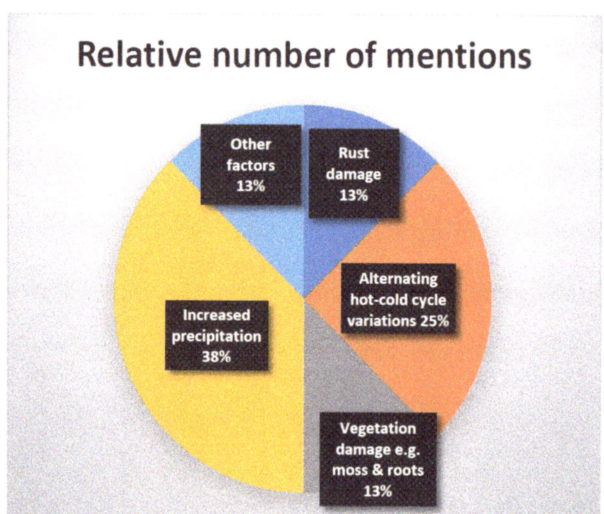

Figure 3. Observations of urban planners dealing with climate change damage factors in the cultural environment.

The respondents were asked to indicate their observations concerning the following factors:

1. Increased precipitation
2. Increased stormwater amounts
3. Increased flooding
4. Increased loads from snow
5. Alternating hot–cold cycle variations
6. Increased wind
7. Rust damage
8. Damage from new fauna species
9. Vegetation damage, e.g., from moss and roots
10. Any other factors, what are they?

4.1. Effects of Climate Change in Areas with Valuable Cultural Heritage

The respondents were asked their opinions about what impacts of climate change need to be considered in areas that have valuable cultural heritage (Question 11). The planners and other experts gave varied responses, but the predominant observation was connected with stormwater management and flood protection. The following excerpts contains both direct quotations and freely reported comments (translated from the Finnish where necessary).

In one response actions at governmental level were expected as a matter of urgency:

In the short term, climate adaptation measures should be identified, and action taken to protect, as far as possible, important assets from climate change. In Belfast, this particularly relates to flood risk and sea level rise.

The other factors mentioned, accounting for a total of 13% of all factors mentioned in the questionnaire, were: sea-flooding, wind and storms. Eight out of ten of the responses included a mention of stormwater management, while four out of ten mentioned flood protection specifically. Plans for flood protection and stormwater infiltration are needed and implemented (Porvoo). Stormwater management should consider better foundation drainage systems and water drainage especially in area development (Jyväskylä). Where the urban area is situated next to a shoreline, protective structures are needed to prevent flood damage (Lappeenranta). It was noticed that there are definitely cultural environments threatened by flooding. In these locations the situation is particularly problematic, because the relocation of buildings interrupts the pattern of linking of the site to the environment and landscape (Vihti).

For the most part, the same kinds of factors we mentioned for the Finnish cities and the Irish and British ones, except for some additional factors such as sandstone structures and failed gutters and downpipes in Glasgow and ivy growth damaging walls in Limerick. There are not many sandstone structures in Finland, and the plant ivy does not grow in the wild in its climate zone. Most Finnish traditional buildings are either made of wood, or of brick with plaster on the surface. In Figure 4, the challenge of having major cultural heritage buildings and areas in an area regularly exposed to potential water damage is highlighted.

Figure 4. The Old Town part of the City of Porvoo is situated along the river and not far from the sea, where the risk of flooding is a current and relevant issue. Photograph by author Eeva Aarrevaara.

Other responses considered both stormwater management and vulnerable qualities of traditional buildings at the same time, and thereby the relationship between the building and the environment. Special care should be paid to preventing the sealing of the ground in cultural environments and to providing adequate tilting away from buildings; additionally, the air should be allowed to circulate around the building (Lahti, Vihti). It was also mentioned that carbon-neutral forms of energy should be available in urban areas to mitigate climate change impact (Porvoo).

Improvements to buildings were suggested to ensure their survival and resistance against increasing rain, wind and snow (Porvoo). It was also noticed that in the repair of buildings, the introduction of new construction standards, e.g., thermal insulation, usually causes damage to old structures and should not be demanded (Heinola). A suggestion was made to be able to replace a part of a valuable building and build a replica in its place, in a case where the building is considered significant and valuable. In general, the identity of cultural heritage and its local representatives should be considered more deeply than before (Hyvinkää). For the cultural environments it is essential that the buildings remain in use and are maintained (Vihti). Some respondents identified the need for extra funding to be made available to maintain the cultural heritage, and also in cases to raise the awareness of historical values. Extra funding could be applied in the form of grants, which might deal with treatment for dampness or for additional rethatching of roofs (since thatching does not last as long as before) would help (Limerick). A summary of the main cultural heritage sites in each city and the perceived threats is given in Table 4 below.

Table 4. Details of relevant cultural heritage and threats. This table introduces the target cities which the respondents represent, as well as the characteristics of the cultural heritage in each city. Additionally, the table summarises the observations of the respondents dealing with climate change impacts and damage in their city area.

City, Country	Types of Cultural Heritage Sites	Climatic Factors	Damage Caused
Porvoo, Finland	Medieval church and town plan, wooden old town	Increased sea- and river-flooding, stormwater	Damage to the buildings near water level
Jyväskylä, Finland	Town centre with church park and university area, industrial environments	Lake flooding, stormwater	Damage by stormwater
Heinola, Finland	Governor's residence park, old teacher's college, wooden town buildings	Stormwater	
Hyvinkää, Finland	Railway station, industrial environments	Stormwater	
Lappeenranta, Finland	Old fortress and garrison, town hall and Saimaa canal	Lake flooding, stormwater	Damage by stormwater
Lahti, Finland	Town hall, market square and ceremony axis (church -town hall), railway station, garrison	Stormwater	Damage by stormwater, building basements
Vihti, Finland	Railway stations, village landscapes	Stormwater	
Limerick, Ireland	Medieval castle, town walls, Georgian Quarter	Increased wind, increased precipitation	Ivy damage to walls, vegetation damage from moss and weeds, stormwater increase
Glasgow, Scotland	Cathedral and several historic buildings	Increased precipitation	Sandstone decay, failing gutters and downpipes, damage from moss and roots
Belfast, N. Ireland	Industrial archeological sites, vernacular sites, historic parks and gardens	Flood risk	Sea level rise (damage not specified)

4.2. The Opportunities of Town Planning to Preserve and Reduce the Impact of Climate Change

The opportunities of town planning to influence the preservation of buildings in terms of reducing the impact of climate change were asked about at the end of the inquiry (Question 12). The responses received considered the different options of zoning and its contents, as well as aspects dealing with stormwater management. It was possible to identify different factors and phases in the town planning process through the responses. In general, the requirements for a sustainable planning process and its goals were described. Additionally, the importance of qualified background research and reports was mentioned. It was noticeable that communication and participation were not highlighted in the material, but only mentioned in passing. This could be interpreted as meaning that the climate change impact on cultural heritage is considered to require professional-based knowledge and understanding.

The usefulness of searching for different options and alternatives in the planning process was recognised as a positive feature. A good plan was described as being one that contains explicit regulations for preserving cultural heritage. Clear planning instructions for dealing with stormwater in urban areas were also required on a large scale. The importance of identifying the connection between a town plan and building permits was also mentioned. In the end, some uncertainties were identified, such as changes of legislation (in the case of Finland), and political decision-making processes and their impact on planning. Table 5 summarises the respondents' findings concerning the planning process and its contents. It can be noticed that the same observation is also shared by as many as 5 respondents connected with stormwater management solutions.

Table 5. The material from responses is grouped into different levels according to the general to detailed nature of the comments.

PROCESS LEVEL	Zoning aims to promote a sustainable community structure, which at best tackles the progress of climate change (Vihti, Porvoo, Belfast). Adaptation measures include reducing the overall need to travel, promoting active travel, sustainable design and reuse of materials. Adaptation measures include flood protection and green design (Belfast).
BACKGROUND STUDIES, SEPARATE PROGRAMMES	Separate studies and plans related to the zoning process, (e.g., stormwater management, building condition surveys) contribute to the preparation of zoning regulations as well as to the further planning and implementation of the immediate environment, supporting the preservation (Jyväskylä, Heinola, Porvoo).
COMMUNICATION	Adequate discussion and involvement (Porvoo).
ALTERNATIVE SOLUTIONS	Development of options and impact assessment (Porvoo). Instead of demolition it's possible to find solutions for protection together with suitable additional and supplementary construction (Porvoo).
FEATURES OF A GOOD TOWN PLAN (Connected with Preservation and Climate Change Adaptation)	Binding planning regulations for other environments and the building stock (Jyväskylä). Various protection regulations play a key role (Jyväskylä). Structural prevention of heavy rain and flood damage, flood route control, stormwater planning and planned space allocations (Lappeenranta, Lahti, Vihti, Limerick, Glasgow).
CONNECTION TO BUILDING PERMITS	Building permit processes could be managed better when there is already a plan for applying for a permit or notification procedure (Hyvinkää).
UNCERTAINTIES	Renewal process of the existing legislation will cause a new situation—uncertainty about the opportunities for building protection (Hyvinkää). Political decision-making process in zoning—changes during the process always possible (Hyvinkää).

4.3. Discussion Dealing with Themes of the Questionnaire

The questions were also reflected on by the respondents with different connections to larger discussions in society. One city planner (from a coastal city along a river) critically evaluated the starting point of the questionnaire as follows: The influences of climate change are clearly visible—can we already say this? But sea, river and stormwater floods are nowadays more common. This statement describes the situation in which the impacts of climate change might not be clearly visible or considered as the most urgent problem in urban planning. Another respondent identified a connection between preservation and mitigation: Preserving buildings seems to be a principal means of approaching climate change mitigation in general (City planner, a coastal city in Finland).

The importance of background studies in planning processes was highlighted and several benefits were detected in a thorough preparation to the actual plan. Separate studies and plans related to the zoning process, such as stormwater management, building condition surveys etc., also contribute to the preparation of zoning regulations as well as to the further planning and implementation of the immediate environment, supporting the preservation of the built cultural environment and mitigating climate change. (City planner, a lakeshore city in Finland.) Additionally, the connection with sustainable communities, urban planning and climate change was identified in a response: On a larger scale, zoning aims to promote a sustainable community structure, which at best tackles the progress of climate change (City planner, an inland town in Finland).

5. Conclusions and Discussion

Urban planning has traditionally involved accounting for present needs and anticipating future ones; in the past, this has meant integrating the potential for growth in the population and the resulting greater demands on the urban infrastructure, such as in the volume of traffic. In a time of climate change there is a new set of factors to be considered by urban planners, with corresponding new risks to be considered. Even with the ambitious scenario according to the Paris agreement of 2015, the climate will continue to change for some time, at least to a certain extent, while there is the danger that this goal of a temperature rise of only 1.5 °C will not be reached, and there could be larger changes [43]. These uncertainties provide big challenges for urban planners in such matters as stormwater control, flood control, snow loads on buildings, and catering for high-speed winds. Cultural heritage is not usually the first priority when it comes to the safety of the population in emergency situations, where drinking water, sanitation and other basic services take priority. Nevertheless, cultural heritage buildings and sites that were constructed in a pre-climate-change period and have been preserved up until now for the common good will now need to be able to withstand additional stresses from the impacts of climate change.

This paper sets out mainly to identify how urban planners and heritage authorities perceive these challenges and how they take them into account in their work, or intend to do so in the future. The literature review provided mostly circumstantial information and findings, with these being related to international activities and trends. To address the specific question of how urban planners consider and implement climate change impacts on cultural heritage in particular, the authors carried out a small-scale survey, the results of which are provided above. As a summary, it can be stated that there is good knowledge of the general and often specific threats involved, even when these urban planners are not in a position to prioritise heritage buildings over other urban aspects in their work. However, they usually need to pay special attention to cultural heritage as a part of urban planning processes, and also to justify any actions caried out in environments containing traditional buildings. In the responses, the position of stormwater management and flooding threats was a dominant aspect of the contents. It can be stated that there is a growing need to discuss other threats that climate change is evidently causing to cultural heritage in European countries, and to facilitate for a more detailed understanding of the phenomenon among urban planners. Differences between national-level strategies were also discovered: the Republic of Ireland, Northern Ireland and Scotland have updated strategies for cultural heritage under climate change, while Finland had a more limited strategy for cultural heritage, not taking climate change into account. The authorities working in museums and other institutions responsible for cultural heritage possess specific knowledge, which should be better considered in urban planning. This need was already recognised in the introduction and literature review of this article [3,30]. Updated national strategies can also help in awareness raising among different professional groups.

The urban planners, and specifically employees responsible for urban heritage, who answered the survey were well aware of how well cultural heritage matters were prioritised in their cities and had realistic and constructive suggestions for what would need to be done to protect the buildings and sites in question in the future. Although limited in scope geographically (to North and West Europe) and in number of respondents, the views expressed can be regarded as being well indicative of those of urban planners as a whole. The direct quotes provided give perhaps the most valuable results of this questionnaire (free translations from the Finnish are those of the authors), offering a range of perceived needs and proposed solutions. The question about their own perceptions of attitudes is only of limited value, but it does give some indicative background information about how they reflect on the urban heritage environment in which they operate.

Summarising the findings in the literature and this exploration in the field, one can confirm the statements from the previous studies raising the awareness of the gaps arising when attempting to deal simultaneously with climate adaptation, urban planning and cultural heritage. The existence of different threatening phenomena is recognised as such, but the interdependence between them needs more research and interdisciplinary discussion to improve the connectivity between urban planning and climate adaptation in the perspective of cultural heritage.

Author Contributions: Conceptualization, E.A. and P.C.; methodology, E.A. and P.C.; investigation, E.A. and P.C. writing—original draft preparation, E.A. and P.C.; writing—review and editing, E.A. and P.C. Both authors have read and agreed to the published version of the manuscript.

Funding: This research received no external funding, but was enabled through the participation of the organisation in Finland, where the two authors are employed; LAB University of Applied Sciences, Faculty of Technology.

Institutional Review Board Statement: Not applicable.

Informed Consent Statement: Not applicable.

Data Availability Statement: Not applicable.

Conflicts of Interest: The authors declare no conflict of interest.

References

1. Fatorić, S.; Seekamp, E. Are cultural heritage and resources threatened by climate change? A systematic literature review. *Clim. Chang.* **2017**, *142*, 227–254. [CrossRef]
2. UNESCO, 2017 Earth Sciences; UNESCO Global Geoparks. Available online: http://www.unesco.org/new/en/natural-sciences/environment/earth-sciences/unesco-global-geoparks/ (accessed on 30 August 2018).
3. Janssen, J.; Luiten, E.; Renes, H.; Stegmeijer, E. Heritage as sector, factor and vector: Conceptualizing the shifting relationship between heritage management and spatial planning. *Eur. Plan. Stud.* **2017**, *25*, 1654–1672. [CrossRef]
4. Calthorpe, P. *Urbanism in the Age of Climate Change*; Island Press: Washington, DC, USA, 2010.
5. Hunt, A.; Watkiss, P. Climate change impacts and adaptation in cities—A review of the literature. *Clim. Chang.* **2010**, *104*, 13–49. [CrossRef]
6. Bigio, A.G. Historic Cities and Climate Change. In *Reconnecting the City*; van Oers, R., Bandarin, F., Eds.; Wiley Blackwell: Hoboken, NJ, USA, 2015; pp. 113–128.
7. Oke, T.R.; Mills, G.; Christen, A.; Voogt, J.A. (Eds.) *Urban Climates. Cambridge*; Cambridge University Press: Cambridge, UK, 2017.
8. Jiang, Y.; Hou, L.; Shi, T.; Gui, Q. A Review of Urban Planning Research for Climate Change. *Sustainability* **2017**, *9*, 2224. [CrossRef]
9. Rumbach, A.; Bierbrauer, A.; Follingstad, G. Are We Protecting Our History? A Municipal-Scale Analysis of Historic Preservation, Flood Hazards, and Planning. *J. Plan. Educ. Res.* **2020**, 1–14. [CrossRef]
10. Sesana, E.; Gagnon, A.S.; Bertolin, C.; Hughes, J. Adapting Cultural Heritage to Climate Change Risks: Perspectives of Cultural Heritage Experts in Europe. *Geosciences* **2018**, *8*, 305. [CrossRef]
11. UNESCO. *Case Studies on Climate Change and World Heritage*; UNESCO World Heritage Center: Paris, France, 2007.
12. Haugen, A. Preparations for climate change's influences on cultural heritage. *Int. J. Clim. Chang. Strateg. Manag.* **2011**, *3*, 386–401. [CrossRef]
13. Sabbioni, C.; Cassar, M.; Brimblecombe, P.; Lefevre, R.A. Vulnerability of Cultural Heritage to Climate Change. Council of Europe. 2008. Available online: https://www.coe.int/t/dg4/majorhazards/ressources/Apcat2008/2008_44_culturalHeritage_EN.pdf (accessed on 24 August 2018).
14. Bertollin, C. Preservation of Cultural Heritage and Resources Threatened by Climate Change. *Geosciences* **2018**, *9*, 250. [CrossRef]
15. Vallega, A. The coastal cultural heritage facing coastal management. *J. Cult. Herit.* **2003**, *4*, 5–24. [CrossRef]
16. Fatorić, S.; Egberts, L. Realising the potential of cultural heritae to achieve climate change actions in the Netherlands. *J. Environ. Manag.* **2020**, *274*, 1–9. [CrossRef] [PubMed]
17. Phillips, H. Adaptation to climate change at UK World Heritage sites: Progress and challenges. *Hist. Environ. Policy Pract.* **2014**, *5*, 288–299. [CrossRef]
18. Hall, M.; Baird, T.; James, M.; Ram, Y. Climate change and cultural heritage: Conservation and heritage tourism in the Anthropocene. *J. Herit. Tour.* **2016**, *11*, 10–24. [CrossRef]
19. UN DESA (United Nations Department of Economic and Social Affairs, Division for Sustainable Development). Climate Change Mitigation and Sustainable Development. 2007. Available online: https://sustainabledevelopment.un.org/content/documents/1489mitigation_paper.pdf (accessed on 2 February 2020).
20. Leus, M.; Verhelst, W. Sustainability Assessment of Urban Heritage Sites. *Buildings* **2018**, *8*, 107. [CrossRef]
21. ICOMOS (International Council on Monuments and Sites) Resolution at the International Workshop on Impact of Climate Change on Cultural Heritage New Delhi. 2007. Available online: https://www.icomos.org/climatechange/pdf/New_Delhi_Resolution_EN.pdf (accessed on 2 February 2020).

22. UNESCO. List of National Cultural Heritage Laws. 2021. Available online: https://en.unesco.org/cultnatlaws/list (accessed on 17 May 2021).

23. IPCC. *Climate Change 2014: Synthesis Report. Contribution of Working Groups I, II and III to the Fifth Assessment Report of the Intergovernmental Panel on Climate Change*; Core Writing Team, Pachauri, R.K., Meyer, L.A., Eds.; IPCC: Geneva, Switzerland, 2021; p. 151.

24. Howard, A.J. Managing global heritage in the face of future climate change: The importance of understanding geological and geomorphological processes and hazards. *Int. J. Herit. Stud.* **2013**, *19*, 632–658. [CrossRef]

25. Jabareen, Y. City planning deficiencies & climate change—The situation in developed and developing cities. *Geoforum* **2015**, *63*, 40.

26. Hasse, J. From vision to action: Roadmapping as a strategic method and tool to implement climate change adaptation—the example of the roadmap 'water sensitive urban design 2020'. *Water Sci. Technol.* **2016**, *73*, 2251–2259. [CrossRef]

27. Carmichael, B.; Wilson, G.; Namarnjilk, I.; Daly, C. Testing the scoping phase of a bottom-up planning guide designed to support Australian Indigenous rangers manage the impacts of climate change on cultural heritage sites. *Local Environ.* **2017**, *22*, 1197–1216. [CrossRef]

28. Adger, W.N.; Barnett, J.; Chapin, F.S., III; Ellemor, F. This must be the place: Underrepresentation of identity and meaning in climate change decision making. *Glob. Environ. Politics* **2011**, *11*, 1–25. [CrossRef]

29. UNESCO. What is Meant by Cultural Heritage? 2020. Available online: http://www.unesco.org/new/en/culture/themes/illicit-trafficking-of-cultural-property/unesco-database-of-national-cultural-heritage-laws/frequently-asked-questions/definition-of-the-cultural-heritage/ (accessed on 15 February 2021).

30. Gandini, A.; Garmendia, L.; Prieto, I.; Alvarez, I.; San-José, J. A holistic and multi-stakeholder methodology for vulnerability assessment of cities to flooding and extreme precipitation events. *Sustain. Cities Soc.* **2020**, *63*, 102437. [CrossRef]

31. Government of Ireland, Department of Culture, Heritage and the Gaeltacht. Built and Archaeological Heritage, Climate Change Sectoral Adaptation Plan. 2019. Available online: https://www.chg.gov.ie/heritage/climate-change/the-built-and-archaeological-heritage-climate-change-sectoral-adaptation-plan/ (accessed on 26 November 2020).

32. A Climate Change Action Plan for Historic Scotland 2012–2017. Available online: https://www.historicenvironment.scot/media/2611/climate-change-plan-2012.pdf (accessed on 17 May 2021).

33. Historic Environment Scotland. Climate Action Plan 2020-25. 2020. Available online: https://www.historicenvironment.scot/archives-and-research/publications/publication/?publicationId=94dd22c9-5d32-4e91-9a46-ab6600b6c1dd (accessed on 19 May 2021).

34. Northern Ireland Climate Change Adaptation Programme 2019–2024. Climate Change Unit, Environmental Policy Division, Environment, Marine and Fisheries Group, Department of Agriculture, Environment and Rural Affairs (DAERA). Available online: https://www.daera-ni.gov.uk/publications/northern-ireland-climate-change-adaptation-programme-2019-2024 (accessed on 19 May 2021).

35. Kulttuuriympäristöstrategia 2014–2020. Opetus- Ja Kulttuuriministeriö, Ympäristöministeriö. 2014. Available online: https://helda.helsinki.fi/bitstream/handle/10138/43197/Kulttuuriymp%C3%A4rist%C3%B6strategia_2014.pdf?sequence=1 (accessed on 17 May 2021). (In Finnish).

36. Kulttuuriympäristöstrategian (2014–2020) Arviointi. Valtioneuvoston Julkaisuja 2021:3. Available online: https://julkaisut.valtioneuvosto.fi/bitstream/handle/10024/162680/VN_2021_3.pdf?sequence=1&isAllowed=y (accessed on 17 May 2021). (In Finnish).

37. Carroll, P.; Aarrevaara, E. Review of Potential Risk Factors of Cultural Heritage Sites and Initial Modelling for adaptation to Climate Change. *Geosciences* **2018**, *8*, 322. [CrossRef]

38. Brimblecombe, P. Refining climate change threats to heritage. *J. Inst. Conserv.* **2014**, *37*, 85–93. [CrossRef]

39. Forino, G.; MacKee, J.; von Meding, J. A proposed assessment index for climate change-related risk for cultural heritage protection in Newcastle (Australia). *Inter. J. Disaster Risk Reduct.* **2016**, *19*, 235–248. [CrossRef]

40. Kaslegard, A. *Climate Change and Cultural Heritage in the Nordic Countries*; TemaNord 2010:599; Nordic Council of Ministers: Copenhagen, Denmark, 2011; pp. 9–18.

41. European Commission. Safeguarding cultural heritage from natural and man-made disasters. In *A Comparative Analysis of Risk Management in the EU*; EU Publications: Luxembourg, 2018; pp. 54–57.

42. Finnish Environmental Institute (FEI). Suomen Tulvariskien Ennakoidaan Kasvavan Tulevaisuudessa. 2020. Available online: https://www.syke.fi/fi-FI/Ajankohtaista/Suomen_tulvariskien_ennakoidaan_kasvavan (accessed on 30 September 2019).

43. UNEP Emissions Gap Report 2019 Global Progress Report on Climate Action. 2019. Available online: https://www.unenvironment.org/interactive/emissions-gap-report/2019/ (accessed on 30 August 2018).

Article

Retrofit Strategies for Energy Efficiency of Historic Urban Fabric in Mediterranean Climate

Meltem Ulu [1] and Zeynep Durmuş Arsan [2,*]

1   Department of Architecture, Erciyes University, 38280 Kayseri, Turkey; meltemulu@erciyes.edu.tr
2   Department of Architecture, Izmir Institute of Technology, 35430 Izmir, Turkey
*   Correspondence: zeynepdurmus@iyte.edu.tr

Received: 31 May 2020; Accepted: 7 July 2020; Published: 13 July 2020

Abstract: Energy-efficient retrofitting of historic housing stock requires methodical approach, in-depth analysis and case-specific regulatory system, yet only limited efforts have been realized. In large scale rehabilitation projects, it is essential to develop a retrofit strategy on how to decide energy-efficient solutions for buildings providing the most energy saving in a short time. This paper presents a pilot study conducted at a neighborhood scale, consisting of 22 pre-, early-republican and contemporary residential buildings in a historic urban fabric in the Mediterranean climate. This study aims to develop an integrated approach to describe case-specific solutions for larger scale historic urban fabric. It covers the building performance simulation (BPS) model and numerical analysis to determine the most related design parameters affecting annual energy consumption. All the case buildings were classified into three main groups to propose appropriate retrofit solutions in different impact categories. Retrofit solutions were gathered into two retrofit packages, Package 1 and 2, and separately, three individual operational solutions were determined, considering a five-levelled assessment criteria of EN 16883:2017 Standard. Energy classes of case buildings were calculated based on National Building Energy Regulations. Changes in building classes were evaluated considering pre- and post-retrofit status of the buildings. For the integrated approach, the most related design parameters on annual energy consumption were specified through Pearson correlation analysis. The approach indicated that three buildings, representing each building group, can initially be retrofitted. For all buildings, while maximum energy saving was provided by Package 2 with 48.57%, minimum energy saving was obtained from Package 1 with 19.8%.

Keywords: energy-efficient retrofit; historic residential buildings; energy consumption prediction

1. Introduction

Many countries have introduced numerous policy measures and strategies on energy efficiency depending on national circumstances and political goals, together with increasing risks of climate change and global warming, rapidly depleting natural sources and rising energy demand/consumption [1]. There is an urgent need to implement energy-oriented solutions for buildings, since the building sector comprises the largest portion of energy saving potential. It is explicit that the building and construction sectors are the highest final energy-consumers, being responsible for 36% and 39% of energy- and process-related emissions at global level in 2018, respectively [2]. Particularly, residential buildings account for 22% [2] and 27.2% [3] of final energy consumption in 2018 in world and the EU-28 countries, respectively. In Turkey, the residential sector has the highest share of total energy consumption, with 24.5% in 2016 [4].

While the current attention is towards the upgrading of energy-efficiency policies and retrofit efforts on existing building stock, historic buildings are also a non-negligible contributor, since historic buildings constitute over 25% of total buildings [5] and more than 40% of residential buildings in

Europe were built before the 1960s [6]. In Turkey, the percentage of historic buildings built before 1945 is 6.1% [7]. Moreover, the largest share within officially registered immovable cultural properties belongs to residential buildings with 63% in 2019 [8].

Differently from other existing buildings, historic buildings form distinctive architectural and aesthetical characteristics in many urban areas, as well as keeping intangible elements, such as associations of historic people, events and aspects of social history, within cultural heritage values. They also comprise inherently sustainable characteristics in terms of material use, construction type, spatial decisions and topographic unity [9]. Energy-efficient retrofit of historic buildings is undoubtedly vital for ensuring their proper re-use to meet modern-day requirements, keeping away from desolation and demolition, enhancing comfort conditions, i.e., thermal and visual, and maintaining distinctive characteristics and heritage values.

Building conservation and energy efficiency are both key aspects for sustainable development which covers social, economic and environmental requirements and balances them in harmony. It is possible to improve the energy efficiency of historic buildings without compromising their historic fabric and distinctive characteristics [10]. The fact remains that the energy efficiency of historic buildings requires a special concern in comparison to existing one. This issue should be addressed in an interdisciplinary approach to find a convenient balance between conservation principles and energy-efficient retrofits.

Energy-efficient retrofit of existing buildings has become a prominent policy argument at both the national and international levels, specifically over the last 20 years. The European Commission (EC) enacted a series of policies and regulations addressing new and existing buildings to make them more energy efficient and reduce CO_2 emissions in the EU countries through Energy Performance of Building Directives (EPBDs) [11–13] and Energy Efficiency Directive (EED) [14]. The EC initially stipulated three key targets: reducing GHG emissions by 20%, increasing energy efficiency by 20% and increasing the share of renewable energy sources in energy consumption by 20% by 2020 [15]. Beyond the 2020 strategy, The EU drew longer-term low carbon economy and energy roadmaps in 2011. The low carbon economy roadmap set out target levels to reach the 2050 goal in a cost-effective way, a 40% reduction in emissions by 2030 and a 60% reduction by 2040, as well as a 25% reduction by 2020 [16]. The energy roadmap emphasizes that improving energy efficiency is a key driver in all decarbonization scenarios [17]. Recently, the recast EPBD also stipulates strong long-term renovation strategies aiming at achieving an energy-efficient and decarbonized European building stock, with indicative milestones for 2030, 2040 and 2050 [13].

As a candidate country for the EU, Turkey is upgrading its legislative efforts to be compatible with the policies of the EU. It prefaced with the first national standard, namely, the TS 825 Thermal Insulation Requirements for Buildings, which established the rules for thermal insulation in the buildings of Turkey in 2000 [18]. The Energy Efficiency Law, established in 2007, was promulgated aiming to increase energy efficiency, minimizing energy costs and ensuring the use of energy sources for a clean environment. Then, Energy Performance Regulation on Buildings was published in 2008. It obligates the building energy certification scheme, which includes information about energy classification categories and the minimum energy requirements of existing buildings for their renovation [19].

Although the EU Directives addresses the major aspects of the renovation of existing building stock and retrofitting technical elements and systems, there is no specific statement about retrofitting historic buildings. In the first EPBD 2002/91/EC, the EU left the decision about implementing minimum energy performance requirements for historic buildings to its member states [11]. The following directives revised and rearticulated this statement [12,14]. In other respects, the recast EPBD [13] promotes researching and testing of new solutions to improve the energy performance of historic buildings and sites, while safeguarding and preserving heritage value.

In Turkey, the Energy Performance Regulation on Buildings refers to a similar statement as specified in the first EPBD 2002/91/EC. The Article 2 (ç) in the Regulation indicates that energy-efficient interventions on the buildings that are officially registered as a cultural asset should be conducted by receiving the consultancy of competent authorities in a way not to affect building fabric and appearance [19]. In its 2023 projections, Turkey points out the necessity of energy improvements of existing buildings; however, there is no specific expression covering historic buildings [20].

Recently launched by the European Committee in 2017, the Standard EN 16883 Conservation of Cultural Heritage-Guidelines Improving the Energy Performance of Historic Buildings directly focuses on the energy efficiency of historic buildings. EN 16883: 2017 covers historically, architecturally or culturally valuable buildings, regardless of whether they are officially registered or not. It presents a systematic procedure about the identification of objectives for refurbishment based on various assessment categories, such as energy saving, heritage significance, economic viability, compatibility, selecting and evaluating interventions and deciding on the most appropriate ones while respecting heritage significance of buildings [21].

Consequently, there is no legal certainty about how to improve the energy efficiency of historic buildings while preserving their function, quality or character in building regulations of the EU and Turkey. Another concern is the lack of any protection procedure covering historic buildings, even if not legally protected. This situation poses a risk when it comes to physical alterations for those buildings constituting large part of historic urban centers [22].

The review of recent literature emphasizes that the lack of a specific protocol on the energy efficiency of historical buildings at an individual building level comes to the fore, as well as this lack being even more noticeable for urban scale approach [23]. Nevertheless, research on the energy efficiency of historic urban stock affirms that a certain number of publications have accelerated after 2010 in the European Countries. The course of studies is discussed in terms of diversity of the research topics, methodologies, focus groups and level of retrofit. Research topics are grouped under five sub-topics, consisting of energy efficiency, thermal comfort, environmental impact, economic impact and heritage value. Energy-efficient retrofit solutions addressing both individual cases and building stock are categorized in the surveyed publications. While individual building level solutions are related to building systems, equipment and building envelope, such as walls, floors, roofs, windows, doors and shutters, integration renewable energy sources and district heating are included in district scale solutions. Table 1 summarizes research topics and energy-efficient retrofit solutions for both building and district scale.

Improving energy efficiency while protecting the heritage value of historic buildings is an essential purpose for all studies. Additionally, some studies aim at improving thermal comfort [24–29], achieving carbon emissions reductions and assessing energy-efficient measures via life-cycle approach [30,31] and increasing economic performance with regard to cost-effectiveness [25,26,29–31] (Table 1).

District-level retrofit solutions have multiscale approach, comprising of the building scale and district scale. Regarding the building scale, the majority of studies deals with retrofit solutions on building envelope covering insulation of walls, floors and roofs and repairing or replacing door and window systems. Interior insulation of the external walls and improving windows are the most preferable ones. It is followed by improvements of the HVAC systems and equipment, while the integration of renewable energy systems has been in the minority due to concerns on building appearance and heritage value. Moreover, using weather stripping is the cheapest way to improve energy performance among the studies whilst still being effective (Table 1). Almost all studies combine single retrofit solutions, and then use these as multiple retrofit packages. Considering the district scale, Broström et al. (2014) [26] and Sugár et al. (2020) [32] propose the installation of a district heating system as a prominent solution. Bonomo and Benardinis (2014) also concentrate on the integration of solar PV technologies not only into the building but also the urban and landscape scale in a historical settlement in Italy [33].

Table 1. Categorization and main characteristics of the retrofit-related publications by a systematically reviewed literature.

| References | Date | Country | Research Topics | | | | | Energy-Efficient Retrofit Solutions | | | | | | | | | | | | |
| | | | | | | | | Building Scale | | | | | | | | | | | District Scale | |
			EE	TC	ENI	ECI	HV	EIW	IIW	CWI	IF	IR	IAF	WS	RRWD	IS	IHVAC	IEA	IRES	IRES DH
Boarin and Davoli	2013	Italy	●		●		●	√	√		√	√			√					
De Berardinis et al.	2014	Italy	●	●		●	●	√	√						√					
Arumägi and Kalamees	2014	Estonia	●	●		●	●	√	√		√		√		√		√			
Alev et al.	2014	Estonia, Finland and Sweden	●	●			●	√			√		√		√		√	√		
Bonomo and De Benardinis	2014	Italy	●			●	●												√	√
Broström et al.	2014	Sweden	●	●		●	●	√	√				√	√	√		√			√
Eriksson et al.	2014	Sweden					●	√												
Giombini and Pinchi	2015	Italy	●				●	√	√				√		√					
Tadeu et al.	2015	Portugal	●		●	●	●	√			√	√			√		√			
Egusquiza et al.	2015	Spain	●			●	●		√			√			√		√	√		
Arumägi et al.	2015	Estonia	●	●			●	√			√	√			√		√			
Fatiguso et al.	2015	Italy	●				●			√	√	√	√			√				
Egusquiza et al.	2018	Spain	●			●	●	√	√	√	√	√		√	√					√
Belpoliti et al.	2018	Italy	●				●	√	√		√	√			√		√			
Blumberga et al.	2020	Latvia	●		●	●	●										√		√	
Sugar et al.	2020	Hungary	●				●	√	√		√	√			√		√			√
Caro and Sendra	2020	Spain	●	●	●		●						√			√	√			

An energy-efficient retrofit approach at district scale differs from the individual building level. Various studies have developed a neighborhood scale approach, instead of one-by-one approach, to assess the energy-efficient retrofit potential of historic building stock. They identify buildings as representative or archetype cases which characterize a specific building group to evaluate the outcomes on these cases by extrapolating to wider scale. The reason is to speed up the decision-making process in determining retrofit solutions and provide a higher level of energy-efficient improvements [23,24,27–30,34]. Several studies also present overall methodologies on deciding and evaluating the effects of energy-efficient retrofit solutions while avoiding the potential risks for historic building stock [23,31,32,35]. Eriksson et al. (2014) presented a methodology developed for EU Historic Districts' Sustainability (EFFESUS) research project to analyze the impacts of energy-efficient measures on heritage significance in a historic district in Visby, Sweden [35]. Egusquiza et al. (2018) suggested a method that provides early-stage suitability assessment of energy conservation measures (ECM) for historic urban areas within a multi-scale approach through ICOMOS guidance on Heritage Impact Assessment [23]. ECMs are evaluated to decide their impact on heritage value and Santiago de Compostela, Spain was selected to test this method, as supported by 3D models. In the study of Sugar et al. (2020), the heritage-respecting energetic retrofit methodology was developed for the historic building stock of Budapest, Hungary, based on the EPBD [32]. Blumberga et al. (2020) presented a decarbonization strategy of urban block in the historic center of Riga, Latvia by discussing the impact of energy efficiency measures on historic heritage values [31].

When planning large-scale retrofit process, extensive data collection and energy investigations are inevitable to define building characteristics, energy behavior and energy saving potential of building stock [23,31,32,34,36,37]. Building categorization is another significant indicator to determine urban typologies and, therefore, appropriate solutions are to be applied in accordance with district scale. Multiple studies list various categorization criteria, such as type of building, purpose of use, number of floors, building geometry, construction year, building system and remarkable architectural characteristics and degree of protection, as well as heritage value. The most notable one is the heritage value [23,26,29,32,38,39].

It has been seen that building performance simulation (BPS) tools are widely used not only in earlier stages of a design process for sustainability but also in analyzing existing building performance and evaluating potential retrofit solutions to attain more energy-efficient historic buildings. They are useful for obtaining fast and actual results in a short time, especially in larger scale studies [24,27,29,39].

The above-mentioned studies make explicit that energy efficiency and heritage value protection are hot topics in discussed publications. Energy-efficient retrofit of historic buildings and urban areas is a delicate matter that needs to be considered in an interdisciplinary way. Therefore, the retrofit process of historic buildings requires a distinctive roadmap in comparison to the other existing buildings. All retrofit works on historic buildings are specific in their context. Before implementing a retrofit solution, in attempt to improve energy efficiency of historic buildings, a number of principles should be thoroughly considered: intervention should be kept at a minimum level and retrofits should be reversible, compatible and respectful to the original fabric, distinctive characteristics and heritage value of buildings.

Historic districts are well-defined areas by distinctive characteristics in urban areas, in terms of size, fabric, form, construction, material used and density, as well as heritage value, integrity, memory and perception in modern urban environment. It is necessary to turn historic urban areas into an energy-efficient model for sustainable development in communities by balancing between the conservation of historic buildings and sustainability requirements to ensure their continuity for future generations while protecting their heritage value. The fact remains that, in large scale rehabilitation projects, the requirement of developing a retrofit strategy is crucial regarding the question of how to decide solutions for buildings which provide the most energy saving in a short time. Since large scale retrofit studies require extensive data collection to define building characteristics, field survey takes a long time, and the economic impact of this is high.

In historic urban fabrics of Turkey, street and/or façade rehabilitation projects are generally conducted at a neighborhood scale. They are limited to various efforts such as improving the physical appearances of buildings and façade components, painting of buildings' façades and fixing street furniture and decoration elements by protecting fabric and the distinctive characteristics of the buildings and streets. The key innovation of this study is to expand this approach from the energy-efficient point of view. The main aim of the study is to develop an integrated approach to identify case-specific energy-efficient solutions toward retrofit strategies for larger scale historic urban fabric.

The present study expands upon the current literature by bringing a distinctive decision support methodology about how to decide energy-efficient retrofit solutions at a neighborhood scale, consisting of both historic and contemporary residential buildings, in a short time and with a limited budget. It conveys an integrated roadmap to speed up the decision-making process in determining more precise and context-specific retrofit solutions for larger scale historic urban fabric. Moreover, this study will be the first in Turkey which considers historic building retrofit from an energy-efficiency point of view at the urban scale.

2. Case Study

The study has been carried out in the neighborhood located in Basmane District, the quite old Ottoman residential area of Izmir, Turkey. The city is situated on the west coast of the country, next to the Aegean Sea, and thus has a Mediterranean climate; summers are hot and humid, while winters are mild and rainy. The Basmane District constitutes a considerable part of the Kemeraltı Urban Historical Site within the historic urban residential texture. 1273 Street, as the selected neighborhood, is in a residential zone which hosts qualified historic buildings within Basmane District. It lies on the east–west direction with a 38.25° N latitude and a 27.08° E longitude and is 12 m elevation above the sea level. The neighborhood has a key position due to its proximity to Basmane Historic Train Station, Agios Voukolos Church and the Altınpark Archaeological Excavation Area of the antique Smyrna City. In the last decade, it has been mostly populated by transboundary migrants, especially Syrian refugees, as the temporary residential area. The historic urban fabric of neighborhood has been damaged.

Izmir Metropolitan Municipality prepared the Façade Rehabilitation Project for 1273 Street in 2013 to regenerate the neighborhood, requiring intervention strategies for improving security and rehabilitating living conditions. In addition, the municipal boards wanted to interfere with the existing conditions, at least, to improve pedestrian routes for local and foreign tourists. However, any energy-efficient approach was not considered during the rehabilitation process. Therefore, this study has been prepared as a proposal for Izmir Metropolitan Municipality to provide local, applicable and quick retrofit solutions with the most energy saving potential within the limited project budget.

A total of 22 buildings, covering historic and contemporary buildings which lie on 1273 Street, are investigated (Figure 1). There are 4 solely commercial and 18 residential buildings, 3 of which have shops on their ground floors. Both historic and contemporary buildings coexist in the street, situated in adjacently. Of the 22 buildings, 13 are historic ones, in total: 11 of them are officially registered, while the remaining 2 are determined as non-registered in character. The rest are the contemporary buildings. The number of floors varies between one and three. A large majority of extant historic buildings were built between the end of 19th and the first quarter of 20th centuries. A total of 20 buildings were constructed with stone or a brick masonry system. There are only two reinforced concrete buildings. The oriels on the second floors designed as a protrusion with wooden or iron structure, ornamentations on iron doors and stone wall order are typical periodic characteristics of historic buildings (Figure 2).

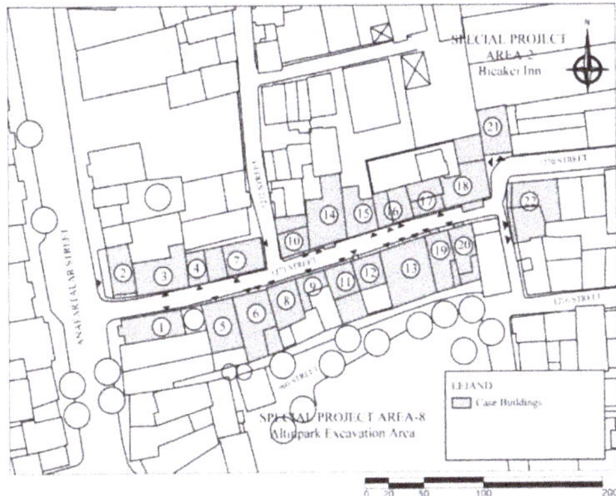

Figure 1. Numbered case buildings in 1273 Street. Source: modified from the drawings of Konak Municipality.

Figure 2. View from 1273 Street.

3. Materials and Methods

This study proposes a method to develop retrofit strategies about the energy efficiency of existing buildings in historic urban district, including both historic and new buildings via appropriate solutions only for the buildings' envelopes. The retrofit strategy considers the impact assessment criteria and scale of retrofit measures for historic buildings, presented by a five-level assessment scale of EN 16883:2017 [21].

The identification process of retrofit strategies is composed of seven main stages: data collecting, data processing, creating possible retrofit solutions, categorizing buildings, assessing retrofit solutions, analyzing data and presenting results (Figure 3). This method starts with the quick field survey, i.e., data collection conducted in several levels. First, documents about the case area are obtained to get preliminary information. The data sheets are created to characterize of the buildings in the case area. Then, on-site measurements on the buildings' envelope are carried out through data sheets. Additionally, it is attempted to get information about the buildings in use from the users of buildings.

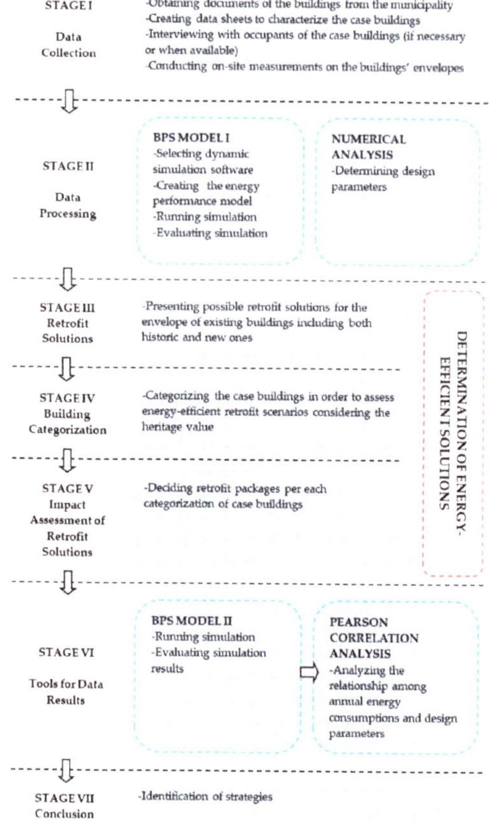

Figure 3. Identification process of energy-efficient retrofit strategies.

The second stage continues with the characterization of the tool selected to put in the process of the method and explanation of how they work. The case area is modelled in dynamic simulation software to calculate the energy consumption of the case buildings in existing conditions. This building performance simulation (BPS) model represents the base model of case buildings. The next three stages lay emphasis on the energy-efficient retrofit strategy for the case buildings. Accordingly, the third stage discusses and presents possible energy-efficient solutions for the retrofitting of existing buildings, covering new and historic ones. It primarily concerns the retrofit solutions of the components of buildings' envelopes, including external walls, floors, roofs, oriels, windows and doors. The fourth stage goes forward with the categorization of case buildings considering the heritage value.

The fifth stage aims at deciding possible packages of retrofit solutions to provide energy savings. It presents the most appropriate solutions and eliminates the inappropriate ones for the categorized case buildings. Therefore, a five-level impact assessment scale for retrofit solutions is investigated for historic buildings. In the sixth stage, new building performance models for retrofit solutions are created and simulated for each retrofit package. After a comparative study between the base case simulation results and retrofitted ones, the relationships between annual energy consumptions and design parameters are presented. In the seventh stage, possible retrofit strategies for the case buildings are introduced.

3.1. Data Collection

Data collection, aiming to gather the required adequate and reliable information about buildings' envelope through a quick survey, is composed of two steps: pre-study and field survey. The former includes the collection of any research and official documentation about case area and buildings. The latter is mainly grouped under site investigations: creating data sheets, conducting on-site measurements for components of buildings' envelopes and interviews with the buildings' users. Data collection was held in two separate time periods of 20–25 June 2014 and 10–15 February 2016.

Documentation about the case area and its surrounding were obtained from the Street Rehabilitation Project Proposal of Izmir-Konak Municipality held in 2013. Particularly for historic buildings, inventory forms of 11 officially registered buildings were provided by the Izmir Metropolitan Municipality Directorate of Historic Environment and Cultural Properties. These forms provide information about the degree of protection status of historic buildings; they do not include any construction drawings and details. Except for three registered buildings, the architectural drawings, such as floor plans and sections of the buildings, were not reached during the site investigations.

Data sheets, composed of a double-sided page in A4 format, were prepared for each building in the case area. The aim was to characterize the current status (historic/old/new) of the case buildings, e.g., physical (envelope) qualities, construction details and surrounding features. The type of data about building envelope and surrounding required by BPS model were determined. Observation, measurements and photography techniques were used in collecting required data for the data sheets.

On-site measurements were carried out for external walls, floors, roofs, oriels, windows, external doors and shutters to characterize construction materials with simple sections, elevations and plan drawings. Dimensions of the structural components were identified via a laser distance meter and then noted on the relevant section in the data sheets. Moreover, the height and width of surrounding buildings were measured by laser distance meter to identify the adiabatic surfaces of the adjacent neighboring buildings and their shading effect. Through the measurements, the following specifications about the envelope are clarified and corrected:

- width length–height of the external walls;
- width–length of the external floors (external floor below the oriels and external floor over the entrances designed as a door niche);
- width–length–height of the windows and their position on the external wall surfaces;
- width–length–height of external doors and their position on the external wall surfaces;
- width–length–height of the oriels;
- width–length–height of the shutters;
- width and length of the eaves of the roofs.

Moreover, traditional building material samples collected from immediate environment were tested to determine their thermal properties. The thermal conductivity (W/mK) of various stone and solid brick samples was measured by the Quick Thermal Conductivity Meter (KEM Q500 with a measuring range of 0.023 to 12 W/mK and a precision of ±5% reading value per reference plate) [40] in the Geothermal Energy Research and Application Centre of Izmir Institute of Technology (IZTECH JEOMER).

Short interviews with the occupants of several buildings were conducted in order to obtain adequate information about the case buildings, i.e., the purpose of use, number of occupants/users, user profile, construction date, how the buildings are heated and cooled and type of fuel used.

3.2. Data Processing

3.2.1. BPS Model

A reliable and verified building performance simulation (BPS) software, DesignBuilder v.5.2, was used for model creation, energy simulation and analysis, as well as decision-making processes [41]. Model creation of the case buildings was prepared according to the most recent field survey which was completed in 2016 (Figure 4). Through data processing, the base case models, which indicate the real status of each case building, and then the retrofitted case models to assess the energy-efficient retrofit solutions were prepared. Seasonal energy consumption for heating and cooling and annual energy consumption were analyzed. The model geometry of case buildings was simplified in line with the purpose of the study: identification of strategies for energy-efficient retrofit via quick field survey. The model abstraction was conducted for both the façades and layout plans of buildings.

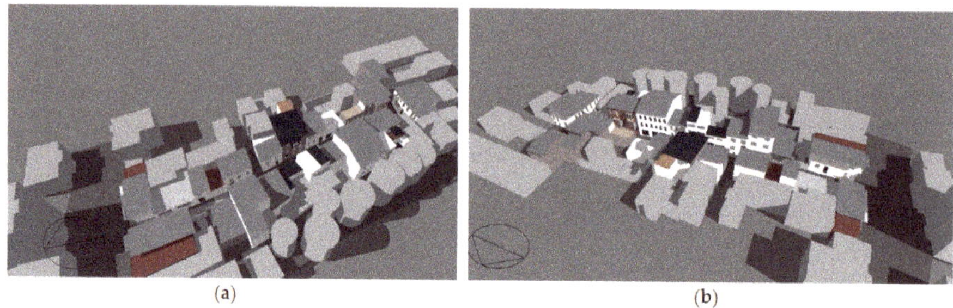

(a)   (b)

Figure 4. Model of case buildings and their surrounding from the (**a**) south and (**b**) north direction in DesignBuilder Software.

3.2.2. Numerical Analysis

This study addresses a number of design parameters to focus on prominent geometric variables regarding building form, i.e., the envelope characteristics of case buildings. It discusses total seven design parameters:

- DP1: Total surface area (m^2) to conditioned volume (m^3) ratio (S/V);
- DP2: Total window area (m^2) to total wall area (m^2) ratio;
- DP3: Window area (m^2) to wall area (m^2) ratio (main façades of buildings face to 1273 Street);
- DP4: Shape factor (building length (m) to depth (m));
- DP5: Usable ground floor area (m^2) to conditioned volume (m^3);
- DP6: Total usable floor area (m^2) to conditioned volume (m^3);
- DP7: Building height (m) to plan depth (m).

Bivariate Pearson correlation coefficient analysis was selected to understand empirically the relation between the design parameters and annual energy consumption of the case buildings. It is a statistical analysis method used to determine whether there is a linear relationship between two numerical variables, and, if any, the degree of the relationship. The Pearson correlation coefficient is expressed in 'r'. It can take a range of values from +1 to −1, depending on whether the relationship is positive or negative, respectively [42] (Table 2):

- r = −1, a perfect negative linear relationship. One variable increases, the other decreases or one variable decreases while the other increases;
- r = 1, a perfect positive linear relationship. One variable increases, the other increases or one variable decreases while the other decreases;
- r = 0, no relationship. There is no relationship between two variables.

Table 2. Range of values and relation of Pearson correlation coefficient.

R	Relation
0.00–0.25	Very low
0.26–0.49	Low
0.50–0.69	Medium
0.70–0.89	High
0.90–1.00	Perfect

3.3. Determination of Energy-Efficient Solutions

In order to determine energy-efficient solutions for historic urban fabric and find the most appropriate solutions for building envelope, a set of actions were organized, starting with a preliminary analysis of possible retrofit solutions by taking into account a literature survey, including guidelines, standards and publications. This section continues with the categorization step. A total of 22 buildings were classified and characterized by number of qualitative and quantitative data. Afterwards, a pre-assessment was conducted to find the best retrofit solutions by excluding inappropriate ones and identify a series of acceptable measures. After this process, retrofit solutions were grouped under retrofit packages by combining the best solutions. This step served the purpose of revealing which packages were most appropriate toward the targets of the study by evaluating and comparing different retrofit scenarios with each other and the base case. The final step was composed of the decision and presentation of retrofit packages (Figure 5).

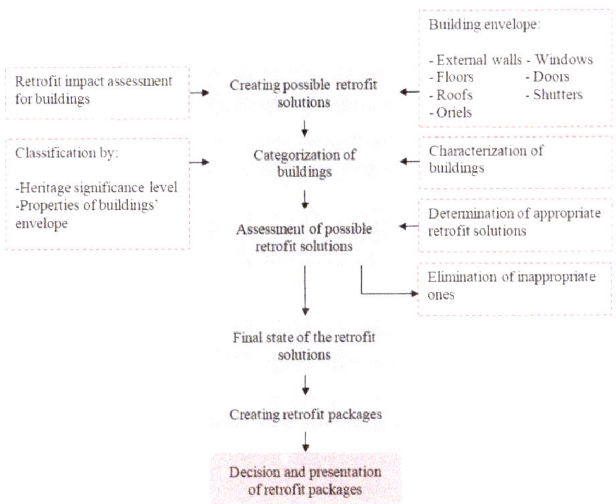

Figure 5. Determination process of energy-efficient solutions of the study.

3.3.1. Retrofit Impact Assessment for Historic Buildings

Historic buildings differentiate in two major ways that can affect energy retrofits in comparison with other building categories. The first one is physical characteristics, such as the complexity of geometry, method of construction, used materials and existence of inherently passive climatic strategies. The second one is conservation principles, since historic buildings are held to account for established conservation principles to preserve their historic fabric and distinguishing characters [43].

It is essential to point out the need for providing a convenient balance between building conservation principles and energy-efficient improvements. Implication of a well-understood energy-efficient retrofit approach protects architectural, aesthetic and heritage values, reduces energy bills and improves comfort conditions and health of occupants, as well as increases the value and prolongs the life of historic buildings.

As such, in existing non-historic and contemporary buildings, it is possible to develop energy-efficient strategies for historic building envelope and system and equipment. However, a standardized retrofit process cannot be conducted, and accordingly, a method for retrofit impact assessment becomes inevitable in determining the best retrofit solutions that can be implemented to historic buildings. Therefore, energy-efficient retrofitting of historic buildings requires an interdisciplinary approach [44].

The retrofit impact assessment has an approach based on certain criteria for each type of existing building. These criteria can be grouped under different topics, such as energy aspect, i.e., energy saving, embodied and operational energy, and economic aspect, indoor and outdoor environment and hygrothermal performance, i.e., durability, moisture risk and thermal transmittance. However, the criteria and process of retrofit differ by conservation principles for historic buildings [42]. Thus, heritage value protection commonly plays a leading role in the assessment of historic buildings [26,35,43,45]. The criteria for retrofit impact assessment specific to historic buildings can be scrutinized under specific subtopics: retrofit effects on building envelope, i.e., visual and spatial effects from the interior and exterior, and properties of retrofit materials, i.e., reversibility, damage potential and fabric compatibility [35,46] (Table 3).

Table 3. Criteria for retrofit impact assessment.

Criteria for Retrofit Impact Assessment				Criteria for Heritage Value Impact Assessment	
Şahin et al., 2015 [46]	Eriksson et al., 2014 [35]	Broström et al., 2014 [26]	Webb, 2017 [43]	Grytli et al., 2012 [46]	Eriksson et al., 2014 [35]
Energy saving	Indoor environment	Energy savings	Global environment	Reversibility	Visual
Cultural heritage values	Fabric compatibility	Economic aspect	Building fabric	Visibility	Physical
Durability	Heritage significance	Heritage values	Indoor environment	Effects on the interior or the exterior	Spatial
Economic return	Embodied energy	Moisture	Economics		
Moisture	Operational energy	Indoor environment			
Indoor environment	Economy				

An overview about the retrofit impact assessment of various scientific studies and guidelines are presented based on two major criteria, including heritage value protection and energy saving. The assessment was conducted for all building components covering walls, floors, roofs, oriels, openings and shutters and a range of retrofit solutions in accordance with these building components. Moreover, the retrofit assessment of sources was interpreted by utilizing the five-level assessment criteria introduced by EN 16883:2017 (Figure 6 and Table 4).

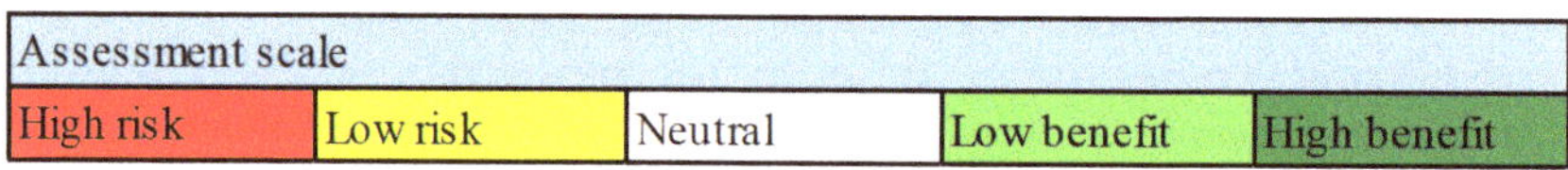

Figure 6. Five-level assessment scale for retrofit impact assessment [21].

Table 4. Retrofit solutions for the building envelope based on retrofit impact assessment (red: high risk; yellow: low risk; white: neutral; light green: low benefit; dark green: high benefit) (Sources: 1, [46]; 2, [47]; 3, [48]; 4, [49]; 5, [50]; 6, [46]; 7, [26]; 8, [51]; 9, [10]; 10, [52]).

Components of Building Envelope	Type of Retrofit	Sources	Retrofit Impact Assessment	
			Heritage Value Protection	Energy Saving
All components	Weather stripping/draught proofing	1, 5, 6, 7, 8, 9, 10	yellow	light green, dark green
External walls	External insulation	1, 3, 4, 6, 7, 8, 9	red, yellow	dark green
	Internal insulation	2, 3, 4, 6, 7, 8, 9, 10	yellow	dark green
Floors	Insulation of basement floor	1	yellow	light green
	Insulation of ground floor	1, 6, 8, 9	red, yellow	light green
	Insulation of attic floor	1, 6, 7, 8, 9	yellow	light green
Roofs	Insulation of roof at rafter level	1, 4, 5, 6, 8, 9	yellow	dark green
	Insulation of flat roof	4, 9	yellow	dark green
Windows, doors and shutters	Adding secondary glazing on existing windows	2, 5, 7, 9, 10	yellow	dark green
	Changing windows with double/triple glasses	2, 6, 7, 9	red, yellow	dark green
	Changing windows with low-e double/triple glasses	2	yellow	dark green
	Changing/improving shutters	8, 9, 10	red, yellow	light green
	Changing/improving doors	8	red, yellow	light green
	Shutter control			light green
	Nighttime ventilation			light green
Oriels	Use of oriels as sun space			light green

According to the retrofit assessment, external insulation of walls, ground floor insulation, changing/improving windows, doors and shutters predominantly result in high risk on the heritage value and historic building character, while they provide substantial energy savings. Internal insulation of walls, basement floor insulation, attic floor insulation, flat roof insulation and adding a secondary glazing on existing windows have less risk on the heritage value, while they provide low benefit for energy efficiency. Implementation of weather stripping and roof insulation at rafter level have no risk on the heritage value and building appearance, as well as presenting moderate energy savings. Finally, shading control, night-time ventilation and use of oriels as sun space can be considered as the retrofit solutions without risk for the heritage value, because they do not cause any change on buildings' envelopes.

3.3.2. Possible Retrofit Solutions for Building Envelope

A list of possible energy-efficient solutions was addressed to develop the retrofit strategies for case buildings, including both historic and contemporary ones. Among a wide range of possible energy efficient retrofit solutions based on the literature survey, only 17 envelope-related retrofit solutions were selected (Table 5):

Table 5. Possible retrofit solutions for building envelope based on the literature survey.

For All Heated Zones	
Draught Proofing/Weather Stripping	
Walls	**Roofs**
External insulation of walls	Insulation of flat roof
Internal insulation of walls	Insulation of pitched roof
Floors	**Windows and doors**
Insulation of basement floor	Changing windows
Insulation of ground floor	
Insulation of attic floor	Adding a secondary glazing to existing windows
Insulation of external floors	
Insulation of oriels' ground floor	Changing doors
Insulation of oriels' attic floor	Use of oriels as a sunspace

3.4. Categorization of Buildings

In this study, a categorization process was conducted for the case buildings by characterizing according to their heritage values and architectural characteristics. This process aims at ensuring the most appropriate energy-efficient solutions by properly matching the retrofit packages with each building category.

Categorization starts with the heritage significance level, which is of top priority because of the most decisive and distinctive criteria at the first stage of categorization. It continues with characterizing the architectural components of building envelopes, i.e., walls, floors, roofs and oriels affecting the number and type of retrofit solutions produced for each building and how they work and are applied to building structure. The availability of basement floors and oriels in the case buildings were initially selected as criteria after the selection of the heritage significance level.

In accordance with the categorization process, the case buildings were gathered under three main groups based on the heritage significance level of buildings. These groups are officially registered historic buildings named Group 1 buildings, with non-registered historic buildings named Group 2 buildings and contemporary (non-historic) buildings named Group 3 buildings (Figure 7)

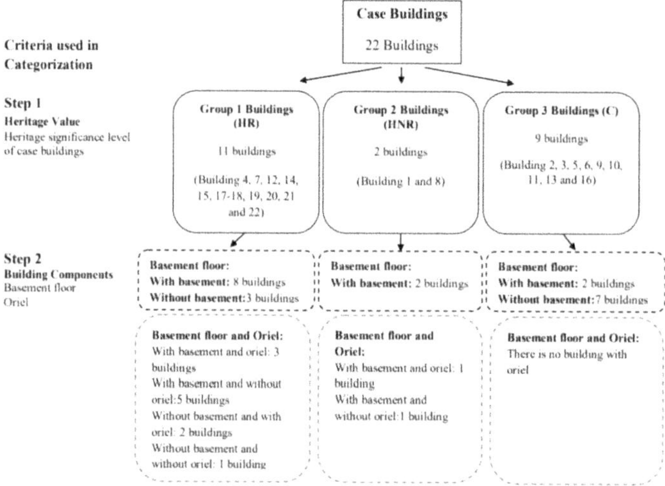

Figure 7. Categorization of buildings.

Group 1 buildings were solely composed of 11 officially registered buildings, encoded as "historic registered (HR)". Two buildings, representing Group 2 "historic non-registered (HNR)"—were not officially registered buildings but they were in harmony with the officially registered ones in terms of the physical, visual and material characteristics of the historic buildings' envelopes. Group 3 buildings were non-historic ones, consisting of nine buildings and characterized as "contemporary (C)" buildings.

3.5. Impact Assessment of Possible Retrofit Solutions for Building Groups

3.5.1. Impact Assessment of Possible Retrofit Solutions for Group 1 Buildings

First, defining the retrofit targets is of importance to properly assess the energy-efficient retrofit solutions and decide the appropriate ones for buildings. Several targets are specified for Group 1 buildings:

- to produce the energy-efficient retrofit solutions primarily considering the heritage value of case buildings;
- to provide as much energy saving as possible while protecting the heritage value of case buildings;
- to select the retrofit solutions as natural, breathable, reversible and compatible with the historic fabric, character and façade components of case buildings;

- to select the insulation materials to meet TS 825 Thermal Insulation Requirements, although there is no description about officially registered historic buildings in the Energy Performance Regulation on Buildings of Turkey [18].

After the definition of retrofit targets, 18 possible energy-efficient retrofit solutions for Group 1 buildings were evaluated based on the retrofit impact assessment through utilizing the five-level assessment criteria introduced by EN 16883:2017 (see Figure 6 in Section 3.3.1). As a result of the assessment, all retrofit solutions were gathered under three risk groups, based on the heritage value: high-risk solutions (red), low-risk solutions (yellow) and neutral solutions (white). Additionally, grey colored boxes show that there is no solution defined for the case buildings (Table 6).

Initially, three retrofit solutions, including the external insulation of external walls, changing windows and doors, were specified the high-risk solutions for the heritage value after the assessment. Therefore, they were excluded from the scope of the solutions for Group 1 buildings.

Six retrofit solutions were determined as the low-risk solutions after the heritage value impact assessment. These solutions are considered to have less impact on the heritage value and the buildings' appearances while causing changes on the buildings' constructions. The retrofit solutions with low risk were:

- internal insulation of external walls;
- internal insulation of oriel wall;
- insulation of oriels' ground floor;
- insulation of oriels' attic floor;
- insulation of the external floors (floor of protrusion and floor above entrance);
- adding secondary glazing to existing windows.

Considering the heritage value impact assessment, six retrofit solutions were determined as the neutral solutions causing physical change on buildings' envelope. These solutions improve the energy efficiency of Group 1 buildings without damaging the heritage value. The neutral retrofit solutions were:

- weather stripping to improve air-tightness of the building envelope;
- insulation of ground floor;
- insulation of attic floor;
- insulation of flat roof;
- insulation of oriels' roof;
- insulation of pitched roof.

The remaining three retrofit solutions were the passive solutions related to building operation. They were also entitled as the neutral solutions which do not cause any physical change on the buildings' envelopes (white color shown in building operation section of Table 6. These retrofit solutions were the following:

- use of oriels as a sunspace;
- night-time ventilation;
- shading control.

After the determination of appropriate retrofit solutions based on the heritage value impact assessment for Group 1 buildings, some of the retrofit solutions were grouped under the retrofit packages while some of were individually assessed. The neutral retrofit solutions causing physical changes on buildings' envelopes were included in a package named Package 1, separately simulated for Group 1 Buildings. Low-risk solutions for Group 1 buildings were not being separately evaluated, since they are not a foremost option in terms of preference and application priority. Therefore, Package 2 is generated by adding the low-risk retrofit solutions to Package 1 solutions and then evaluated. Moreover, operational solutions were not grouped, in order to observe their individual effects on the buildings' envelopes. Table 7 presents all retrofit solutions and packages determined for Group 1 buildings.

Table 6. Heritage value impact assessment for 18 energy-efficient retrofit solutions of Group 1 buildings (red: high risk; yellow: low risk; white: neutral; gray: no solution).

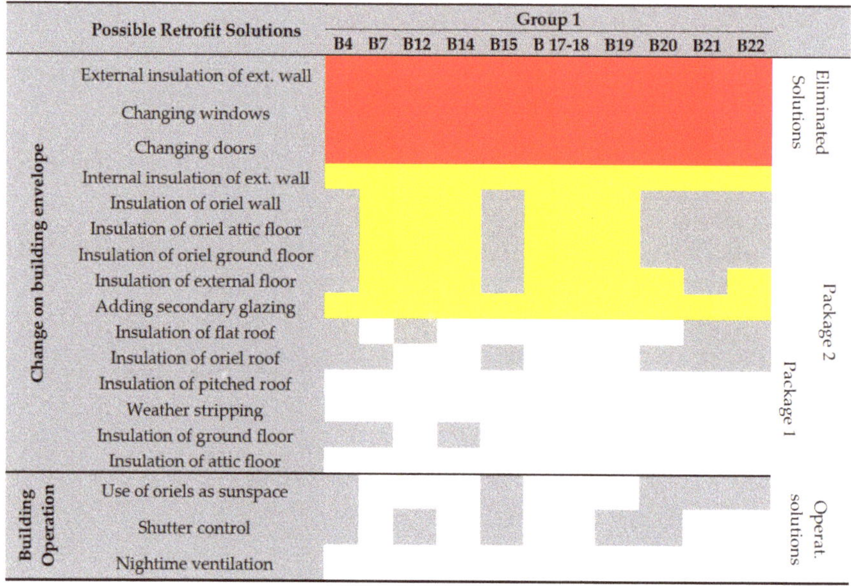

Table 7. Determined retrofit solutions and packages for Group 1 buildings.

Group 1 Buildings		
Package 1 (Neutral Related to Building Envelope)	**Package 2** (Combination of Neutral and Low-risk)	**Individual Operational Solutions** (Neutral Related to Building Operation)
Insulation of flat roof	Internal insulation of external wall	Use of oriels as sunspace
Insulation of oriel roof	Insulation of oriel wall	Shutter control
Insulation of pitched roof	Insulation of oriel attic floor	Nightime ventilation
Weather stripping	Insulation of oriel ground floor	
Insulation of ground floor	Insulation of external floor	
Insulation of attic floor	Adding secondary glazing	
	Insulation of flat roof	
	Insulation of oriel roof	
	Insulation of pitched roof	
	Weather stripping	
	Insulation of ground floor	
	Insulation of attic floor	

Specifications about Package 1 for Group 1 buildings: Package 1 aspired to enhance the energy efficiency of Group 1 buildings without damaging the heritage value. The package contained an implementation of weather stripping to improve air-tightness of the building envelope, insulation of attic floor, insulation of flat roof, insulation of oriels' roof, insulation of pitched roof and insulation of ground floor. Considering air-tightness improvements, the air exchange rate (ACH) was assumed to have improved from 0.7 h^{-1} to 0.4 h^{-1} in heated zones and from 0.9 h^{-1} to 0.7 h^{-1} in unheated zones to repair the cracks and holes on the building envelope. For floors, ground floor and attic floor insulation existed in Package 1. On the other hand, regarding the ground floors of some buildings, it was decided that they should not undergo a change during the retrofit interventions due to their structural composition and historic material properties.

Specifications about Package 2 for Group 1 buildings: Package 2 was intended to reveal the effects of both neutral and low risk retrofit solutions by providing as much as energy saving for Group 1 buildings as possible. Package 2 presented the combination of the low-risk and neutral solutions named (Package 1). The content of the low-risk solutions was composed of six retrofit solutions, including internal insulation of external wall, insulation of oriel wall, insulation of oriels' attic floor and oriel ground floor and roof, insulation of the external floors (floor of protrusion and floor above entrance) and adding secondary glazing to existing windows.

Considering the walls, internal insulation of external wall and oriel wall were implemented for heated spaces. If there was a case building with a gable roof, the gable walls were also insulated from inside. Moreover, external wall surfaces, as an adiabatic, were not insulated.

Specifications about Insulation Materials for Group 1 buildings: Determination of insulation material carries importance to protect the heritage value and fabric of historic buildings. Therefore, the use of natural, breathable and reversible materials was beneficial for minimizing the risks, i.e., moisture generation, on historic building construction and components. Wood fiber board and sheep wool were selected as the internal insulation material for Group 1 buildings.

Installation of secondary glazing was selected to provide an effective insulation for historic windows and limit draughts without changing any components of windows and damaging their character and heritage values. However, changing windows (glazing and frame) was envisaged on the façades of some Group 1 buildings. As for floors, the ground and attic floor of the oriels and external floors such as ground floor of protrusions and floors above the buildings' entrances were also included in low-risk solutions. Overall heat transfer coefficient targets to meet TS 825 Thermal Insulation Requirements for Buildings were achieved after the retrofits.

3.5.2. Impact Assessment of Possible Retrofit Solutions for Group 2 Buildings

Identified retrofit targets for Group 2 buildings:

- to select and evaluate the energy-efficient retrofit solutions primarily considering the historic character and façade constituents of case buildings;
- to provide as much energy savings as possible without damaging the historic character and façade constituents of case buildings;
- to select effective insulation materials compatible with case building character;
- to meet TS 825 Thermal Insulation Requirements for the components of buildings' envelopes.

A total of 17 possible energy-efficient retrofit solutions were evaluated considering the five-level assessment criteria introduced by EN 16883: 2017. All retrofit solutions were divided into three risk groups, including high-risk solutions, low-risk solutions and neutral solutions (Table 8).

First, external insulation of external walls and changing doors were eliminated, since they carry a risk for Group 2 buildings according to the heritage value impact assessment. Thus, these solutions were left out of the scope of appropriate solutions for Group 2 buildings.

Five retrofit solutions, the yellow colored boxes shown in Table 8, were specified as low-risk solutions which cause less impact on the buildings' façade characters and appearance while causing change on the buildings' components. The retrofit solutions with low risk were:

- internal insulation of external walls;
- insulation of oriel wall;
- insulation of oriels' ground floor;
- insulation of external floor;
- changing windows.

Seven retrofit solutions were specified as the neutral solutions causing physical change on the buildings' envelopes, but without damaging the historic character of buildings according to the heritage value impact assessment. The neutral retrofit solutions for Group 2 buildings were:

Atmosphere **2020**, 11, 742

- weather stripping to improve air-tightness of the building envelope;
- insulation of basement floor;
- insulation of ground floor;
- insulation of attic floor;
- insulation of oriel attic floor;
- insulation of oriel roof;
- insulation of pitched roof.

Table 8. Heritage value impact assessment for 17 energy-efficient retrofit solutions of Group 2 buildings (red: high risk; yellow: low risk; white: neutral; gray: no solution).

Category	Possible Retrofit Solutions	Group 2 B1	Group 2 B8	Grouping
Change on building envelope	External insulation of external wall	red	red	Eliminated Solutions
	Changing doors	red	red	Eliminated Solutions
	Internal insulation of external wall	yellow	yellow	Package 2
	Insulation of oriel wall	yellow	gray	Package 2
	Insulation of oriel ground floor	yellow	gray	Package 2
	Insulation of external floor	yellow	yellow	Package 2
	Changing windows	yellow	yellow	Package 2
	Insulation of oriel attic floor	white	gray	Package 1 / Package 2
	Weather stripping	white	white	Package 1 / Package 2
	Insulation of basement floor	white	white	Package 1 / Package 2
	Insulation of attic floor	white	white	Package 1 / Package 2
	Insulation of oriel roof	white	gray	Package 1 / Package 2
	Insulation of pitched roof	white	white	Package 1 / Package 2
	Insulation of ground floor	white	white	Package 1 / Package 2
Building Operation	Use of oriel as sunspace	gray	white	Operational Solutions
	Shutter control	white	white	Operational Solutions
	Nightime ventilation	white	white	Operational Solutions

Some of the retrofit solutions were grouped under the retrofit packages while some were individually evaluated. The neutral solutions which cause the physical changes on buildings' envelope were included in a package named as Package 1 and separately simulated for Group 2 buildings. Low-risk solutions for Group 2 buildings were not individually performed because of similar reasons to the Group 1 buildings. Unchanged or less changed façades for this group of buildings are principally desired due to their façade characteristics. Low-risk solutions were combined and simulated with the Package 1 for Group 2 buildings. Thus, Package 2 originated from the combination of solutions with low risk and a neutral effect on the heritage value of Group 2 buildings. Furthermore, operational solutions were not grouped in order to observe their individual effects on the buildings' envelopes (Table 9).

Table 9. Determined retrofit solutions and packages for Group 2 Buildings.

Group 2 Buildings.		
Package 1 (Neutral Related to Building Envelope)	**Package 2** (Combination of Neutral and Low-Risk)	**Individual Operational Solutions** (Neutral Related to Building Operation)
Insulation of oriel attic floor Insulation of basement floor Insulation of attic floor Insulation of oriel roof Insulation of pitched roof Insulation of ground floor	Internal insulation of external wall Insulation of oriel wall Insulation of oriel ground floor Insulation of external floor Changing windows Insulation of oriel attic floor Weather stripping Insulation of basement floor Insulation of attic floor Insulation of oriel roof Insulation of pitched roof Insulation of ground floor	Use of oriel as sunspace Shutter control Nightime ventilation

3.5.3. Assessment of Possible Retrofit Solutions for Group 3 Buildings

The retrofit targets determined for Group 3 buildings were:

- to provide as much as energy savings as possible by implementing the appropriate retrofit solutions;
- to select the retrofit materials compatible with the buildings' envelopes;
- to ensure the buildings' components meet TS 825 Thermal Insulation Requirements after retrofit.

The heritage value impact assessment for Group 3 buildings was not performed, because they consisted of contemporary buildings which did not present any historic character and heritage value. Therefore, a list of 11 possible retrofit solutions was prepared. Then, the retrofit solutions were grouped according to their vertical and horizontal building components. The solutions applied for the vertical building components, i.e., walls, windows and doors, were determined as Package 1. Package 2 consisted of the combination of retrofit solutions applied for both the vertical and horizontal building components. Then, the operational solutions were individually assessed for Group 3 buildings (Table 10). Package 1 included the retrofit solutions which were thought to be preferable in terms of priority and ease of implementation.

Table 10. Determined 11 retrofit solutions and packages for Group 3 buildings.

Group 3 Buildings		
Package 1 (Vertical Components)	**Package 2** (Horizontal Components)	**Individual Operational Solutions**
External wall insulation changing windows Changing doors	Insulation of attic floor Insulation of flat roof Insulation of pitched roof insulation of external floors Insulation of basement floor Insulatation of ground floor	Shutter control Nightime ventilation

Package 2 aimed to observe the effect of the combined solutions produced for both vertical and horizontal components of Group 3 buildings, except for the operational solutions. Therefore, following specifications on the solutions for horizontal building, components were presented in addition to the above-mentioned specifications about the vertical components within Package 1. When insulating all the floor types for Group 3 buildings, it was considered whether to directly implement insulation materials to the upper level of the existing floor without excavating and then place floor covering material on it.

4. Results

The results of the determined retrofit packages and individual retrofit solutions for each building category, i.e., Group 1, Group 2, and Group 3, are presented in separate subsections. All retrofit proposals were simulated by using DesignBuilder BPS software version 5.2.003 and 5.5.0.012. The comparison among retrofit packages and base case conditions of the buildings are illustrated in the next sections.

4.1. Results of Retrofit Solutions Belonging to Group 1 Buildings (HR)

4.1.1. Results of Retrofit Packages for Group 1 Buildings (HR)

Regarding total energy consumption for heating, the amount of energy saving for Group 1 buildings ranged from 9.0% (Building 15) to 17.89% (Building 19) with Package 1. This rate changed from 34.64% (Building 12) to 60.6% (Building 22) through Package 2 (Table 11). For the total energy consumption for cooling, Package 1 significantly enabled reductions on energy consumption of almost all the buildings. However, there exists an increase in cooling consumption of 25.92% and 41.7% in Building 15 and 21, respectively. The minimum energy saving achieved by Package 1 was 2.45% in Building 12, while the maximum energy saving was 85.87% for Building 22. Package 2 provided a minimum energy saving of 28.11% in Building 15 and a maximum of 91.15% in Building 22, in comparison to the base case of buildings (Table 11).

Table 11. Change rates in annual energy consumption compared to base case of Group 1 buildings through Package 1 and Package 2.

Group 1 Buildings	Package 1 (%)		Package 1 Total (%)	Package 2 (%)		Package 2 Total (%)
	Heating	Cooling		Heating	Cooling	
Building 4	−9.09	−6.74	−8.99	−38.84	−35.24	−38.70
Building 7	−13.31	−18.93	−13.52	−52.00	−66.50	−52.52
Building 12	−11.92	−2.45	−11.71	−34.64	−36.08	−34.67
Building 14	−10.25	−12.38	−10.32	−52.47	−46.65	−52.27
Building 15	−9.00	25.92	−8.12	−41.51	−28.11	−41.17
Building 17–18	−13.12	−3.99	−12.77	−59.68	−30.25	−58.55
Building 19	−17.89	−19.02	−17.92	−43.71	−43.38	−43.70
Building 20	−15.97	−5.31	−15.68	−37.84	−33.61	−37.72
Building 21	−11.19	41.70	−9.38	−60.60	−36.96	−59.79
Building 22	−11.30	−85.87	−31.43	−56.42	−91.15	−65.80

The results of annual energy consumption indicated that the minimum energy saving obtained from Package 1 was 8.12% (Building 15) and the maximum was 31.43% (Building 22). Through Package 2, the energy saving rate increased by a minimum of 34.67% (Building 12) and a maximum of 65.8% (Building 22) in proportion to base case (Table 11).

4.1.2. Results of Operational Solutions for Group 1 Buildings (HR)

According to the results of individual operational solutions, it could be remarked that the night-time ventilation slightly differed from other operational solutions, in terms of providing more energy saving specifically for the cooling season and applicability to all existing windows of Group 1 buildings. The use of oriels as a sunspace and shading control did not completely perform for all case buildings because of the use of the existing shutters and the fact that not all buildings had an oriel. These solutions implemented to the existing building components had close results as compared with the night-time ventilation strategy (Table 12).

Table 12. Change rates in annual energy consumption compared to base case of Group 1 buildings through operational solutions.

Group 1 Buildings	Use of Orielsas a Sunspace (%)			Nighttime Ventilation (%)			Shading Control (%)		
	Heating	Cooling	Total	Heating	Cooling	Total	Heating	Cooling	Total
Building 4	-	-	-	−3.2	−0.9	−3.1	-	-	-
Building 7	−4.0	1.1	−3.8	−4.0	−23.7	−4.7	−3.9	−16.3	−4.4
Building 12	−1.5	−4.2	−1.5	−1.5	−37.6	−2.2	−1.6	−3.1	−1.6
Building 14	−1.6	6.0	−1.4	−1.7	−23.1	−2.4	−1.6	5.5	−1.4
Building 15	-	-	-	−2.8	−37.7	−3.7	-	-	-
Building 17–18	−0.3	8.9	0.1	−0.3	−13.5	−0.8	−0.3	−1.0	−0.3
Building 19	−1.9	4.5	−1.7	−1.9	−28.2	−2.6	−1.9	1.2	−1.8
Building 20	-	-	-	0.0	−0.3	0.0	−2.5	−8.9	−2.6
Building 21	-	-	-	0.7	4.1	0.9	-	-	-
Building 22	-	-	-	0.0	−23.3	−22.5	0.0	−4.8	−21.9

4.2. Results of Retrofit Solutions Belonging to Group 2 Buildings (HNR)

4.2.1. Results of Retrofit Packages for Group 2 Buildings (HNR)

The energy consumption results for heating season indicates that there is a remarkable difference between Package 1 and Package 2. Through Package 2, the highest reduction occurred in Building 8, by 47.34%, and Building 1, by 42.63%. Package 1 provides less energy saving, by 17.56% in Building 1 and 10.06% in Building 8, compared to the base case (Table 13).

Table 13. Change rates in annual energy consumption compared to base case of Group 2 buildings through Package 1 and Package 2.

Group 2 Buildings	Package 1 (%)		Package 1 Total (%)	Package 2 (%)		Package 2 Total (%)
	Heating	Cooling		Heating	Cooling	
Building 1	−17.56	60.13	−15.38	−47.34	48.69	−44.65
Building 8	−10.06	34.99	−9.41	−42.63	−23.98	−42.36

The results of energy consumption for cooling show that Package 1 and Package 2 unexpectedly increased the energy consumption of two buildings while only Package 2 provided a reduction of 23.98% for Building 8. The increase rate was 34.99% for Building 1 and reached 13% for Building 8 through Package 1 (Table 13).

Considering the total annual energy saving rates, there was a significant difference varying from 30% to 35% between Package 1 and Package 2. The maximum saving obtained from Package 2 was 44.65% for Building 1 and 42.36% for Building 8. The minimum energy saving rate was 9.41% for Building 8 and 17.56% for Building 1 through Package 1 (Table 13).

4.2.2. Results of Operational Solutions for Group 2 Buildings (HNR)

Individual operational solutions provided minor energy savings for Group 2 buildings. Nevertheless, these solutions can be considered as favorable since they did not damage on historic buildings' envelopes and appearances, although Group 2 Buildings were not officially registered. Among the individual solutions, the strategy of the use of oriels as a sunspace was conducted for only Building 1. All operational solutions indicated the same amount of saving, i.e., 1.7%, for the heating season, compared with energy consumption for cooling. It can be concluded that the night-time ventilation strategy is more effective for reducing energy for cooling (Table 14).

Table 14. Change rates in annual energy consumption compared to base case of Group 2 buildings through operational Solutions.

Group 2 Buildings	Use of Orielsas a Sunspace (%)			Nighttime Ventilation (%)			Shading Control (%)		
	Heating	Cooling	Total	Heating	Cooling	Total	Heating	Cooling	Total
Building 1	−1.7	2.6	−1.6	−1.7	−19.4	−2.2	−1.7	−20.1	−2.2
Building 8	-	-	-	−2.9	−34.9	−3.4	−2.9	2.0	−2.8

4.3. Results of Retrofit Solutions Belonging to Group 3 Buildings (C)

4.3.1. Results of Retrofit Packages for Group 3 Buildings (C)

Regarding the total energy consumption for heating, both packages revealed significant reductions for most of Group 3 buildings. The highest energy reduction for heating occurred in Building 11, with 67.43%, through Package 2, while the lowest result was in Building 16, with 9.3%, through Package 1 (Table 15).

Table 15. Change rates in annual energy consumption compared to base case of Group 3 buildings through Package 1 and Package 2.

Group 3 Buildings	Package 1 (%)		Package 1 Total (%)	Package 2 (%)		Package 2 Total (%)
	Heating	Cooling		Heating	Cooling	
Building 2	−51.74	−35.14	−50.91	−52.49	−39.01	−51.82
Building 3	−9.31	−13.82	−9.83	−35.13	−72.95	−39.48
Building 5	−26.12	−49.91	−26.88	−27.01	−54.75	−27.90
Building 6	−36.25	−43.70	−36.41	−52.15	−16.40	−51.41
Building 9	−15.50	−41.25	−16.22	−25.29	−46.88	−25.89
Building 10	−18.95	−54.58	−19.90	−43.83	−26.65	−43.37
Building 11	−44.93	−26.32	−44.06	−67.43	−59.27	−67.05
Building 13	−27.56	−40.40	−27.90	−36.14	−45.66	−36.39
Building 16	−9.30	−8.77	−9.26	−45.51	−83.59	−48.42

The energy consumption for cooling also decreased with both packages, yet Package 2 provided higher energy saving rates. The highest reduction occurred in Building 16 (83.59%), through Package 2, while Package 1 resulted in the lowest reduction rate of 8.77%, in Building 16 (Table 15).

Regarding annual energy consumption, although Package 2 enables more energy saving, both retrofit packages provide significant energy conservation. The maximum rate was 67.05% for Building 11, through Package 2, while the minimum was 9.26% in Building 16, through Package 1 (Table 15).

4.3.2. Results of Operational Solutions for Group 3 Buildings (C)

Individual operational solutions create more energy savings compared to other building groups. Both strategies including night-time ventilation and shading control with shutters providing close results for energy consumption for heating and cooling. The use of shutters resulted in higher energy saving rates for Building 2 and 3 with a marked difference. The operation of the night-time ventilation strategy was effective on Building 11 and 13, compared to shading control for cooling season (Table 16).

Table 16. Change rates in annual energy consumption compared to base case of Group 3 buildings through operational solutions.

Group 3 Buildings	Nighttime Ventilation (%)			Shading Control (%)		
	Heating	Cooling	Total	Heating	Cooling	Total
Building 2	−0.9	−46.4	−3.2	−0.9	−60.9	−3.9
Building 3	−30.2	−67.8	−34.5	−30.2	−80.5	−36.0
Building 5	12.3	10.0	12.2	12.3	6.8	12.1
Building 6	−8.9	−14.9	−9.1	−8.9	−14.9	−9.1
Building 9	0.0	12.3	0.3	0.3	−4.8	0.1
Building 10	−0.5	−22.3	−1.0	−0.5	−21.1	−1.0
Building 11	−13.9	−43.7	−15.3	−13.9	−7.9	−13.6
Building 13	−0.3	−33.4	−1.1	−0.3	−4.3	−0.4
Building 16	4.6	3.0	4.5	-	-	-

5. Discussion

5.1. Evaluation among Building Categorizations

For Group 1 buildings, Package 1 (the neutral solutions) provided an average of 15.11% of energy saving, while the highest energy savings was 50.90% obtained from Package 2 (the combination of low risk and neutral solutions). Considering the Group 2 buildings, 11.93% of energy saving was achieved by implementing Package 1 (the neutral solutions) while Package 2 (the combination of low risk and neutral solutions) provided an energy saving of 43.33%. Regarding Group 3 buildings, Package 1 (the retrofit solutions for vertical building components) provided a 30.11% energy saving for Group 3 buildings. Package 2 (the combination of solutions for both vertical and horizontal building components) resulted in 50.90% of the energy saving rate (Table 17).

Table 17. Evaluation of annual energy consumption results among building categories.

Case Buildings	Retrofit Packages (%)		Individual Operational Solutions (%)		
	Package 1	Package 2	Use of Oriels as Sunspace	Nighttime Ventilation	Shading Control
Group 1 Buildings	−15.11	−50.90	−1.45	−4.96	−5.25
Group 2 Buildings	−11.93	−43.33	−4.49	−4.40	−4.35
Group 3 Buildings	−30.11	−45.83	-	−6.81	−6.38

Among all the building categories, Package 2 achieved the maximum energy saving of 50.90%, while the minimum energy saving rate of 11.93% was obtained from Package 1. It was deduced that for all building groups, Package 2 enabled the saving of considerably more energy than Package 1. Only for Group 3 buildings, there occurred a close energy saving result between Package 1 and Package 2, compared to the other building groups (Table 17).

Individual operational solutions provided minor energy savings, in comparison to the solution included in packages for case building groups. Although these individual solutions did not provide as effective energy savings as the packages, they did not cause any changes on buildings' envelopes; therefore, these solutions have an importance in terms of protecting the heritage values and the characteristics of the Group 1 and Group 2 buildings. The night-time ventilation strategy saved the highest energy of 6.81% for the Group 3 buildings. A minimum energy saving rate of 1.45% was obtained from the strategy of the use of oriels as sunspace for Group 1 buildings. Night-time ventilation was the most effective solution in terms of energy saving among individual operational solutions (Table 17).

5.2. Evaluation of All Case Buildings

Out of all case buildings, the maximum energy saving was provided by Package 2, with 48.57%, while the minimum energy saving was obtained from Package 1, with 19.8% (Table 18). Among the individual operational solutions, night-time ventilation and shading control provided similar energy savings. The energy saving rates were 5.2% and 5.4%, for night-time ventilation and shading control, respectively. The use of oriels as sunspace resulted in a minor energy saving rate of 1.5% for all case buildings (Table 19).

Table 18. Evaluation of annual energy consumption results for retrofit packages in all case buildings.

All Case Buildings	Annual Energy Consumption		
	Base Case	Package 1	Package 2
Total (kWh)	506,924	406,568	260,735

Table 19. Evaluation of annual energy consumption results for individual operational solutions in all case buildings.

All Case Buildings	Annual Energy Consumption			
	Base Case	Use of Oriels as Sunspace	Nighttime Ventilation	Shading Control
Total (kWh)	506,924	191,837	480,556	420,493

5.3. Evaluation of Building Groups Based on Energy Classes

In this section, all the building groups were evaluated according to the energy classes for energy consumption. The energy class of the buildings was determined by calculating the annual primary energy consumption per unit occupied floor area. For new and existing buildings, the Energy Performance Regulation on Buildings of Turkey stipulates the preparing of a Building Energy Certificate that includes a classification of energy performance varying between A (the best) and G (the worst). According to the regulation, new buildings are required to have a rating of class C or higher [37]. Although there is no restriction about the energy class of historic buildings, all building groups were included in this evaluation.

Among Group 1 buildings (HR), two base case buildings (Building 14 and 20), three buildings with Package 1 (Building 14, 20 and 21) and seven buildings with Package 2 (Building 4, 7, 14, 17–18, 20, 21, and 22) met the minimum energy class of C and above, according to the regulation. Building 12, 15 and 19 did not meet the minimum energy class of C and above in any cases, before or after retrofit. Package 1 provided the highest change rate on energy class for Building 22 (from F to C), compared to base case. There was no change in energy class for Building 4, 12 and 15 by implementing Package 1. Through Package 2, the highest change rate on energy class occurred in Building 17–18 and 22 (from F to B) compared to the base cases (Table 20).

Among Group 2 buildings (HNR), the energy class results for the base case and all the packages were the same as energy class B for Building 1. For Building 8, the energy classes of both base case and Package 1 were found to be energy class F, which is not acceptable, according to the Energy Performance Regulation on Buildings of Turkey. Through Package 2, it achieves minimum energy class (C) (Table 20).

Table 20. Primary energy consumption and energy classes of the base and retrofitted cases for all case buildings (light green: minimum energy class C; green; energy class B; dark green: energy class A).

Building Groups		Base Case Total (kWh/m^2)	Base Case Energy Class	Package 1 (kWh/m^2)	Package 1 Energy Class	Package 2 (kWh/m^2)	Package 2 Energy Class
GROUP 1	Building 4	215.05	E	195.95	E	132.20	C
	Building 7	202.25	E	174.40	D	94.71	B
	Building 12	278.61	F	246.71	F	181.91	D
	Building 14	143.52	C	128.58	B	68.85	B
	Building 15	281.06	F	261.41	F	166.57	D
	Building 17–18	247.14	F	216.66	E	105.92	B
	Building 19	294.44	G	241.56	F	165.81	D
	Building 20	149.90	C	126.95	B	93.57	B
	Building 21	197.56	E	183.52	D	81.44	B
	Building 22	256.66	F	138.46	C	70.31	B
GROUP 2	Building 1	130.58	B	114.12	B	76.75	B
	Building 8	257.70	F	235.66	F	149.44	C
GROUP 3	Building 2	426.99	G	209.61	C	205.74	C
	Building 3	293.56	E	264.69	D	177.65	B
	Building 5	389.43	F	237.58	C	234.28	C
	Building 6	117.72	B	74.63	B	58.33	A
	Building 9	325.13	E	272.39	D	240.94	D
	Building 10	159.92	C	126.16	B	91.49	B
	Building 11	203.05	D	115.73	B	67.85	A
	Building 13	156.28	C	111.99	B	98.89	B
	Building 16	506.16	G	459.27	G	261.06	D

Among Group 3 buildings (C), the highest change rate in energy class (from G to C) was provided in Building 2 through Package 1. There was no change in the energy class of Building 16 with Package 1. Through Package 2, energy classes of most buildings were class B. The highest change rate occurred in Building 2 (from G to C), followed by Building 3 (from E to B), Building 5 (from F to C), Building 11 (from D to A), Building 16 (from G to D), Building 6 (from B to A), Building 13 (from C to B) and Building 9 (from E to D) (Table 20).

For all building groups, Package 2 provided the highest improvements on energy classes compared to Package 1. Moreover, Group 3 buildings indicated better performance in energy classes in comparison to the other building groups.

5.4. Evaluation of Relationship between Design Parameters and Building Energy Consumption

The numerical analysis was conducted by using Pearson correlation analysis. Seven design parameters, based on the geometric variables of building form, mentioned in the Section 3.2.2, were investigated for each case building to find the most influential parameters on building energy consumption.

The calculated values of parameters belonging to each case building are presented in (Table 21), while the results of Pearson correlation analysis (R values) are presented in Table 22. The outcomes convey that DP1 (total surface area to conditioned volume ratio (S/V)) and DP5 (usable ground floor area (m^2) to conditioned volume (m^3)) are negatively and significantly related to the annual energy consumption of buildings. In other words, buildings that have lower levels of S/V and usable ground floor area (m^2) to conditioned volume (m^3) are likely to have higher annual energy consumption per m^2. P4 (length to depth), DP6 (total usable floor area to conditioned volume) and building DP7 (height to depth), on the other hand, are the variables are positively related to energy consumption, although the correlation is statistically insignificant (Table 22).

Table 21. Numerical analysis results for design parameters.

Case Buildings	Total Annual Energy Consumption (kWh)	DP1	DP2	DP3	DP4	DP5	DP6	DP7
Building 1	18,546.45	0.89	14.83	16.86	2.26	0.18	0.37	1.27
Building 2	6980.19	1.36	10.82	3.48	0.51	0.21	0.21	0.52
Building 3	10,072.96	1.28	12.6	31.55	1.40	0.34	0.34	0.45
Building 4	9750.90	1.51	1.54	7.79	0.57	0.18	0.36	0.78
Building 5	8030.71	1.40	9.17	46.4	0.67	0.29	0.29	0.37
Building 6	32,205.86	0.80	5.58	20.14	0.58	0.14	0.36	0.74
Building 7	27,693.98	1.16	7.08	11.19	1.84	0.25	0.41	1.10
Building 8	25,414.80	1.08	4.25	19.04	0.74	0.16	0.31	0.78
Building 9	6670.72	1.56	4.83	18.86	0.84	0.18	0.36	0.85
Building 10	13,831.17	1.02	7.96	12.88	0.73	0.24	0.38	0.70
Building 11	29,327.03	1.16	12.13	51.91	0.96	0.15	0.45	1.63
Building 12	18,834.18	1.58	6.37	21.11	0.98	0.18	0.38	1.06
Building 13	51,986.99	0.64	5.51	14.63	0.79	0.10	0.31	0.74
Building 14	47,009.47	0.88	3.87	16.66	0.58	0.15	0.49	0.64
Building 15	26,581.31	1.14	3.76	19.84	0.66	0.15	0.29	0.64
Building 16	8538.26	1.96	-	-	2.58	0.40	0.40	1.50
Building 17–18	47,314.48	1.23	8.33	19.02	2.61	0.18	0.34	1.25
Building 19	35,287.91	1.33	6.28	28.3	0.41	0.18	0.36	0.77
Building 20	23,614.83	1.29	5.56	16.85	0.55	0.22	0.51	0.95
Building 21	17,631.81	1.17	3.41	17.33	0.67	0.17	0.34	0.74
Building 22	41,600.29	0.81	11.74	19.9	1.41	0.18	0.37	0.75

Table 22. Results of Pearson coefficient analysis.

Design Parameters	R Value	Relation
DP1	−0.660320 *	Medium
DP2	−0.100082	Very low
DP3	−0.012648	Very low
DP4	0.219320	Very low
DP5	−0.565679 *	Medium
DP6	0.289929	Very low
DP7	0.260149	Low

Note: * r values closest to −1.

5.5. Integrated Approach to Identify Case-Specific Energy-Efficient Solutions for Retrofit Strategy of Larger Scale Historic District

The need for developing retrofit strategy for larger scale case studies was confronted while deciding which buildings could provide the most energy saving within the given time limitations of the project. In cases with insufficient building data, it was required to focus on accessible data derived from building envelope with a quick field survey. This study introduced an integrated approach to identify case-specific energy efficient solutions for a retrofit strategy of a larger scale historic district.

This approach was composed of eight main steps, the approach starting with determination of the most effective design parameters on annual energy consumption of buildings (Table 23). Then, identified parameters were sorted within themselves and 50% of buildings with more energy consumption were chosen. The same building(s) in each identified design parameter were selected. Then, BPS model of the selected buildings was created, and their annual energy consumption was calculated. Energy classes of the buildings, considering primary energy consumption, were defined. Afterwards, the buildings meeting the minimum energy class (C) and above (B and A) were eliminated.

The retrofit solutions/packages were applied to the rest of the BPS model of the building(s). Finally, it was decided whether the buildings met the minimum energy class (C).

Table 23. Integrated approach of this study for retrofit strategy of larger scale historic district.

1	Determination of the design parameters related to annual energy consumption of buildings
2	Identification of the most-related ones among the design parameters
3	Sorting identified parameters within themselves and determination of 50% of buildings consuming more energy
4	Determination of same building(s) in each identified design parameter
5	Creation of determined building(s)' BPS model and calculation of their annual energy consumption
6	Identification of energy classes based on primary energy consumption of the building(s)
7	Elimination of the building(s) that meet minimum energy class (C) and application of retrofit solutions/packages in the rest of building(s)' BPS model
8	Determination of the building(s) that meet minimum energy class (C)

In this research, P1 (total surface area to conditioned volume ratio (S/V)) and P5 (usable ground floor area (m^2) to conditioned volume (m^3)) were determined as the two most influential parameters. The calculated values of these parameters were sorted from minimum to maximum value. Of the buildings with more annual energy consumption per each design parameter, 50% corresponded to the first 11 case buildings, colored grey in Table 24 (a) and (b). Then, the same buildings in both parameters were determined, as shown in blue in Table 24 (a) and (b). This means that the number of case buildings to work on decreased to eight.

Table 24. Sorted and determined case buildings based on (a) DP1 (total surface area to conditioned volume ratio (S/V)) and (b) DP5 (usable ground floor area (m^2) to conditioned volume (m^3) ratio) (grey: 50% of case buildings; blue: same buildings in both parameters).

Case Buildings	DP1	Case Buildings	DP5
Building 13	0.64	Building 13	0.10
Building 6	0.80	Building 6	0.14
Building 22	0.81	Building 15	0.15
Building 14	0.88	Building 11	0.15
Building 1	0.89	Building 14	0.15
Building 10	1.02	Building 8	0.16
Building 8	1.08	Building 21	0.17
Building 15	1.14	Building 9	0.18
Building 7	1.16	Building 4	0.18
Building 11	1.16	Building 1	0.18
Building 21	1.17	Building 12	0.18
Building 17–18	1.23	Building 22	0.18
Building 3	1.28	Building 17–18	0.18
Building 20	1.29	Building 19	0.18
Building 19	1.33	Building 2	0.21
Building 2	1.36	Building 20	0.22
Building 5	1.40	Building 10	0.24
Building 4	1.51	Building 7	0.25
Building 9	1.56	Building 5	0.29
Building 12	1.58	Building 3	0.34
Building 16	1.96	Building 16	0.40

The energy classes of the identified eight buildings, based on primary energy consumption, are presented. The buildings with minimum energy class (C) and above, i.e., Building 1 (HNR), 6 (C) and 14 (HR), in base cases were eliminated since they already met the requirements of the Energy Performance Regulation on Buildings of Turkey.

Finally, the remaining four buildings, including Building 8 (HNR), 11 (C), 15 (HR) and 21 (HR), were evaluated according to the energy class change of the retrofit packages. Building 15 (HR) was disregarded in the evaluation process, because both Package 1 and Package 2 did not cause any change in the energy class of this building. Consequently, three buildings (Building 8 (HNR), 11 (C) and 21 (HR)), that did not meet minimum energy class (C) were determined as the buildings which could be initially retrofitted (Table 25).

Table 25. Energy classes of identified eight buildings based on primary energy consumption (blue: buildings which do not meet minimum class (C)).

	Base Case Energy Class	Package 1 Energy Class	Package 2 Energy Class
Building 13	C	B	B
Building 6	B	B	A
Building 14	C	B	B
Building 1	B	B	B
Building 8	F	F	C
Building 15	F	F	D
Building 11	D	B	A
Building 21	E	D	B

Package 1 solutions provided an improvement for only Building 11 (C), from D to B. Package 2 solutions provided an improvement for Building 8 (HNR) from F to C and for Building 21 (HR) from E to B and for Building 11 (C) from D to A (Table 25).

Three buildings, which can initially be retrofitted, represent each building group: Building 21 (HR) belonged to officially registered historic buildings (Group 1), Building 8 (HNR) belonged to non-registered but historic buildings (Group 2) and Building 11 (C) belonged to contemporary buildings (Group 3) (Figure 8). Building 8 (HNR) and 21 (HR) were the second most energy saving buildings in their own groups, through Package 2. Package 1 did not provide effective energy saving results for Building 8 (HNR) and 21 (HR). Building 11 (C) had the second most energy saving potential, through Package 1, and the most energy saving potential by Package 2 among Group 3 Buildings.

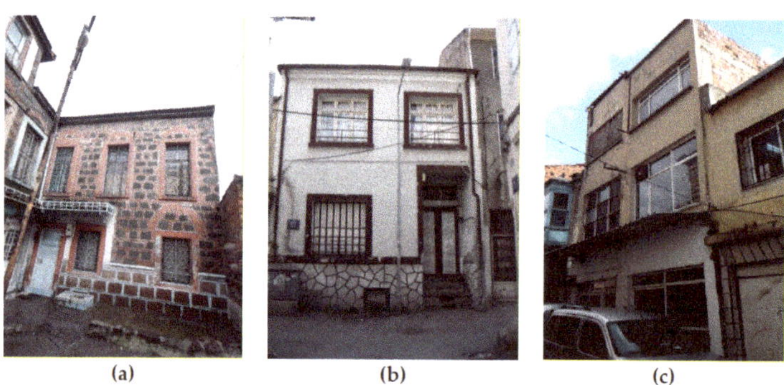

(a) (b) (c)

Figure 8. (**a**) Building 21(HR) (9% energy saving by Package 1; 60% energy saving by Package 2), (**b**) Building 8 (HNR) (9% energy saving by Package 1; 42% energy saving by Package 2) and (**c**) Building 11 (C) (44% energy saving by Package 1; 67% energy saving by Package 2) where retrofit solutions can be applied initially.

6. Conclusions

This paper presents the study at the urban neighborhood, consisting of a total of 22 historic and contemporary buildings with residential and commercial use. It introduces an integrated approach to retrofit buildings in the larger scale historic urban fabric. Improving the energy performance of the buildings' envelopes by proposing energy-efficient retrofit solutions in different impact categories, while protecting and maintaining the heritage value and architectural character of historic buildings, was the primary concern.

As a consequence, following conclusions can be derived:

1. This study indicates the evidence for the possibility of decreasing energy consumption on a neighborhood scale without extensive data collection and in-depth energy audits.
2. The methodology of the research is applied to the historic fabric located in the Mediterranean climate. It is developed as the distinctive roadmap for a rapid reaction required by many historic cities under the thread of rapid transformation and degradation. Therefore, it combines quick survey analysis and statistical assessment with the BPS tool to decide the energy-efficient solutions providing the most energy saving over a short time.
3. There is an ongoing argument between conservation principles and an energy-efficient approach for historic buildings in the restoration practice and previous literature. Proposing energy-efficient retrofit solutions, especially for officially registered buildings, is a major subject that needs to act with deliberation.
4. With respect to conservation principles, a minimum level of intervention is always expected in historic buildings. Accordingly, retrofit solutions should be determined which do not require intervention and/or require minimum intervention for protecting heritage value of buildings.
5. Such interventions which cause changes to building constructions and interior spaces should be considered in detail by the reciprocal communication of architectural restoration and energy conservation specialists, and multifaceted investigation of previous experiences and research.
6. Each historic building necessitates a case-specific approach when the energy-efficient retrofit is the major subject. Each solution may not be appropriate for each building; in other words, the generalization of solutions may inevitably cause conservation risks or energy losses. For instance, a low-risk solution, i.e., internal insulation of external wall, for a historic building may not have any risk at all for another historic building.
7. The implementation of current energy performance regulations for existing and new buildings is a matter of debate for historic buildings. When it comes to energy-efficient improvements, it is controversial whether historic buildings can be treated as other existing buildings or whether they are to be given a specific thermal target, such as certain U-value. In the case that there is no predefined calculation procedure about historic buildings in the Energy Performance Regulation on Buildings of Turkey, this study draws attention to whether the applicability of the TS 825 Thermal Insulation Requirements for Buildings, in which the specific thermal requirements are defined for building components of envelope per each climatic zone, is possible for historic buildings, or not.
8. This study points out the necessity of using the procedure on how to decide and assess energy-efficient solutions for historic buildings. Therefore, the study utilized the assessment criteria and scale of EN 16883: 2017. This standard provided the guidance on sustainability and improvement of the energy performance of historic buildings while respecting their heritage value. It presents a systematic procedure which enables the user to find the best solution for historic buildings with a case by case approach.

This study provides information regarding different retrofit approaches, i.e., the energy-efficient retrofitting of buildings from different categorizations in the same neighborhood. Overall consideration in determining the possible energy-efficient retrofit solutions for urban building stock, hosting both historic and contemporary buildings, enables its usability, in terms of bringing different retrofit approaches together. This contributes to the current literature by developing an integrated approach about how to decide retrofit solutions at a neighborhood scale, consisting of both historic and contemporary buildings, via quick survey analysis without extensive data collection. It points out the retrofit strategies are characterized for the representative neighborhood so that they may be extrapolated to wider scale urban historic fabric of that particular city.

On the other hand, this research indicates the significance of determining case-specific retrofit packages. The findings related to the retrofit solutions and their interpretation cannot be generalized for other studies. However, the approach of the study can serve as a model for historic building stock in the Mediterranean climate, i.e., the determination process of energy-efficient retrofit solutions and packages within several retrofit strategies. Nevertheless, the number and type of retrofit solutions and packages differ in other studies, since historic buildings have different historic and architectural characteristics depending on cultural, social and geographical facts.

Author Contributions: Conceptualization, M.U. and Z.D.A.; methodology, M.U. and Z.D.A.; software, M.U.; formal analysis, M.U. and Z.D.A.; investigation, M.U. and Z.D.A.; data curation, M.U.; writing—original draft preparation, M.U. and Z.D.A.; writing—review and editing, M.U. and Z.D.A.; visualization, M.U.; supervision, Z.D.A. All authors have read and agreed to the published version of the manuscript.

Funding: This research received no external funding.

Acknowledgments: The authors would like to thank Izmir Metropolitan Municipality Directorate of Historic Environment and Cultural Properties and Konak Municipality for documents and information shared. The authors wish to express special thanks to H. Engin Duran from the Department of City and Regional Planning at Izmir Institute of Technology for his supervision on statistical analysis of this study. Lastly, many thanks should be given to Geothermal Energy Research and Application Centre of Izmir Institute of Technology (IZTECH JEOMER) for the support on measurement of thermal properties of local building materials.

Conflicts of Interest: The authors declare no conflict of interest.

Abbreviations

BPS	building performance simulation;
ECM	energy conservation measures;
EE	energy efficiency;
EPBD	Energy Performance Building Directive;
TC	thermal comfort;
ENI	environmental impact;
ECI	economic impact;
HV	heritage value;
EIW	external insulation of walls;
IIW	internal insulation of walls;
CWI	cavity wall insulation;
IF	insulation of floors;
IR	insulation of roofs;
IAF	insulation of attic floors;
WS	weather stripping;
RRWD	repairing or replacing of windows and doors;
IS	improving shutters;
IHVAC	improving HVAC systems;
IEA	improving of electrical appliances;
IRES	integration of renewable energy systems;
DH	District Heating.

References

1. *World Energy Perspective, Energy Efficiency Policies: What Works and What Does Not*; World Energy Council: London, UK, 2013.

2. Global Alliance for Buildings and Construction, International Energy Agency and the United Nations Environment Programme: 2019 Global Status Report for Buildings and Construction: Towards a Zero–Emission, Efficient and Resilient Buildings and Construction Sector. 2019. Available online: https://wedocs.unep.org/bitstream/handle/20.500.11822/30950/2019GSR.pdf?sequence=1&isAllowed=y (accessed on 29 April 2020).
3. Eurostat. Energy, Transport and Environment Statistics: 2019 Edition. Available online: https://ec.europa.eu/eurostat/documents/3217494/10165279/KS-DK-19-001-EN-N.pdf/76651a29-b817-eed4-f9f2-92bf692e1ed9 (accessed on 29 April 2020).
4. *Environmental Indicator 2016*; Ministry of Environment and Urbanization: Ankara, Turkey, 2018.
5. Moran, F.; Blight, T.; Natarajan, S.; Shea, A. The use of passive house planning package to reduce energy use and CO2 emissions in historic dwellings. *Energy Build.* **2014**, *75*, 216–227.35. [CrossRef]
6. *Europe's Buildings under the Microscope, a Country-by-Country Review of the Energy Performance of Buildings*; Buildings Performance Institute Europe: Brussels, Belgium, 2011.
7. EFFESUS. Energy Efficiency for EU Historic Districts Sustainability. European Building and Urban Stock Data Collection. 2013. Available online: http://www.effesus.eu/wp-content/uploads/2016/01/D-1.1_European-building-and-urban-stock-data-collection.pdf (accessed on 5 April 2017).
8. Kültür Varlıkları ve Müzeler Genel Müdürlüğü. [General Directorate of Cultural Heritage and Museums]. 2019. Available online: https://kvmgm.ktb.gov.tr/TR-44798/turkiye-geneli-korunmasi-gerekli-tasinmaz-kultur-varlig-.html (accessed on 29 March 2020).
9. English Heritage. *Energy Efficiency and Historic Buildings: Application of Part L of the Building Regulations to Historic and Traditionally Constructed Buildings*; English Heritage: London, UK, 2012.
10. ChangeWorks. *A Guide to Improving Energy Efficiency in Traditional and Historic Homes*; English Heritage: London, UK, 2008.
11. Directive 2002/91/EC of the European Parliament and of the Council of 16 December 2002 on the Energy Performance of Buildings. Available online: https://eur-lex.europa.eu/LexUriServ/LexUriServ.do?uri=OJ:L:2003:001:0065:0071:EN:PDF (accessed on 29 March 2020).
12. Directive 2010/31/EU of the European Parliament and of the Council of 19 May 2010 on the Energy Performance of Buildings (Recast). Available online: https://eur-lex.europa.eu/legal-content/EN/TXT/PDF/?uri=CELEX:32010L0031&from=EN (accessed on 29 March 2020).
13. Directive 2018/844/EU of the European Parliament and of the Council. Available online: https://eur-lex.europa.eu/legal-content/EN/TXT/PDF/?uri=CELEX:32018L0844&from=EN (accessed on 29 March 2020).
14. Directive 2012/27/EU of the European Parliament and of the Council. Available online: https://eur-lex.europa.eu/legal-content/EN/TXT/?uri=celex%3A32012L0027 (accessed on 29 March 2020).
15. Europe 2020 A Strategy for Smart, Sustainable and Inclusive Growth. Available online: https://ec.europa.eu/eu2020/pdf/COMPLET%20EN%20BARROSO%20%20%20007%20-%20Europe%202020%20-%20EN%20version.pdf (accessed on 30 March 2020).
16. A Roadmap for Moving to a Competitive Low Carbon Economy in 2050. Available online: https://ec.europa.eu/clima/sites/clima/files/strategies/2050/docs/roadmap_fact_sheet_en.pdf (accessed on 30 March 2020).
17. Energy Roadmap 2050. [COM (2011) 885 Final]. Available online: https://www.buildup.eu/en/practices/publications/com2011-885-final-energy-roadmap-2050 (accessed on 30 March 2020).
18. Türk Standartları Enstitüsü. [Turkish Standards Institution]. TS 825—Binalarda ısı yalıtım kuralları (Revize) [Thermal Insulation Requirements for Buildings (Revised)]. Ankara, Turkey. 2008. Available online: http://www1.mmo.org.tr/resimler/dosya_ekler/cf3e258fbdf3eb7_ek.pdf (accessed on 30 March 2020).
19. Bayındırlık ve İskân Bakanlığı. [Ministry of Public Works and Settlement]. Binalarda Enerji Performansı Yönetmeliği [Energy Performance Regulation on Buildings]. Ankara, Turkey. 2008. Available online: https://www.resmigazete.gov.tr/eskiler/2008/12/20081205-9.htm (accessed on 30 March 2020).
20. Ministry of Energy and Natural Resources. Energy Efficiency Strategy Paper 2012–2023. Available online: http://www.yegm.gov.tr/verimlilik/document/Energy_Efficiency_Strategy_Paper.pdf (accessed on 30 March 2020).
21. European Committee for Standardizations (CEN). *EN 16883: Conservation of Cultural Heritage. Guidelines for Improving the Energy Performance of Historic Buildings*; CEN: Brussels, Belgium, 2017.
22. Mazzarella, L. Energy retrofit of historic and existing buildings. The legislative and regulatory point of view. *Energy Build.* **2015**, *95*, 23–31. [CrossRef]

23. Egusquiza, A.; Prieto, I.; Izkara, J.L.; Béjar, R. Multi-scale urban data models for early-stage suitability assessment of energy conservation measures in historic urban areas. *Energy Build.* **2018**, *164*, 87–98. [CrossRef]

24. Arumägi, E.; Kalamees, T. Analysis of energy economic renovation for historic wooden apartment buildings in cold climates. *Appl. Energy* **2014**, *115*, 540–548. [CrossRef]

25. De Berardinis, P.; Rotilio, M.; Marchionni, C.; Friedman, A. Improving the energy-efficiency of historic masonry buildings. A case study: A minor centre in the Abruzzo region, Italy. *Energy Build.* **2014**, *80*, 415–423. [CrossRef]

26. Broström, T.; Eriksson, P.; Liu, L.; Rohdin, P.; Ståhl, F.; Moshfegh, B. A method to assess the potential for and consequences of energy retrofits in Swedish historic buildings. *Hist. Environ. Policy Pract.* **2014**, *5*, 150–166. [CrossRef]

27. Alev, Ü.; Eskola, L.; Arumägi, E.; Jokisalo, J.; Donarelli, A.; Siren, K.; Kalamees, T. Renovation alternatives to improve energy performance of historic rural houses in the Baltic Sea region. *Energy Build.* **2014**, *77*, 58–66. [CrossRef]

28. Arumägi, E.; Mändel, M.; Kalamees, T. Method for Assessment of energy retrofit measures in milieu valuable buildings. *Energy Procedia* **2015**, *78*, 1027–1032. [CrossRef]

29. Caro, R.; Sendra, J.J. Evaluation of indoor environment and energy performance of dwellings in heritage buildings. The case of hot summers in historic cities in Mediterranean Europe. *Sustain. Cities Soc.* **2020**, *52*, 101798. [CrossRef]

30. Tadeu, S.; Rodrigues, C.; Tadeu, A.; Freire, F.; Simões, N. Energy retrofit of historic buildings: Environmental assessment of cost-optimal solutions. *J. Build. Eng.* **2015**, *4*, 167–176. [CrossRef]

31. Blumberga, A.; Vanaga, R.; Freimanis, R.; Blumberga, D.; Antužs, J.; Krastiņš, A.; Jankovskis, I.; Bondars, E.; Treija, S. Transition from traditional historic urban block to positive energy block. *Energy* **2020**, *202*, 117485. [CrossRef]

32. Sugár, V.; Talamon, A.; Horkai, A.; Kita, M. Energy saving retrofit in a heritage district: The case of the Budapest. *J. Build. Eng.* **2020**, *27*, 100982. [CrossRef]

33. Bonomo, P.; De Berardinis, P. PV integration in minor historical centers: Proposal of guide-criteria in post-earthquake reconstruction planning. *Energy Procedia* **2014**, *48*, 1549–1558. [CrossRef]

34. Gregório, V.; Seixas, J. Energy savings potential in urban rehabilitation: A spatial-based methodology applied to historic centres. *Energy Build.* **2017**, *152*, 11–23. [CrossRef]

35. Eriksson, P.; Hermann, C.; Hrabovszky-Horváth, S.; Rodwell, D. EFFESUS methodology for assessing the impacts of energy-related retrofit measures on heritage significance. *Hist. Environ. Policy Pract.* **2014**, *5*, 132–149. [CrossRef]

36. Boarin, P.; Davoli, P. A systemic approach for preliminary proposals of sustainable retrofit in historic settlements-the case study of villages hit by earthquake. In Proceedings of the European Conference on Sustainability, Energy and the Environment 2013, Brighton, UK, 4–7 July 2013.

37. Ascione, F.; De Masi, R.F.; De Rossi, F.; Fistola, R.; Sasso, M.; Vanoli, G.P. Analysis and diagnosis of the energy performance of buildings and districts: Methodology, validation and development of urban energy maps. *Cities* **2015**, *35*, 270–283. [CrossRef]

38. Fatiguso, F.; De Fino, M.; Cantatore, E.; Scioti, A.; De Tommasi, G. Energy models towards the retrofitting of the historic built heritage. *WIT Trans. Built Environ.* **2015**, *153*, 159–170.

39. Belpoliti, V.; Bizzarri, G.; Boarin, P.; Calzolari, M.; Davoli, P. A parametric method to assess the energy performance of historical urban settlements. Evaluation of the current energy performance and simulation of retrofit strategies for an Italian case study. *J. Cult. Herit.* **2018**, *30*, 155–167. [CrossRef]

40. KEM QTM 500-Quick Thermal Conductivity Meter. Available online: https://www.kyoto-kem.com/en/pdf/catalog/QTM-500.pdf (accessed on 30 June 2020).

41. EnergyPlus. Available online: https://energyplus.net/ (accessed on 7 December 2018).

42. Kalaycı, Ş. *SPSS uygulamalı çok değişkenli istatistik teknikleri*; Volume 5; Asil Yayın Dağıtım: Ankara, Turkey, 2016.

43. Webb, A.L. Energy retrofits in historic and traditional buildings: A review of problems and methods. *Renew. Sust. Energ. Rev.* **2017**, *77*, 748–759. [CrossRef]

44. Troi, A.; Bastian, Z. Energy Efficiency Solutions for Historic Buildings. A Handbook, Berlin, Basel: Birkhäuser. Available online: https://www.degruyter.com/view/product/429524 (accessed on 10 September 2017).

45. Şahin, C.D.; Arsan, Z.D.; Tuncoku, S.S.; Broström, T.; Akkurt, G.G. A transdisciplinary approach on the energy efficient retrofitting of a historic building in the Aegean Region of Turkey. *Energy Build.* **2015**, *96*, 128–139. [CrossRef]

46. Grytli, E.; Kværness, L.; Rokseth, L.S.; Ygre, K.F. The impact of energy improvement measures on heritage buildings. *J. Archit. Conserv.* **2012**, *18*, 89–106. [CrossRef]

47. Morelli, M.; Rønby, L.; Mikkelsen, S.E.; Minzari, M.G.; Kildemoes, T.; Tommerup, H.M. Energy retrofitting of a typical old Danish multi-family building to a "nearly-zero" energy building based on experiences from a test apartment. *Energy Build.* **2012**, *54*, 395–406. [CrossRef]

48. Aste, N.; Adhikari, R.S.; Buzzetti, M. Energy retrofit of historical buildings: An Italian case study. *J. Green Build.* **2012**, *7*, 144–165. [CrossRef]

49. Zagorskas, J.; Zavadskas, E.K.; Turskis, Z.; Burinskienė, M.; Blumberga, A.; Blumberga, D. Thermal insulation alternatives of historic brick buildings in Baltic Sea Region. *Energy Build.* **2014**, *78*, 35–42. [CrossRef]

50. Ben, H.; Steemers, K. Energy retrofit and occupant behaviour in protected housing: A case study of the Brunswick Centre in London. *Energy Build.* **2014**, *80*, 120–130. [CrossRef]

51. *Short Guide 1: Fabric Improvements for Energy Efficiency in Traditional Buildings*; Historic Scotland: Edinburgh, UK, 2013.

52. The American Society of Heating and Refrigeration Engineers. [ASHRAE]. ASHRAE 34P Energy Guideline for Historical Buildings. Second Public Review Draft. 2016. Available online: https://osr.ashrae.org/Public%20Review%20Draft%20Standards%20Lib/GPC%2034%20Guideline%20Final%20Draft%202015%2008%2024_2ndPPRDraft_chair_approved.pdf (accessed on 30 March 2020).

 *atmosphere*

Article

Outdoor Thermal Comfort and Building Energy Use Potential in Different Land-Use Areas in Tropical Cities: Case of Kuala Lumpur

Yasemin D. Aktas [1,*], Kai Wang [1], Yu Zhou [1], Murnira Othman [2], Jenny Stocker [3], Mark Jackson [3], Christina Hood [3], David Carruthers [3], Mohd Talib Latif [4], Dina D'Ayala [1] and Julian Hunt [5]

[1] Department of Civil, Environmental and Geomatic Engineering (CEGE), University College London (UCL), London WC1E 6BT, UK; kai.wang@ucl.ac.uk (K.W.); ucesyz5@alumni.ucl.ac.uk (Y.Z.); d.dayala@ucl.ac.uk (D.D.)

[2] Institute for Environment and Development (LESTARI), Universiti Kebangsaan Malaysia (UKM), Bangi 43600, Malaysia; murnira@ukm.edu.my

[3] Cambridge Environmental Research Consultants (CERC), Cambridge CB2 1SJ, UK; jstocker@cerc.co.uk (J.S.); mjackson@cerc.co.uk (M.J.); christina@cerc.co.uk (C.H.); David.Carruthers@cerc.co.uk (D.C.)

[4] Department of Earth Sciences and Environment, Universiti Kebangsaan Malaysia (UKM), Bangi 43600, Malaysia; talib@ukm.edu.my

[5] University College London (UCL), Earth Sciences, London WC1E 6BT, UK; julian.hunt@ucl.ac.uk

* Correspondence: y.aktas@ucl.ac.uk; Tel.: +44-(0)-2076791566

Received: 1 May 2020; Accepted: 17 June 2020; Published: 19 June 2020

Abstract: High air temperature and high humidity, combined with low wind speeds, are common trends in the tropical urban climates, which collectively govern heat-induced health risks and outdoor thermal comfort under the given hygrothermal conditions. The impact of different urban land-uses on air temperatures is well-documented by many studies focusing on the urban heat island phenomenon; however, an integrated study of air temperature and humidity, i.e., the human-perceived temperatures, in different land-use areas is essential to understand the impact of hot and humid tropical urban climates on the thermal comfort of urban dwellers for an appraisal of potential health risks and the associated building energy use potential. In this study, we show through near-surface monitoring how these factors vary in distinct land-use areas of Kuala Lumpur city, characterized by different morphological features (high-rise vs. low-rise; compact vs. open), level of anthropogenic heating and evapotranspiration (built-up vs. green areas), and building materials (concrete buildings vs. traditional Malay homes in timber) based on the calculated heat index (HI), apparent temperature (T_{App}) and equivalent temperature (T_E) values in wet and dry seasons. The results show that the felt-like temperatures are almost always higher than the air temperatures in all land-use areas, and this difference is highest in daytime temperatures in green areas during the dry season, by up to about 8 °C (HI)/5 °C (T_{App}). The T_E values are also up to 9% higher in these areas than in built-up areas. We conclude that tackling urban heat island without compromising thermal comfort levels, hence encouraging energy use reduction in buildings to cope with outdoor conditions requires a careful management of humidity levels, as well as a careful selection of building morphology and materials.

Keywords: thermal comfort; land-use; tropics; urban microclimate

---

1. Introduction

An urban heat island (UHI) is often defined as the significant temperature differential between urban and surrounding rural areas [1]. This is primarily attributable to the reduced evapotranspiration due to less greenery, construction materials with higher thermal admittance, lower ventilation due to high surface roughness, and higher anthropogenic heat sources in cities, such as traffic and waste heat from air-conditioning (AC) systems [2]. Quantification of UHI and its implications on heat induced risks on health and wellbeing has gained considerable prominence in academic literature despite the fact that temperature has been shown to be an insufficient metric by a tremendous number of epidemiological studies to appraise heat stress and mortality, which are more meaningfully linked with the combined impact of temperature and humidity [3–10]. This combined impact of temperature and humidity is better indicative of the physiological experience of heat, and the higher the humidity, the higher the perceived temperature, the higher the potential health risks, the poorer the thermal comfort, and hence, the higher the building energy use to keep the indoor conditions at favorable levels. Despite recent sporadic studies that demonstrated that humid heat is increasingly a global trend [11] and highlighted urban moisture as a prevalent issue and an aggravator of heat island impact [12–15] with important implications on health and energy use [16], the scholarly discourse on urban microclimate under current and future climatic trends is still heavily dominated by air temperatures alone.

In this respect, our presuppositions with regards to the thermal comfort in different land-use areas need also to be revisited, especially in tropical and subtropical cities (compared to mid-latitude cities), because (1) high air temperatures and UHI in tropical and subtropical cities is an almost year-round critical phenomenon, (2) humidity levels are rather high due to frequent and intense precipitation, and (3) ventilation is low due to overall lower wind speed values, further exacerbating the adverse impact of humidity on perceived temperatures.

The aim of this study was to demonstrate how urban microclimate and hence thermal comfort and building energy use potential vary in different land-use areas in Kuala Lumpur in wet and dry seasons, with specific emphasis on heritage and green areas, through multiple metrics by using on-site monitoring data. Kuala Lumpur is located in West Malaysia over the tropics at 3°09′35″ N 101°42′00″ E. The urban climatology of (sub)tropical megacities is in general a relatively sparsely studied field [17]; however, the evolution and progression of urban heat island in Kuala Lumpur (see [18] for an extensive review of some of these previous studies) and the health implications [19,20] have been examined and documented rather extensively: Greater Kuala Lumpur, with a population expected to exceed 10 million in 2020, similar to other Asian megacities, suffers from substantial urban heating [21], which is attributable mainly to the rate of urbanization (and conurbation), which is among the highest in Southeast Asia [22], subsequent changes in the cityscape and land-use, and increasingly higher anthropogenic heating due to steep population rise.

2. On-Site Monitoring

2.1. Monitoring Equipment and Locations

In order to identify the trends in near-surface air temperature (T) and relative humidity (RH) in different land-use areas, ibutton DS 1923 Hygrocon sensors (Model DS1923F5, Maxim Integrated, San Jose, CA, USA) (sensor resolution and accuracy: (T) 0.0625 °C, <0.5 °C; (RH) 0.04%, <5%) were used along with radiation shields (HOBO, model RS3, Onset Computer, MA, USA), and readings were logged every half an hour. The sensors were set up at approximately 2.5 m height, so that the readings were representative of the physiological hygrothermal experience of the urban dwellers while also ensuring the safety of the sensors.

Using this setup, a rigorous on-site monitoring work was carried out in 11 locations within central Kuala Lumpur, selected mainly based on their morphological/constructive features, and built-up/green differentiation. These included the following built up areas: (1) Intercontinental Hotel, (2) Jalan P Ramlee, (3) Chow Kit, (4) Malaysia Tourism Centre, (5) Kuala Lumpur City Centre (KLCC) Park, (6) Jalan Ampang-Jalan Tun Razak intersection, and (7 and 8) two locations in Kampung Baru. Kampung Baru is an urban heritage site characterized by traditional Malay homes made of timber, in contrast to the mainly concrete building stock in all other built-up measurement sites (see Section 2.2

for more information). In addition to these built-up areas, (9–11) three green areas, i.e., Perdana Botanic Gardens, Tugu Negara, and Eco Park, were included in the monitoring campaign.

In order to correctly categorize the land-uses the built-up monitoring locations represent, the three-dimensional (3D) building data was used to calculate the average level of compactness and average building height within the 200 m around the sensor locations (Figure 1). The assessment regarding the level of compactness was done based on the λ_p parameter, which was calculated as the ratio of the planar area within the grid cell occupied by the buildings to the total grid area [23].

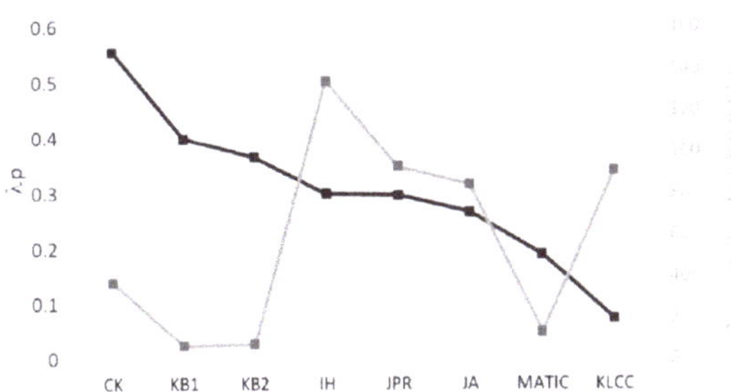

Figure 1. Average λ_p and building heights for the monitoring locations within 200 m distance.

The locations with an average λ_p below and above 0.3 were labelled as open and compact, respectively. Similarly, those with the average building height below and above 40 m were labelled as low- (inclusive of medium-rise) and high-rise, respectively.

The three green areas monitored here, the Perdana Botanic Gardens (BG), Tugu Negara (TN), and Eco Park (ECO), each have varying degrees of tree cover: BG is very open with almost no trees, TN is open with few trees, and ECO is thickly covered by high trees. Therefore, we had: Compact High-Rise areas (CHR) [n = 2, Intercontinental Hotel (IH), Jalan P Ramlee (JPR)], Compact Low-Rise (CLR) [n = 3, two locations within Kampung Baru (KB1 and KB2) and one in Chow Kit (CK)], Open High-Rise (OHR) [n = 2, Kuala Lumpur City Centre Park (KLCC) and Jalan Amp (JA)], Open Low-Rise (OHR) [n = 1, Malaysia Tourism Centre (MATIC)], Open Low-Green (OLG) [n = 2, Perdana Botanic Gardens (BG), Tugu Negara (TN)] and Compact Tall-Green (CTG) [n = 1, Eco Park (ECO)] (Figure 2 and Table 1).

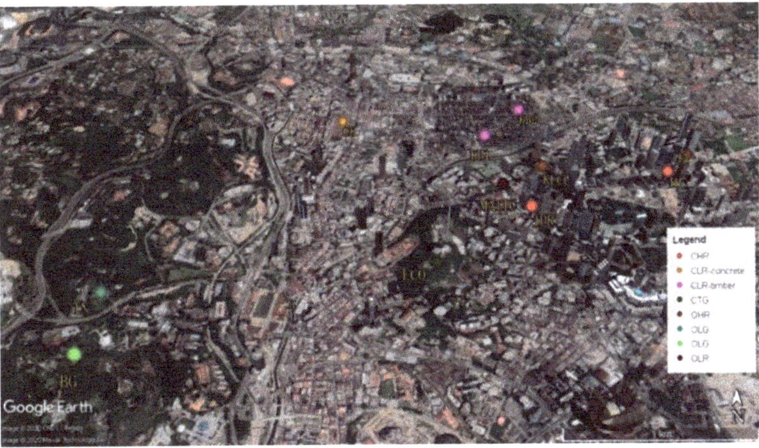

Figure 2. T and RH monitoring locations in Kuala Lumpur city center (drawn on Google Earth).

Table 1. On-site monitoring locations (photos taken by K. Wang and M. Othman in 2017).

Open High-Rise (OHR)	**Jalan Ampang Near KL City Centre Park (KLCC)**	**Jalan Ampang—Jalan Tun Razak Intersection (JA)**	**Green (OLG and CTG)**	**Botanic Gardens (BG)**	**Tugu Negara (TN)**	**Eco Park (ECO)**	**Open Low- Rise (OLR)**	**Malaysia Tourism Centre (MATIC)**

Compact High-Rise (CHR)	**Intercontinental Hotel (IH)**	**Jalan P Ramlee (JPR)**	**Compact Low-Rise (CLR)**	**Kampung Baru (KB1)**	**Kampung Baru (KB2)**	**Chow Kit (CK)**		

2.2. Kampung Baru (KB)

Kampung Baru (also spelled as Kampong Bharu (KB), see Fujita, 2010; meaning "New Village") is an urban heritage site located between Chow Kit and KLCC, i.e., right at the heart of Kuala Lumpur's business and financial center, surrounded by major roads. KB has a total area equal to 110 hectares with a population of 45,000 [24], making this low-rise settlement one of the most densely populated in the city.

Initially a mining area, KB was founded in 1899 as a 91 ha Malay Agriculture Settlement (MAS) to allow Malays a rural life in Kuala Lumpur, although it gradually expanded beyond the MAS, with relatively new-built, non-traditional housing, until it reached its current state. KB is to this day a wholly Malay area and is still governed by a Board of Management in line with the initial management plan set up by the British; however, it is now a completely residential area as opposed to its original multifunctional use combining residential and agricultural purposes [25]. The protocols initially set up to define the land title within the MAS area in KB are still in place, which greatly contributed to remaining almost untouched to the present day. However, echoing a desire ongoing since the 1970s [26], the government in their 2020 development plan states that "many of the original buildings (. . .) are no longer compatible with their surroundings" and mentions the potential to developing it into a "modern commercial area" because of its proximity within the city center [24], which attracted fierce criticism both from the residents themselves, and the public in and outside Malaysia (e.g., [27,28]). The political and economic pressures pushed residents to join forces under various organizations in order to facilitate efficient negotiations with the government to ensure sustainable (re)development of the neighborhood. However, following many town meetings and the government's ever increasing land price offers, the majority of the residents are reported to have agreed on the development of a "Taman Warisan Melayu" (Malay Heritage Park) in the neighborhood: the project involves the reconstruction/refurbishment of 11 traditional homes and preservation of a few monumental landmarks including the 119 year-old Kampung Baru Mosque, along with the development of 45,000 new houses, an upgrade of the famous Kampung Baru food markets with car parks, and other amendments regarding public transport and pedestrian routes (all with dire implications for the original spatial and social setting of KB), though there are reportedly still major issues around land ownership [29–34]. This entangled state of multi-ownership, as well as the absentee landlords, who are difficult to access to discuss an eventual purchase of their lands, are possibly the main reason for the "development" of KB becoming a reality only some 50 years after it was first proposed [35].

At an intriguing contrast to the skyscrapers visible from within it, at the time of writing this paper KB is still home to some fine examples of traditional timber Malay homes (Figure 3), mainly post and lintel structures with steep roofs, including long roofed typologies, *rumah limas* ("five roofs") with crafted carpentry joinery, representative of elaborate workmanship this building culture flourished on [36]. A significant number of these are on stilts, raising floors high above the ground level to protect the building envelope from dampness, catch high winds and facilitate ventilation, and reduce flood damage potential [37,38]. It has been shown that during the "normal" floods that occur commonly during the north-east monsoon, i.e., the wet season, the flood height does not exceed the stilt heights of traditional housing in a given area [39]. The slope of the roofs facilitates easy discharge of rainwater, while the overhangs and gables provide protection from wind driven rain and the natural materials used to build it (such as attap or reeds) prevent heat absorption. The many windows and the lack of conventional partitioning between various usage areas within homes further encourage air circulation providing passive cooling, preventing stagnation and development of moisture-induced decay in the building envelope [40–42]. Therefore, traditional Malay typologies have been developed to ensure the highest comfort under prevailing hot and humid climatic conditions and to provide the most efficient protection from environmental hazards common to the tropics, having much to offer to modern sustainable and climate-resilient building design, as other vernacular typologies do.

Figure 3. Some examples from the traditional Malay building stock in Kampung Baru (photos taken by Y. D. Aktas in 2016).

2.3. Diurnal Average Temperature and Relative Humidity Variations

The diurnal near surface air T and RH variations in the monitoring locations are shown in Figure 4a–d for a month each in wet and dry seasons in 2018: from 20 February to 18 March, which correspond to the end of the rainy Northeast Monsoon, and from 9 June to 1 July, which is a part of the drier Southwest Monsoon, respectively [18,43]. The selection of these two different months also provides an opportunity to delve into the impact of solar angle on the thermal environment, as Kuala Lumpur is located near the equator the solar radiation is stronger in March, while the solar elevation angle is higher in June.

In order to improve the readability of the graphs, trends under the same land-use area with negligible differences have been averaged, such as KB1 and 2, two open low green spaces, i.e., BG and TN, two CHR locations, i.e., IH and JPR, and two OHR locations, i.e., KLCC and JA.

The results show that the T and RH ranges in wet and dry seasons are quite similar, although they are slightly narrower in the latter case. The highest T and lowest RH appear in CLR locations (CK and KB) during daytime, due to less shading that results in higher solar gain, as well as the anthropogenic heat from the traffic. While CK shows the same, among the highest T, with the lowest RH trends during the nighttime, KB tends to show lower T and higher RH trends, which is indicative of more moisture inducing anthropogenic activity such as outdoor cooking, and less anthropogenic heating (both traffic and waste heat from AC systems). This also demonstrates the impact of building materials: the concrete buildings dominating CK have both higher unit thermal storage and thicker walls (i.e., more material available to absorb heat) than traditional *kampung* houses of KB, made of thinner, timber walls (for a more comprehensive discussion on building materials see [44]). Interestingly, Ts on OHR are similar to CLR Ts during the wet season (March), while they are lower than CLR Ts during the dry season (June). This is considered to be because OHR locations are exposed to higher solar radiation levels due to the higher solar angle of Kuala Lumpur during the wet season. CHR locations, on the other hand, are situated somewhere in the middle of all land-use areas in terms of both daytime and nighttime T and RH values. Daytime Ts in CHR locations were found to be lower than CLR and OHR areas, as expected, due to the shading effect and large heat storage in these areas owed to the building morphology, providing increased surface area and volume that can trap and absorb more solar radiation in the daytime. This is released at nighttime, making the nighttime Ts at CHR locations comparable to those at CLR locations and OHR. The lower radiative cooling due to smaller sky view factor may also play a role in the relatively higher nighttime Ts in CHR locations. The lowest daytime Ts are observed in ECO, as expected due to more shading from the trees. However, the temperatures in ECO are higher than that in BG and TN during nighttime. This is considered to be because of the larger nocturnal cooling in BG and TN, due to lower roughness, which may result in higher wind speeds and convection rates, as well as the larger radiative cooling thanks to a more open morphology.

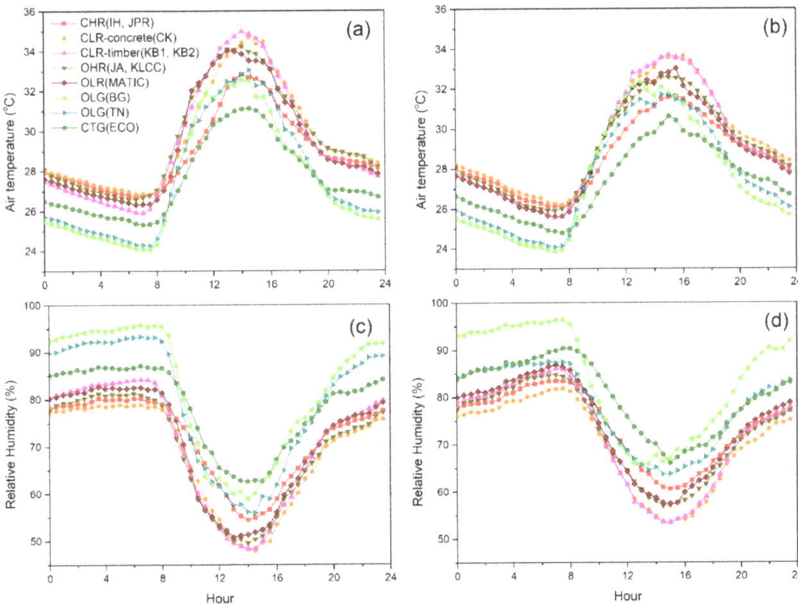

Figure 4. Diurnal variation of air temperature (T) and relative humidity (RH) in wet (**a**,**c**) and dry (**b**,**d**) seasons for different land use areas.

3. Metrics Indicative of Thermal Comfort and Energy Use

To demonstrate the importance of accounting for the impact of humidity while assessing heat induced risks on human health and wellbeing, we first used the monitoring data for an appraisal of mortality likelihood using the risk curves derived by Mora et al. [45] identifying the hygrothermal thresholds for lethal events, using a range of daily climatic data including near-surface air temperature, near-surface relative humidity, solar radiation, and ventilation (Figure 5a,b). The blue curve here indicates the threshold identified by means of Support Vector Machines to best separate lethal and non-lethal conditions based on mean daily surface air T and RH, while the red curve is the 95% probability threshold. As seen, the T and RH values obtained from various locations in Kuala Lumpur indicate that the microclimatic conditions in all land-use areas pose health risks, and green areas are not in any way devoid of risk due to much higher average daily mean relative humidity values than the built-up areas in both seasons.

This clearly shows the need for more comprehensive metrics to assess thermal comfort and heat-induced risk on health and wellbeing. In this paper we used three of these to discuss their suitability for the tropics: heat index, apparent temperature, and equivalent temperature.

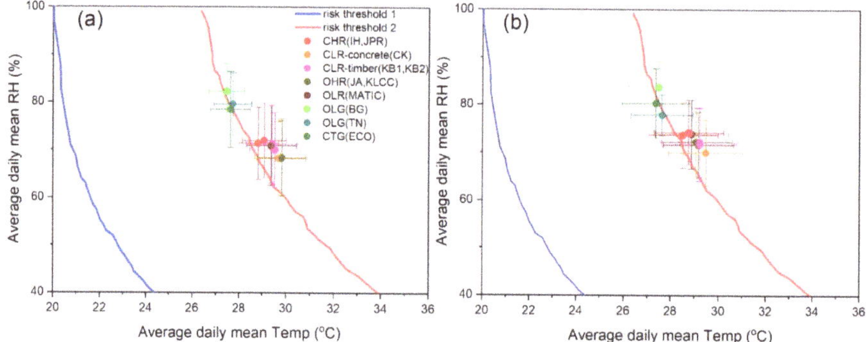

Figure 5. Measured average daily and average daily mean temperature (T) and relative humidity (RH) values in different land-use areas against the risk threshold derived by Mora et al. [45] in wet (**a**) and dry months (**b**).

3.1. Heat Index

Heat index is one of the "simple" indices to determine felt-like or perceived temperature, developed by [46] by multiple regression from the first version of Steadman's apparent temperature model (1979), which is discussed in detail in Section 3.2 [47,48]:

$$
HI = \begin{cases}
\begin{aligned}
& -8.784695 + 1.61139411 \times T + 2.338549 \times RH - 0.14611605 \times T \times RH - \\
& (1.2308094 \times 10^{-2})T^2 - (1.6424828 \times 10^{-2})RH^2 + (2.211732 \times 10^{-3})T^2 \times RH + \\
& (7.2546 \times 10^{-4})T \times RH^2 - (3.582 \times 10^{-6})T^2 \times RH^2, \qquad T \geq 20°C
\end{aligned} \\
\\
T, \qquad T < 20°C
\end{cases}
\tag{1}
$$

where T is air temperature (°C) and RH is the ambient relative humidity (%). The calculated average diurnal heat index variations for different land-use categories are shown in Figure 6. As can be seen, the heat index values are almost always higher than the measured air temperature values. According to the assessment scale of heat index, the index values between 27 °C and 32 °C would be category "caution," where the possible heat disorders for people in high risk groups would include fatigue, possible with prolonged exposure and/or physical activity. The heat indexes between 32°C and 41 °C, on the other hand, would call for "extreme caution," where sunstrokes, muscle cramps, and/or heat exhaustion possible with prolonged exposure and/or physical activity are among possible health risks [47]. As seen in Figure 6, an assessment based only on air temperatures would significantly underestimate heat experience in most of the land-use areas with the maximum temperatures lying on the 34–35 °C range as opposed to maximum heat indices at around 40 °C, which are out of the thermal comfort range by a margin dangerously beyond what is accepted as tolerable, i.e., ±1.1–1.7 °C [49], and fall almost entirely under caution and extreme caution categories. Importantly, the risk categorization used here might not be representative of people native to tropical areas. While there are no conclusive studies regarding how heat-induced health risks vary in the tropics, the natives of tropical areas are reported to have a higher tolerance to elevated temperature and humidity conditions [50–54].

The results indicate that the felt-like temperatures in green areas are up to 5 °C lower than the built-up areas at night and early in the morning, while the daytime temperatures in these locations can be comparable to or even higher than those in the built-up areas. Interestingly, while the daytime temperatures are highest in Kampung Baru among built-up case study monitoring sites, they are lowest at night and in the small hours, which is indicative of the lower heat absorption and storage capacity of the building stock in this neighborhood. The lowest daytime temperatures are observed in CHR locations, as expected, owed to the high thermal admittance capacity.

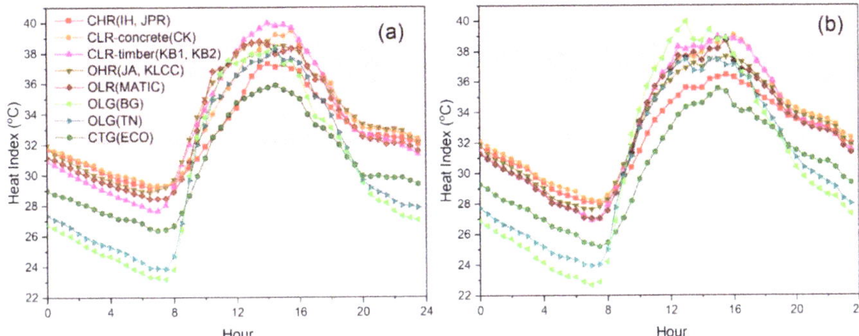

Figure 6. Diurnal variation of calculated heat index in wet (**a**) and dry (**b**) seasons for different land use areas.

3.2. Apparent Temperature

In addition to heat index, we also used Steadman's Universal Apparent Temperature Model [55] for a further appraisal of the felt-like temperatures in different land use areas, accounting also for ventilation as follows:

$$T_{App} = -2.7 + 1.04 \times T + 2.0 \times P - 0.65v \tag{2}$$

where T_{App} is apparent temperature (°C), T is air temperature (°C), P is water vapor pressure (kPa), which was estimated from our monitored temperature and relative humidity using procedures described by Steadman [56], and v is the wind speed at 10 m above ground (m/s).

Meteorological Department data from Subang station show that wind speeds can be as high as 7 m/s, while the average values do not exceed 3.5 m/s (Figure 7). The often-lower median values point out overall very still conditions, as expected in the tropics. Importantly, Subang is in Kuala Lumpur's suburbs in the west, and the wind speed values in the city center are expected to be even lower, due to the increased surface roughness from the buildings. As our monitoring program did not include the measurement of wind speeds, we used wind speed modelling via ADMS-Urban (Atmospheric Dispersion Modelling Software) [57] to estimate wind speeds in different land use to use in the apparent temperature calculations.

ADMS-Urban is a fast local-scale urban climate modelling tool, widely used to calculate the spatiotemporal variation of neighborhood or city scale urban temperatures and dispersion modelling. The parametrization and land-use input data used in the ADMS modelling of Kuala Lumpur to study the urban temperature perturbations was previously described in [58]. In addition to this, in this study a detailed urban canopy model with 200 m resolution was developed to quantify the roughness length in terms of building density and geometry [59] to model hourly wind speeds at 10 m height. The modelled diurnal wind speed trends for 20 February 2018 and 9 June 2018 are shown in Figure 8a,b, respectively. 20 February and 9 June are the first days of our selected wet and dry season windows, respectively, and it was assumed that the overall diurnal wind speed profiles obtained for these two days are representative for the rest of the selected months. The obtained trends clearly demonstrate the contrast of wind speed diurnally between the two green areas, BG and TN, and built up areas for both seasons, with slightly higher values in the dry season.

The diurnal apparent temperature variations calculated as such are shown in Figure 9, which indicates that the apparent temperatures are also always higher than the measured air temperatures shown in Figure 4a,b, especially in the wet season, on green areas and at night, while the ranking of various land-use areas in terms of the thermal comfort they offer does not change. While there are no general risk categories for apparent temperature, similar to the ones we report above for heat index, previous studies suggest that the apparent temperature thresholds for defining levels, beyond which mortality risk increases significantly can be as low as 30.7 °C for Taipei [60], 27 °C for Korea, and 29.4 °C for the Mediterranean basin [6], highlighting the potentially dangerous levels of apparent temperatures in all land-use areas we investigate in Kuala Lumpur. The high temperature values in CK combined with very low wind speed values make this area the one with the highest apparent temperatures, around 4 °C higher than the ambient temperatures in this location at its peak. Because trees have a significant impact on the urban wind flows [61] and as the ADMS modelling is unable to process complex terrain and urban canopy flow simultaneously, and considers only building obstacles, the modelled diurnal wind profile for ECO might not reflect the low ventilation levels prevailing in this very intensely tree covered park. Therefore, we also calculated T_{App} values based on a no-wind case in ECO, which led to around 1.5 °C higher felt-like temperatures in this location than T_{App} estimates if this was a largely open, grass covered park such as BG and TN, indicating thermal comfort levels comparable to a built-up area.

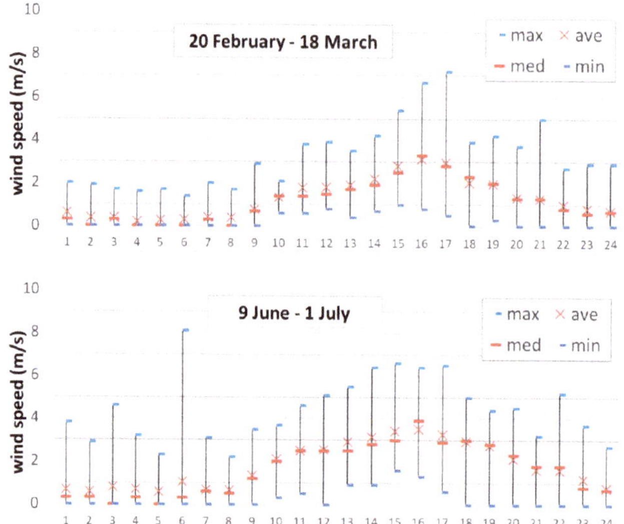

Figure 7. The diurnal wind speed ranges, averages and medians for Subang averaged over the wet and dry periods focused in this study (data from MetMalaysia).

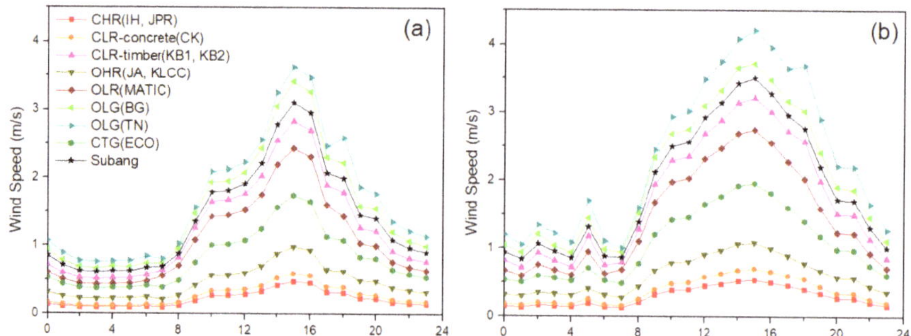

Figure 8. Average diurnal wind speed variation as modelled by ADMS-Urban in wet (**a**) and dry (**b**) seasons for different land use areas.

3.3. Equivalent Temperature

In this study, we also focused on equivalent temperature (T_E), which shows the total enthalpy from both sensible and latent heat [3,62], and is calculated as follows:

$$T_E = \frac{C_p T + Lq}{C_p} \tag{3}$$

where C_p is heat capacity of air, taken as 1.005 kJ/(kg·°C), T is the air temperature (°C) measured at each land-use location, L is the latent heat of vaporization taken as 2.5×10^3 kJ/kg, and q is the specific humidity in kg/kg, which was calculated using the observed RH and air pressure based on the empirical relationship by [63]. As we do not have air pressure measurements at monitoring locations, the air pressure data for Subang Station, which is located at a similar elevation as all the monitoring locations, were used in this study. The data are available from the Integrated Surface Database (ISD) at NOAA (https://www.ncdc.noaa.gov/isd/data-access). Note that the air pressure data are 3-hourly at Subang Station, which were interpolated to a half-hourly dataset in analogy with the monitoring data by Fast Fourier Transform method. The findings can be seen in Figure 10. The results show that open green areas (BG and TN) have the highest equivalent temperature in the daytime, though the air temperature is much lower, due to the very high humidity in these areas. Kampung Baru offers the second highest daytime equivalent temperature, which may be attributed to the high anthropogenic moisture in this area due to a high number of street food vendors. The difference between the equivalent temperature in different land-use areas is otherwise rather small.

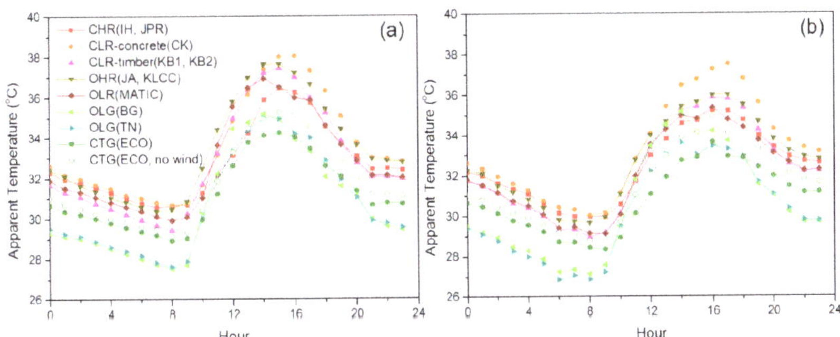

Figure 9. Diurnal variation of calculated apparent temperatures in wet (**a**) and dry (**b**) seasons for different land use areas.

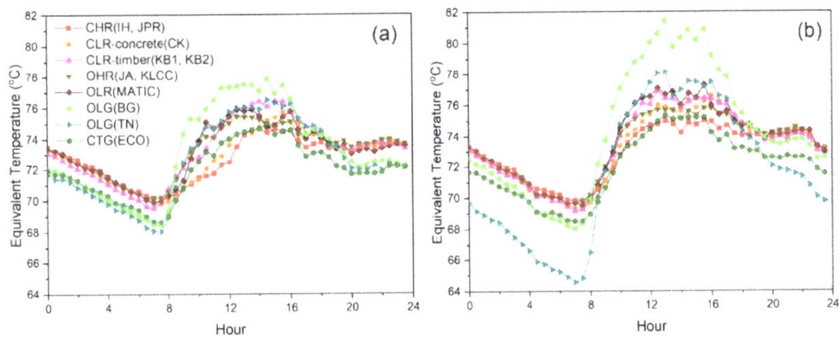

Figure 10. Diurnal variation of calculated equivalent temperature in wet (**a**) and dry (**b**) seasons for different land use areas.

4. Discussion

This study used a near-surface monitoring campaign in 11 different locations in Kuala Lumpur, Malaysia, in order to identify temperature and humidity variations in their associated six different land-use categories in wet and dry seasons, for an appraisal of outdoor thermal comfort, on a comparative basis. The results are briefly discussed below in terms of the role of green areas, morphology, and building materials.

Urban green areas are a major source of evapotranspiration, which is a natural cooling mechanism, and therefore, is considered one of the most obvious ways of mitigating elevated urban temperatures. While greenery is indeed effective at reducing urban temperatures (e.g., [64]) at different levels [65], our results show that temperatures in these areas can still be dangerously high in a tropical city. More critically, the thermal comfort in green areas is further compromised due to high humidity levels and, depending on the local morphology, very low ventilation rates. Our results show that while these areas offer the best thermal comfort in the evening and at night, regardless the season, the daytime thermal comfort in open, grass covered green areas (Botanic Gardens and Tugu Negara) is comparable to built-up areas in the wet season, while it is poorer than the built-up areas in the dry season, especially from early-morning to mid-afternoon, with higher felt-like temperatures of up to 8 °C (heat index)/5 °C (apparent temperatures) than air temperatures measured in these locations. As the general human perception of these spaces is that they are thermally comfortable even when the measurements show that they are not [66], this is important to address while advising about risk areas and times for vulnerable segments of the population (e.g., elderly, very young, those with cardiovascular illnesses), and while regulating the maintenance of these areas (e.g., watering times). The Eco Park, which is a thickly wooded green area, was found to offer the lowest daytime temperatures; however, thermal comfort here can converge to built-up areas, when the humidity levels and very low wind speed values due to high surface roughness are accounted for.

The reduced diurnal temperature values, i.e., cool islands, observed in the Compact High-Rise areas, indicate a strong morphological advantage in the daytime over Open and Low-Rise settings, as expected. The difference in results obtained from two different Open Low-Rise locations (i.e., Chow Kit and Kampung Baru), however, indicate the impact of traffic-induced anthropogenic heating and building materials: use of timber, which has lower thermal inertia and thermal storage capacity, combined with the constructive features of timber housing with thinner walls (see [44]), lead to a smaller thermal mass and higher daytime temperatures in Kampung Baru than in Chow Kit with a building stock with similar morphological characteristics but made in concrete and with thicker walls. This disadvantage in the daytime, however, becomes a major benefit at nighttime: Kampung Baru offers the highest nighttime thermal comfort of all built-up areas.

Previous studies found out that in warm climates/seasons, there is a strong correlation between the outdoor and indoor ambient conditions [67]. Therefore, all three metrics used here can be used as some proxy to assess the building energy use potential under the given outdoor hygrothermal conditions. However, only the equivalent temperature uses absolute humidity, and it is therefore best representative of the cooling load, which is the energy required to remove both sensible and latent heat from an enclosed space through AC systems to maintain a constant indoor dry-bulb air temperature and humidity. Our calculated mean diurnal equivalent temperatures mimic the heat index and apparent temperature findings, though with a more accentuated increases in green areas in the daytime with up to 9% higher values than built-up areas. Of all monitoring locations investigated here, the transferability of outdoor thermal comfort levels to the building energy use potential must be the weakest in Kampung Baru as the building stock here is less reliant on the AC systems to keep the indoor conditions at comfort levels thanks to the constructive features developed over time to tackle hot and humid climatic conditions by encouraging ventilation through the building and limiting heat absorption by materials used and architectural detailing.

Based on these results, humidity and low wind speed values emerge as critical variables governing thermal comfort in tropical areas. Any mitigation method put in place to tackle high urban temperatures, or any urban redevelopment work should consider their impact on humidity and ventilation patterns. Our results do not indicate a significant difference in thermal comfort in dry and wet seasons, at least in the comparative performances of the land-use areas we examine here (cf. [15]). Importantly, it should be noted that within the complexities of an actual urban setting, every area is either a conglomeration of various features that can be defined as low- or high-rise, or open or compact, making labelling them with a given urban morphology class very difficult, or is very closely surrounded by other areas able to be more closely aligned with other, sometimes quite opposing morphologies. To complicate things even further, the traffic-induced heating greatly impacts urban microclimate, especially through the main transport network arteries. Therefore, the temperature and relative humidity values obtained from each site which has been named as a certain land-use area here do not necessarily reflect the ideal hygrothermal characteristics of their associated land-uses. For instance, increasingly heavy traffic surrounding Kampung Baru is known to have made the urban cool island which site once was disappear [68,69].

5. Conclusions

This study shows that our presumptions with regards to thermal comfort levels in different land-use areas should be revisited, especially in a tropical context. In the face of a changing climate and ever-increasing temperatures, mitigating urban heat island without compromising thermal comfort levels and inflating building energy use requires a careful management of humidity levels, as well as a careful selection of building morphology and materials. Urban climate models and a scholarly discourse relying only on-air temperatures will critically underestimate the health, wellbeing, and energy use implications under current and future climates.

Our results, based on mean diurnal variations in different thermal comfort indexes, suggest critically poor thermal comfort levels under normal conditions, even in green areas. A higher risk is expected under increasingly common extreme conditions, such as heatwaves, and absence of background wind, which should be investigated further.

Finally, the pressures that urban heritage sites are facing in growing megacities are not unique to Kampung Baru; rather, it is a global trend with different cultural and legal backdrops against which heritage preservation operates. Vernacular architecture with constructive and architectural features, which have been proven sustainable under a given climatic context do give important clues about thermal comfort and energy efficiency to be considered in modern urban planning, development, and regeneration.

Author Contributions: Conceptualization, Y.D.A.; formal analysis, K.W., Y.Z., J.S., M.J. and C.H.; funding acquisition, Y.D.A., D.D. and J.H.; investigation, Y.D.A.; methodology, Y.D.A. and K.W.; supervision, Y.D.A.; visualization, K.W.; writing—original draft, Y.D.A.; writing—review and editing, K.W., M.O., D.C., M.T.L., D.D. and J.H. All authors have read and agreed to the published version of the manuscript.

Funding: This study was carried out as part of (a) "Future Cities-Disaster Resilient Cities: Forecasting Local Level Climate Extremes and Physical Hazards for Kuala Lumpur," jointly funded by Innovate UK and EPSRC under the Malaysia–UK Research and Innovation Bridges Competition for Collaborative R&D; Y.D.A., K.W. and D.D. by EPSRC Grant ref. EP/P015506/1; J.H. by EPSRC Grant ref. EP/P015476/1; J.S., M.J., C.H. and D.C. by Innovate UK Grant ref. 102718; M.T.L. and M.O. by Newton-Ungku Omar Fund XX-2017-002 and (b) "Modelling Apparent Temperatures in the Tropical City of Kuala Lumpur" funded by UCL EPSRC Impact Acceleration Award; Y.D.A., K.W., D.D. and J.H. for Grant ref. EP/R511638/1.

Acknowledgments: The 3D building dataset is courtesy of the UKM's Southeast Asia Disaster Prevention Research Initiative (SEADPRI-UKM) based on the 2013 LiDAR dataset from the KL City Hall (DBKL). We thank Lim Choun-Sian for facilitating building data acquisition, and the Malaysian Meteorological Department for the weather data. Finally, we are indebted to the Malaysian project leader Joy Pereira for making this collaboration possible.

Conflicts of Interest: The authors declare no conflict of interest.

Abbreviations

T	Air temperature (°C)
RH	Ambient relative humidity (%)
P	Water vapor pressure (kPa)
v	Wind speed at 10 m above ground (m/s)
T_E	Equivalent temperature (°C)
CLR	Compact Low-Rise
OLR	Open Low-Rise
JA	Jalan Ampang
L	Latent heat of vaporization (kJ/kg)
JPR	Jalan P Ramlee
BG	Botanic Gardens
ECO	Eco Park
CHR	Compact High-Rise
OHR	Open High-Rise
CK	Chow Kit
KB	Kampung Baru
IH	Intercontinental Hotel
HI	Heat index (°C)
T_{App}	Apparent temperature (°C)
C_p	Heat capacity of air (kJ/(kg·°C))
q	Specific humidity (kg/kg)
MATIC	Malaysia Tourism Centre
TN	Tugu Negara
KLCC	KL City Centre Park

References

1. Oke, T.R. *Boundary Layer Climates*, 2nd ed.; Routledge: New York, NY, USA, 1987.
2. Hunt, J.C.R.; Aktas, Y.D.; Mahalov, A.; Moustaoui, M.; Salamanca, F.; Georgescu, M.; Palou, F.S. Climate change and growing megacities: Hazards and vulnerability. *Proc. Inst. Civ. Eng. Eng. Sustain.* **2018**, *171*, 314–326. [CrossRef]
3. Matthews, T. Humid heat and climate change. *Prog. Phys. Geogr. Earth Environ.* **2018**, *42*, 391–405. [CrossRef]
4. Wichmann, J.; Andersen, Z.J.; Ketzel, M.; Ellermann, T.; Loft, S. Apparent Temperature and Cause-Specific Emergency Hospital Admissions in Greater Copenhagen, Denmark. *PLoS ONE* **2011**, *6*, e22904. [CrossRef]
5. Baccini, M.; Biggeri, A.; Accetta, G.; Kosatsky, T.; Katsouyanni, K.; Analitis, A.; Anderson, H.R. Heat effects on mortality in 15 European cities. *Epidemiology* **2008**, *19*, 711–719. [CrossRef]
6. Basu, R. High ambient temperature and mortality: A review of epidemiologic studies from 2001 to 2008. *Environ. Health* **2009**, *8*, 40. [CrossRef]
7. Basu, R.; Ostro, B.D. A Multicounty Analysis Identifying the Populations Vulnerable to Mortality Associated with High Ambient Temperature in California. *Am. J. Epidemiol.* **2008**, *168*, 632–637. [CrossRef]
8. Zanobetti, A.; Schwartz, J. Temperature and Mortality in Nine US Cities. *Epidemiology* **2008**, *19*, 563–570. [CrossRef]
9. Bell, M.; O'Neill, M.S.; Ranjit, N.; Borja-Aburto, V.H.; A Cifuentes, L.; Gouveia, N. Vulnerability to heat-related mortality in Latin America: A case-crossover study in São Paulo, Brazil, Santiago, Chile and Mexico City, Mexico. *Int. J. Epidemiol.* **2008**, *37*, 796–804. [CrossRef]
10. Stafoggia, M.; Forastiere, F.; Agostini, D.; Biggeri, A.; Bisanti, L.; Cadum, E.; Caranci, N.; Donato, F.K.D.; De Lisio, S.; De Maria, M.; et al. Vulnerability to Heat-Related Mortality. *Epidemiology* **2006**, *17*, 315–323. [CrossRef]
11. Raymond, C.; Matthews, T.; Horton, R.M. The emergence of heat and humidity too severe for human tolerance. *Sci. Adv.* **2020**, *6*, eaaw1838. [CrossRef] [PubMed]
12. Wang, Z.; Song, J.; Chan, P.W.; Li, Y. The urban moisture island phenomenon and its mechanisms in a high-rise high-density city. *Int. J. Clim.* **2020**. [CrossRef]

13. Doan, Q.-V.; Kusaka, H.; Ho, Q.-B. Impact of future urbanization on temperature and thermal comfort index in a developing tropical city: Ho Chi Minh City. *Urban Clim.* **2016**, *17*, 20–31. [CrossRef]

14. Qaid, A.; Bin Lamit, H.; Ossen, D.R.; Shahminan, R.N.R. Urban heat island and thermal comfort conditions at micro-climate scale in a tropical planned city. *Energy Build.* **2016**, *133*, 577–595. [CrossRef]

15. Chow, W.T.L.; Akbar, S.N. Assyakirin, B.A.; Heng, S.L.; Roth, M. Assessment of measured and perceived microclimates within a tropical urban forest. *Urban For. Urban Green.* **2016**, *16*, 62–75. [CrossRef]

16. Maia-Silva, D.; Kumar, R.; Nateghi, R. The critical role of humidity in modeling summer electricity demand across the United States. *Nat. Commun.* **2020**, *11*, 1–8. [CrossRef]

17. Roth, M. Review of urban climate research in (sub)tropical regions. *Int. J. Clim.* **2007**, *27*, 1859–1873. [CrossRef]

18. Ramakreshnan, L.; Aghamohammadi, N.; Fong, C.S.; GhaffarianHoseini, A.; GhaffarianHoseini, A.; Wong, L.-P.; Hassan, N.; Sulaiman, N.M. A critical review of Urban Heat Island phenomenon in the context of Greater Kuala Lumpur, Malaysia. *Sustain. Cities Soc.* **2018**, *39*, 99–113. [CrossRef]

19. Aghamohammadi, N.; Ramakreshnan, L.; Fong, C.S.; Sulaiman, N.M. Climate-Related Disasters and Health Impact in Malaysia. In *Extreme Weather Events and Human Health: International Case Studies*; Springer: Cham, Switzerland, 2019; pp. 247–264.

20. Wong, L.P.; Alias, H.; Aghamohammadi, N.; Aghazadeh, S.; Lai, S.H. Urban heat island experience, control measures and health impact: A survey among working community in the city of Kuala Lumpur. *Sustain. Cities Soc.* **2017**, *35*, 660–668. [CrossRef]

21. Yatim, A.N.M.; Latif, M.T.; Ahamad, F.; Khan, F.; Nadzir, M.S.M.; Juneng, L. Observed Trends in Extreme Temperature over the Klang Valley, Malaysia. *Adv. Atmos. Sci.* **2019**, *36*, 1355–1370. [CrossRef]

22. Yuen, B.; Kong, L. Climate change and urban planning in Southeast Asia. *SAPIENS Surv. Perspect. Integr. Environ. Soc.* **2009**, *2*, 1–11.

23. CERC. *ADMS Urban Canopy Tool User Guide*; CERC: Cambridge, UK, 2014.

24. DBKL. *Kuala Lumpur Structure Plan 2020*; Kuala Lumpur City Hall: Kuala Lumpur, Malaysia, 2018.

25. Alhabshi, S.M. Urban Renewal of Traditional Settlements in Singapore and Malaysia: The Cases of Geylang Serai and Kampung Bharu. *Asian Surv.* **2010**, *50*, 1135–1161. [CrossRef]

26. Chee, W.T. From Spontaneous Settlement to Rational Land Use Planning: The Case of Kuala Lumpur, Malaysia. *Sojourn J. Soc. Issues Southeast Asia* **1991**, *6*, 240–262.

27. Mayberry, K. A Village Amid Skyscrapers: How Long Can Kuala Lumpur's Enclave Hold out? *The Guardian*. 9 March 2017. Available online: https://www.theguardian.com/cities/2017/mar/09/village-amid-skyscrapers-kuala-lumpur-kampung-bharu (accessed on 14 April 2020).

28. FMT News. Skyscrapers Threaten to Swallow KL's Last Malay Village. *Free Malaysia Today*. 13 April 2018. Available online: https://www.freemalaysiatoday.com/category/nation/2018/04/13/skyscrapers-threaten-to-swallow-kls-last-malay-village/ (accessed on 14 April 2020).

29. Hassan, H. Malaysia's Malay Enclave of Kampung Baru Inches Closer to Redevelopment. *The Straits Times*. 13 January 2020. Available online: https://www.straitstimes.com/asia/se-asia/malaysias-malay-enclave-of-kampung-baru-inches-closer-to-redevelopment (accessed on 18 April 2020).

30. Azis Ngah, O.M. Bandar Moden Simbol Kekuatan Melayu Kota. *BH Online*. 21 April 2019. Available online: https://www.bharian.com.my/berita/nasional/2019/04/555541/bandar-moden-simbol-kekuatan-melayu-kota (accessed on 11 April 2020).

31. Babulal, V.; Ying, T.P. Kampung Baru Land Deal: Final Offer to Be Made by PM. *New Straits Times*. 22 September 2019. Available online: https://www.nst.com.my/news/nation/2019/09/523428/kampung-baru-land-deal-final-offer-be-made-pm (accessed on 3 March 2020).

32. New Straits Times. 2019 Sees a Glimmer of Hope towards Kampung Baru Redevelopment. *New Straits Times*. 26 December 2019. Available online: https://www.nst.com.my/news/nation/2019/12/550875/2019-sees-glimmer-hope-towards-kampung-baru-redevelopment (accessed on 9 March 2020).

33. Rodzi, N.H. PH Govt Makes Fresh Offer to Redevelop Malay Enclave in KL. *The Straits Times*. 22 September 2019. Available online: https://www.straitstimes.com/asia/se-asia/ph-govt-makes-fresh-offer-to-redevelop-malay-enclave-in-kl (accessed on 5 April 2020).

34. Zainol, M.J. Transform Kampung Baru into a Central Park. *New Straits Times*. 3 October 2019. Available online: https://www.nst.com.my/opinion/letters/2019/10/526867/transform-kampung-baru-central-park (accessed on 7 March 2020).

35. Fujita, M.A. Forays into Building Identity. *J. Arch. Educ.* **2010**, *63*, 8–24. [CrossRef]
36. Hashim, W.; Nasir, A.H. *The Traditional Malay House*; Institut Terjemahan Negara Malaysia: Kuala Lumpur, Malaysia, 2011.
37. Yuan, L.J. Under one roof: The traditional Malay house. *IDRC Rep.* **1984**, *12*, 15–16.
38. Chan, N.W.; Parker, D.J. Response to Dynamic Flood Hazard Factors in Peninsular Malaysia. *Geogr. J.* **1996**, *162*, 313. [CrossRef]
39. Chan, N.W. Increasing flood risk in Malaysia: Causes and solutions. *Disaster Prev. Manag. Int. J.* **1997**, *6*, 72–86. [CrossRef]
40. Yuan, L.J. *The Malay House: Rediscovering Malaysia's Indigenous Shelter System*; Institut Masyarakat: Penang, Malaysia, 1987.
41. Kamal, K.S.; Wahab, L.A.; Ahmad, A.C. Climatic design of the traditional Malay house to meet the requirement of modern living. In Proceedings of the 38th International Conference of Architectural Science Association ANZAScA: "Contexts of Architecture", Launceston, Tasmania, 10–12 November 2004; pp. 175–179.
42. Nordin, T.E.; Husin, H.N.; Kamal, K.S. Climatic Design Feature in the Traditional Malay House for Ventilation Purpose. In Proceedings of the International Seminar on Malay Architecture as Lingua Franca, Jakarta, Indonesia, 22–23 June 2005; pp. 41–48.
43. Ooi, M.; Chan, A.; Subramaniam, K.; Morris, K.; Oozeer, M.Y. Interaction of Urban Heating and Local Winds During the Calm Intermonsoon Seasons in the Tropics. *J. Geophys. Res. Atmos.* **2017**, *122*, 11, 499–523. [CrossRef]
44. Aktaş, Y.D.; Stocker, J.; Carruthers, D.; Hunt, J. A Sensitivity Study Relating to Local Urban Climate Modelling within the Built Environment. *Procedia Eng.* **2017**, *198*, 589–599. [CrossRef]
45. Mora, C.; Dousset, B.; Caldwell, I.; Powell, F.E.; Geronimo, R.C.; Bielecki, C.R.; Counsell, C.W.W.; Dietrich, B.S.; Johnston, E.T.; Louis, L.; et al. Global risk of deadly heat. *Nat. Clim. Chang.* **2017**, *7*, 501–506. [CrossRef]
46. Rothfusz, L.P. *The Heat Index Equation*; SR/SSD 90–23; NWS Southern Region Technical Attachment: Fort Worth, TX, USA, 1990.
47. Blazejczyk, K.; Epstein, Y.; Jendritzky, G.; Staiger, H.; Tinz, B. Comparison of UTCI to selected thermal indices. *Int. J. Biometeorol.* **2011**, *56*, 515–535. [CrossRef]
48. Anderson, G.B.; Bell, M.; Peng, R.D. Methods to Calculate the Heat Index as an Exposure Metric in Environmental Health Research. *Environ. Health Perspect.* **2013**, *121*, 1111–1119. [CrossRef] [PubMed]
49. Wang, S.K. *Handbook of Air Conditioning and Refrigeration*; McGraw-Hill: New York, NY, USA, 2001.
50. Saat, M.; Tochihara, Y.; Hashiguchi, N.; Sirisinghe, R.G.; Fujita, M.; Chou, C.M. Effects of Exercise in the Heat on Thermoregulation of Japanese and Malaysian Males. *J. Physiol. Anthr. Appl. Hum. Sci.* **2005**, *24*, 267–275. [CrossRef] [PubMed]
51. Nguyen, M.; Tokura, H. Observations on Normal Body Temperatures in Vietnamese and Japanese in Vietnam. *J. Physiol. Anthr. Appl. Hum. Sci.* **2002**, *21*, 59–65. [CrossRef] [PubMed]
52. Mallick, F.H. Thermal comfort and building design in the tropical climates. *Energy Build.* **1996**, *23*, 161–167. [CrossRef]
53. Nasir, R.A.; Ahmad, S.; Zain-Ahmed, A. Adaptive Outdoor Thermal Comfort at an Urban Park in Malaysia. *Asian J. Behav. Stud.* **2018**, *3*, 13. [CrossRef]
54. Karyono, T.H. Thermal Comfort in the Tropical South East Asia Region. *Arch. Sci. Rev.* **1996**, *39*, 135–139. [CrossRef]
55. Steadman, R.G. A Universal Scale of Apparent Temperature. *J. Clim. Appl. Meteorol.* **1984**, *23*, 1674–1687. [CrossRef]
56. Steadman, R.G. The assessment of sultriness. Part I: A temperature-humidity index based on human physiology and clothing science. *J. Appl. Meteorol.* **1979**, *18*, 861–873. [CrossRef]
57. CERC. *ADMS-Urban Urban Air Quality Management System User Guide*; CERC: Cambridge, UK, 2017.
58. Wang, K.; Aktas, Y.D.; Stocker, J.; Carruthers, D.; Hunt, J.; Malki-Epshtein, L. Urban heat island modelling of a tropical city: Case of Kuala Lumpur. *Geosci. Lett.* **2019**, *6*, 4. [CrossRef]
59. Hood, C.; Carruthers, D.; Seaton, M.; Stocker, J.; Johnson, K. Urban canopy flow field and advanced street canyon modelling in ADMS-Urban. In Proceedings of the 16th International Conference on Harmonisation within Atmospheric Dispersion Modelling for Regulatory Purposes, Varna, Bulgaria, 8–11 September 2014.
60. Chung, J.-Y.; Honda, Y.; Hong, Y.-C.; Pan, X.-C.; Guo, Y.L.; Kim, H. Ambient temperature and mortality: An international study in four capital cities of East Asia. *Sci. Total. Environ.* **2009**, *408*, 390–396. [CrossRef]

61. Salim, M.; Schlünzen, K.H.; Grawe, D. Including trees in the numerical simulations of the wind flow in urban areas: Should we care? *J. Wind. Eng. Ind. Aerodyn.* **2015**, *144*, 84–95. [CrossRef]

62. Pielke, R.A., Sr.; Davey, C.; Morgan, J. Assessing "global warming" with surface heat content. *Eos Trans. Am. Geophys. Union* **2004**, *85*, 210. [CrossRef]

63. Ford, T.W.; Schoof, J.T. Characterizing extreme and oppressive heat waves in Illinois. *J. Geophys. Res. Atmos.* **2017**, *122*, 682–698. [CrossRef]

64. Buyadi, S.N.A.; Mohd, W.M.N.W.; Misni, A. Impact of Land Use Changes on the Surface Temperature Distribution of Area Surrounding the National Botanic Garden, Shah Alam. *Procedia Soc. Behav. Sci.* **2013**, *101*, 516–525. [CrossRef]

65. Richards; Fung, T.; Belcher, R.; Edwards, P. Differential air temperature cooling performance of urban vegetation types in the tropics. *Urban For. Urban Green.* **2020**, *50*, 126651. [CrossRef]

66. Ali, S.B.; Patnaik, S. Thermal comfort in urban open spaces: Objective assessment and subjective perception study in tropical city of Bhopal, India. *Urban Clim.* **2018**, *24*, 954–967. [CrossRef]

67. Nguyen, J.; Schwartz, J.D.; Dockery, D. The relationship between indoor and outdoor temperature, apparent temperature, relative humidity, and absolute humidity. *ISEE Conf. Abstr.* **2013**, *2013*, 4712. [CrossRef]

68. Elsayed, I.S. A study on the urban heat island of the city of Kuala Lumpur, Malaysia. *JKAU Met. Environ. Arid. Land Agric. Sci.* **2012**, *23*, 121–134. [CrossRef]

69. Elsayed, I.S. Mitigation of urban heat island of the city of Kuala Lumpur, Malaysia. *Middle East J. Sci. Res.* **2012**, *11*, 1602–1613.

 *atmosphere*

Article

Exposure Indices of Extreme Wind-Driven Rain Events for Built Heritage

Scott Allan Orr *  and May Cassar

UCL Institute for Sustainable Heritage, University College London, Central House, 14 Upper Woburn Pl, London WC1H 0NN, UK; bseer-communications@ucl.ac.uk
* Correspondence: scott.orr@ucl.ac.uk

Received: 13 December 2019; Accepted: 27 January 2020; Published: 4 February 2020

Abstract: Building performance and material change of cultural heritage in urban areas are negatively impacted by wind-driven rain (WDR). The frequency and intensity of WDR exposure are modified by climate change. Current approaches to exposure assessment emphasise prolonged exposure. Here, we propose indices to represent the exposure of cultural heritage to extreme WDR events. The indices are derived in two stages: (1) time-binning of long-term exposure, and (2) statistical representation of the occurrence of infrequent but intense events by fitting to the Generalised Extreme Value (GEV) distribution. A comparison to an existing exposure assessment procedure demonstrates that the proposed indices better represent shorter, more intense, and more consistent WDR events. Indices developed for seasons had greater statistical confidence than those developed for annual exposure. One index is contextualised within a model of a gutter on a terraced building: this converts the index from a measure of *exposure* to potential *impact*. This evaluation demonstrated the importance of maintenance to reduce the potential impact of WDR events. This work has direct and indirect implications for developing robust assessment procedures for cultural heritage exposure to extreme weather events.

Keywords: historic buildings; risk assessment; WDR; resilience, sustainability; extreme value analysis

1. Introduction

Masonry walls have a high moisture content and are inherently wet to a greater or lesser extent. Historic masonry walls and their mortars have the ability to absorb and evaporate moisture provided foundations are not permanently saturated by rising damp or walls wetted by ineffective guttering. These effects, and the risk from wind-driven rain, were highlighted in a study of Brodick Castle, a 13th century sandstone Category A listed building on the Isle of Arran [1] (pp. 2–3). The study reported on the effects on wall moisture content of the failure of lead guttering and while adequate and well maintained rainwater disposal systems are at the heart of historic building protection, the increase in extreme episodic wind-driven rain events has highlighted the importance of impact indices and risk assessment to effective building maintenance and urban heritage management.

Time-binning is an integral part of assessing exposure to wind-driven rain. A set of rules and conditions are applied to a data series to produce an ordered set of discrete units (events), to which statistical analysis is applied. These units are indices that are intended to represent specific responses of built heritage to the environment. To date, time-binning of wind-driven rain time series have favoured "the average moisture content of exposed building material or when assessing the likely growth of mosses and lichens" [2] (p. 4) and "rain penetration through masonry" [3] (p. 12). These indices are derived from time-binning rules informed by the time scales of associated physical processes.

Time-binning is often combined with extreme value analysis (EVA) [4]. The units (wind-driven rain events) are fitted to an appropriate statistical distribution [5]. This enables the determination of statistical frequency (e.g., return periods), which can be projected far beyond the time scales of the input data. One challenge of EVA is how to interpret the statistical frequency of events. The output of time-binning and EVA primarily represent the intensity of events that built heritage is exposed to. However, what does a once-in-any-given 50-year period event represent in terms of risk to built heritage? The Noah's Ark project—a collaborative EU project—developed and visualised several such indices (referred to as 'Heritage Climate maps') in the context of a changing climate [6]. However, rain events were not parameterised. An open challenge remains to produce representations of *impact* from statistical measures of rain events.

Current indices do not represent exposure to short but intense WDR events [3] (p. v). Primary concerns during these events include rain penetration through building elements (e.g., window frames, cracks, etc.) and the failure of rainwater goods (e.g., overspilling of gutters). Gutter overspill can result in substantial volumes of water to run off the façade of the building which can activate and foster several weathering mechanisms [7]. It is important to note that the relationship between surface run-off and absorption is complicated: for example, recent work [8,9] demonstrates that, after initial uptake, subsequent exposure may not result in significant change of moisture at depth in the walls.

The need for a new index that characterises shorter, more intense, and more consistent wind-driven rain events is evidenced by the damage that can be caused by rapid wetting and drying of the built heritage. For example, wetting dissolves salts presents in building materials and mortars. The speed of drying after a wetting episode can cause salt crystals to form either on or close to the surface of a material, leading to a variety of damage mechanisms such as disfiguring efflorescence, flaking, and spalling.

This paper proposes two new indices for exposure of built heritage to intense wind-driven events. The time-binning procedure is outlined, the output of which is compared to the intensity and temporal characteristics of indices in current use. These indices are derived from extreme value analysis, one of which is contextualised in a model of gutter overflow to represent potential *impact* on built heritage. The discussion that follows emphasises the combined use of these indices and contextualisation as a tool for comparative evaluation of impact assessment.

2. Results

2.1. Regional Case Study: Plymouth

Measured hourly climate data from Plymouth, UK (50.3544°, −4.11986°) from 1986–2015 was used to demonstrate and evaluate the time-binning procedure [10,11]. Plymouth is a port city situated on the south coast of England. Although there are very few pre-war buildings in the city centre due to extensive bombing during WWII [12], the fabric of the city includes extensive post-war rebuilding and terraced housing of various periods in the surrounding urban and peri-urban regions. This urban fabric is a tapestry of built heritage and an important representation of Britain's participation in 20th-century international conflict.

Due in part to its exposed coastal setting, Plymouth experiences intense weather events. Recently, the Met Office recorded that winds reached 83 miles per hour during a storm [13]. Previous work has identified that one once-in-every-three-years wind-driven rain event (based on time-binning rules in ISO 15927-3:2009 [3]) can represent nearly 50% of the average annual exposure [14].

For the purposes of demonstrating the procedure and output of the proposed index for extreme events, a façade oriented to the southwest without any obstructions was used to determine the wind-driven rain exposure from measured climate data.

2.2. A Demonstrative Example

Current indices for wind-driven rain commonly applied to heritage may not represent short periods of intense exposure, while incorporating substantial periods without exposure. These characteristics are demonstrated using one month of exposure (Figure 1). In Plymouth, UK, 122.8 mm of precipitation was recorded in January 2015. For a façade oriented toward the southwest without any obstructions, this equates to a semi-empirical estimation (based on ISO 15927 [3]) of total wind-driven rain intensity equal to approximately 226 L m^{-2}.

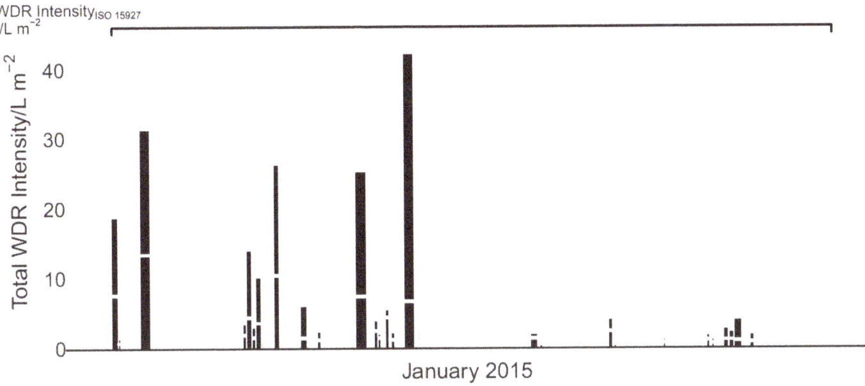

Figure 1. Wind-driven rain intensity during January 2015 in Plymouth, UK. The black bar represents the total exposure during each event under the proposed time-binning rules, while the horizontal white line shows the maximum exposure in a single hour. Under the ISO 15927-3 time-binning rules, the entire month is a single wind-driven rain event. This does not represent the several extreme short exposures that occurred during that period. The thickness of the line is a measure of the duration of the event.

Using the time-binning rules for prolonged events [3], the entire month is binned into one unit: a single wind-driven rain event. Applying a time-binning procedure that applies shorter criteria for the periods between events produces 64 events (1–15 h in duration) within the same month, of which the majority are less than 3 h in duration. These events range in intensity, but several represent more than 10% of the total wind-driven rain exposure for the month in a period of a few hours.

Current indices commonly represent periods of time in which periods of exposure are a small fraction of the total duration. Under the rules for prolonged exposure, approximately 20% of the hours within the event include wind-driven rain exposure. Under the proposed time-binning procedure for extreme events, each event that occurred in January 2015 has wind-driven rain exposure during each hour of their duration.

The maximum hourly intensity within each event varies. In the cases of events that are the duration of the data resolution (1 h), the average intensity is equal to the maximum hourly intensity. As the event duration increases (the maximum was 15 h in January 2015), the maximum hourly intensity as a fraction of the total within the event decreases. However, as can be seen in Figure 1, it commonly represents a third to a half of the total intensity of the event.

Thus, the characteristics of the wind-driven rain exposure within this month demonstrate that indices in current common use for prolonged events do not represent intense, short, and more consistent wind-driven rain events. In contrast, time-binning rules based on shorter periods between events characterises these phenomena when applied to hourly wind-driven rain exposure.

2.3. Temporal Characteristics of the Time-Binned Events

2.3.1. Intensity, Duration, and Maxima

The proposed time-binning rules produce shorter wind-driven rain events. Figure 2 shows that the durations of events under the proposed time-binning rules are, both on average and in their extreme, at least one order of magnitude less than those produced from rules for prolonged events. The former produce events up to 40 h in duration, although 95% of the events are less than 7 h in duration. In contrast, the prolonged events have a median duration of 146 h, with outliers (of the median + 1.5 × IQR, the inter-quartile range) between 850 and 2500 h in duration.

Events characterised by these sets of rules experience similarly-proportioned wind-driven rain exposure (Figure 2, upper row). The short events are primarily characterised by near-zero total (summed) intensities, with outlying events ranging from 0 to 130 L m^{-2}. The prolonged events primarily range between 0 and 200 L m^{-2}, but can reach upwards of 700 L m^{-2}. In both cases, there is a general proportionality between duration and intensity.

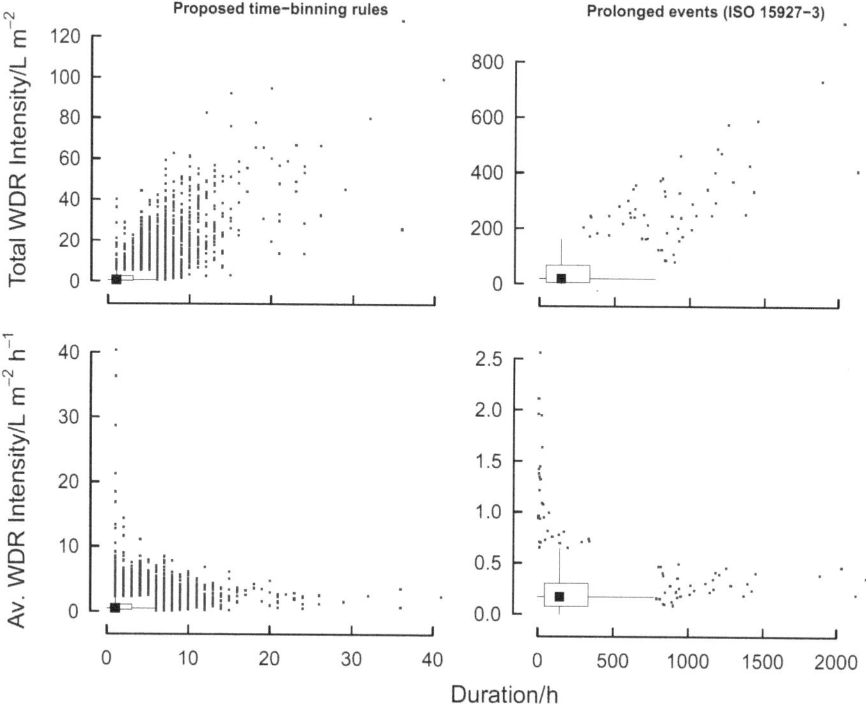

Figure 2. Boxplots of wind-driven rain intensities and average intensities derived from two time-binning rules during 1986–2015 for Plymouth, UK.

A different picture emerges when observing average intensity (Figure 2, lower row). In most cases, both sets of time-binning rules produce a dense cluster of events short durations and low average intensities. Both sets of rules produce an important clusters (within their respective range of durations): short events with high average intensities. The clusters of high duration have similar intensities to most of the shorter events. Although the index for prolonged events yields high total WDR intensities, they are spread over long durations. These do not represent potential impact on gutter performance and the frequency and consequences of over spilling on the historic built fabric.

Some of the events that pose the greatest risk to gutter performance are present in Figure 2 (lower-left panel, outliers with durations less than approximately 5 h). However, characterising events by the maximum hourly intensity demonstrates several additional events that represent similar short-term exposures (Figure 3). Using the maximum hourly exposure during each event means that the index represents hourly periods of very heavy exposure that have the potential to wreak havoc on the function of rainwater goods and induce rapid cycles of physical and chemical weathering mechanisms.

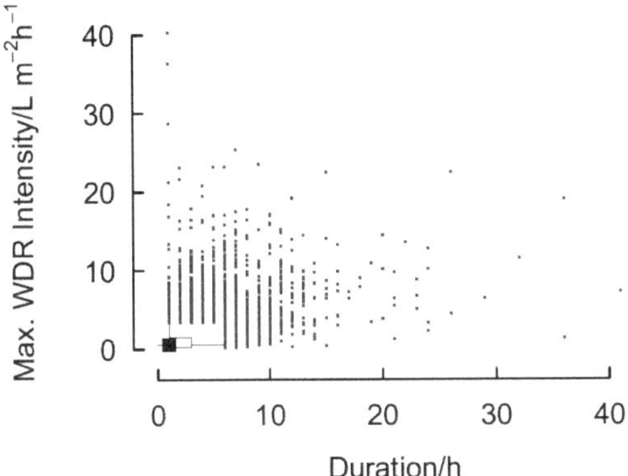

Figure 3. Boxplots of the maximum wind-driven rain intensities derived from the proposed time-binning rules for an extreme event index between 1986–2015 in Plymouth, UK.

2.3.2. Consistency

The proposed time-binning rules yield events that are more consistent: for the Plymouth time series, 99.8% events (all but 21 during the 30-year study period) had a consistent fraction of 1 (wind-driven rain within every hour), while the remainder included wind-driven rain exposure during 90% or greater of the hours within them. In contrast, Figure 4 shows the rules for prolonged events produced short (<10 h duration) spells that have consistent exposure (fraction = 1). This is followed by a decrease in the consistent fraction with increasing duration until it levels out at approximately 0.2 (i.e., there is wind-driven rain exposure for 20% of the hours within the event).

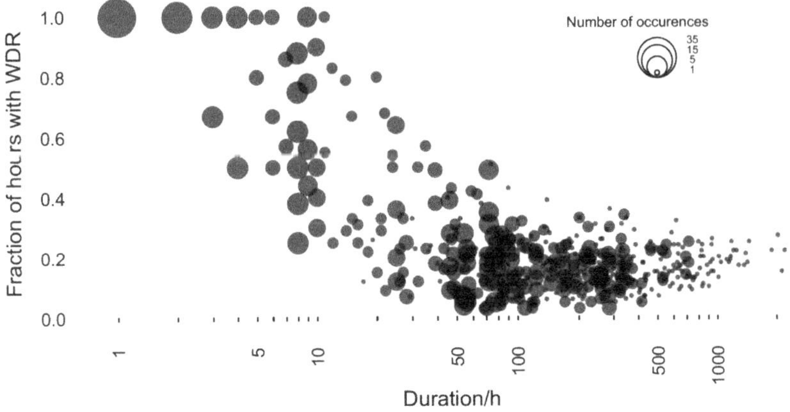

Figure 4. A plot of the fraction of hours in which there was active rain for events time binned according to the ISO 15927-3 [3] rules for prolonged exposure.

2.3.3. Seasonal Characteristics

Under the proposed time-binning rules for extreme events, an average of 353 events occur per year. These do not occur evenly over calendar seasons. Previous work has demonstrated that climate change scenarios result in a polarising in the intensities of seasonal characteristics [15]. In Plymouth, UK, wind-driven rain events are most common during the autumn and winter months

($n = \{96, 113\}$, respectively). In contrast, the spring and summer months experience fewer events ($n = \{73, 71\}$, respectively).

Table 1. Average annual frequency of wind-driven rain events derived from two time-binning rules for Plymouth, UK from 1986–2015.

Season	Proposed Time-Binning Rules	ISO15927-3 (Prolonged Exposure)
Winter (DJF)	113	4
Spring (MAM)	73	5
Summer (JJA)	71	5
Autumn (SON)	96	4
Total	353	18

In contrast, the time-binning rules for prolonged events produce an average of 18 events per year. These are more homogeneous across the year (varying by only one event between calendar seasons), and occur more frequently during the spring and summer months.

2.4. Intensity of Extreme Events

The Extreme Event Index (EEI) and Extreme Event Maxima Index (EEMI) can be determined for return periods of interest. They represent the average and maximum hourly wind-driven rain intensity, respectively, likely to occur once in the specified period.

2.4.1. GEV Fitting

The modelled Generalised Extreme Value (GEV) distribution reproduced the distribution of empirical annual indices. The modelled distributions were evaluated based on the quality of fit of the kernel densities of the annual indices (Table 2). The GEV distributions for seasonal indices produce better fits than the annual indices. This is likely due to the seasonal variation of extreme event occurrence and intensity.

Table 2. Squares of the Pearson correlation coefficient (R^2) of the kernel densities for modelled GEV distributions compared to the empirical annual indices.

Season	Extreme Event Index (EEI)	Extreme Event Maxima Index (EEMI)
Winter (DJJ)	0.90	0.97
Spring (MAM)	0.98	0.99
Summer (JJA)	0.95	0.96
Autumn (SON)	0.91	0.99
Annual	0.89	0.93

It is due to the manner in which the return periods are modelled that causes the annual EEI to be greater than that derived for any season. Since they are determined from the maxima that occurs in each year (or seasonal subset), the annual GEV distribution is fitted to a set of events that occur in varying seasons. Thus, the similarity between the winter and annual EEI suggests that most of the annual maxima occur in this season, but that they are otherwise also occurring in the spring and summer months.

2.4.2. Extreme Event Index (EEI)

The Extreme Event Index (EEI) is derived from the average hourly intensity for each event. It represents a generalised indication of the intensity of these events.

Figure 5 shows the EEI determined for the case study for four calendar seasons, which demonstrate varying characteristics. The winter and autumn months experience the highest and second-highest intensities, respectively, across the range of return periods. The spring months have the lowest intensities of extreme events. Despite the overall low intensities of WDR in Plymouth during summer months [15], the index demonstrates that extreme wind-driven rain events occur with intensities greater than those of the spring months.

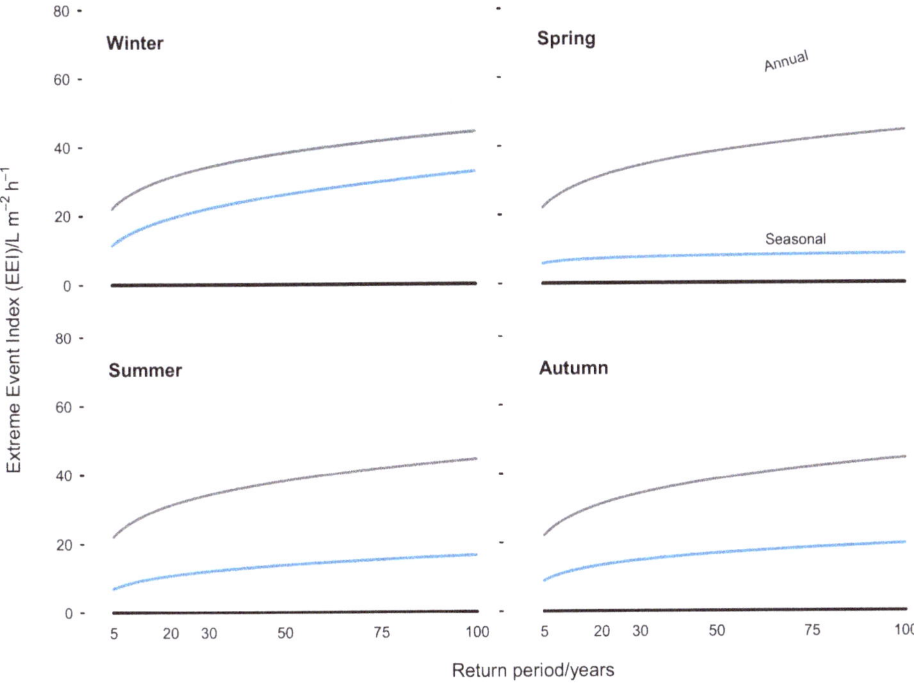

Figure 5. The Extreme Event Index (EEI) for calendar seasons based on semi-empirical wind-driven rain exposure in Plymouth UK between 1986 and 2015. Shaded areas represent 95% confidence intervals.

Another important outcome is that separating the index by seasons increases the confidence interval. This is observed in Figure 5 by the relative sizes of the gray and blue areas across all seasons. Due to the duration of the input data (30 years), the 95% confidence interval for greater return periods increases significantly.

2.4.3. Extreme Event Maximum Index (EEMI)

The Extreme Event Maxima Index (EEMI) is derived from the maximum hourly intensity for each event (Figure 6). The EEMI is useful to evaluate particular impacts on buildings, such as the performance of rainwater goods. Similar to the EEI, the annual EEMI is greater than that of any season since the most extreme annual maxima occurs occurs within a mix of different seasons (see Section 2.4.2) for a more detailed explanation).

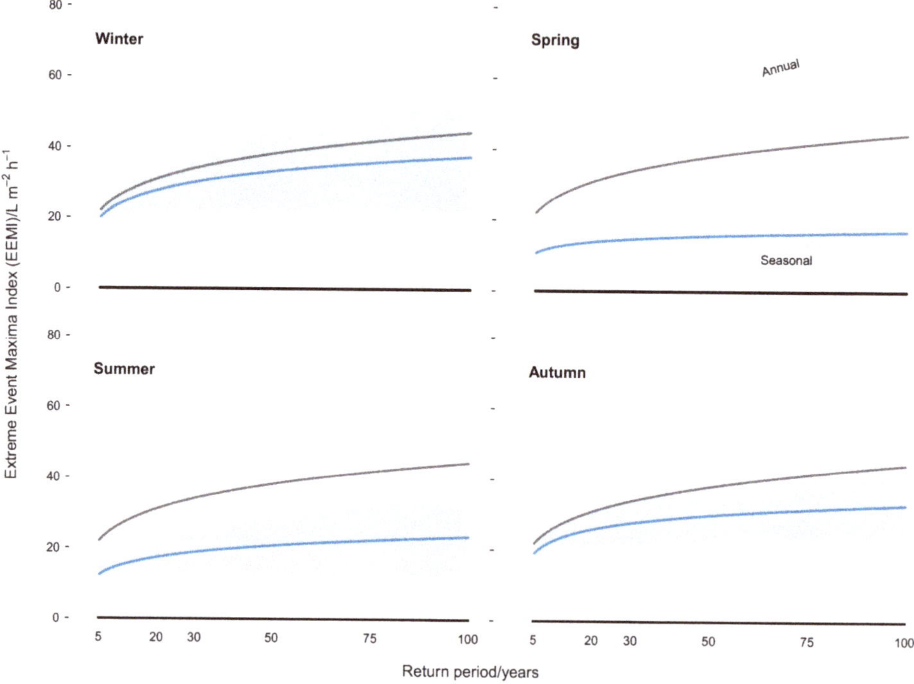

Figure 6. The Extreme Event Maxima Index (EEMI) for calendar seasons based on semi-empirical wind-driven rain exposure in Plymouth UK between 1986 and 2015. Shaded areas represent 95% confidence intervals.

2.5. Threshold Assessment

The EEMI can be contextualised by how rainwater goods function on part of a historic structure oriented toward the southwest. The maximum hourly exposure represents the period of time during the event which poses the greatest risk of gutter overspill. This extends its use beyond a relative indicator of the intensity of exposure to an indicator of potential impact. To demonstrate this, a model of a gutter on a terraced house is used.

Figure 7 shows the same index for the winter calendar months as shown in a panel in Figure 6. The index is contextualised in a model of a standard gutter. The model demonstrates that, if the gutter is properly functioning, no extreme hourly exposure likely to occur once in a hundred years or less should cause the gutter to spill over. However, if the downpipe area is compromised (e.g., detritus blocks part of the drain area), the index falls within the range of possible extreme hourly intensities.

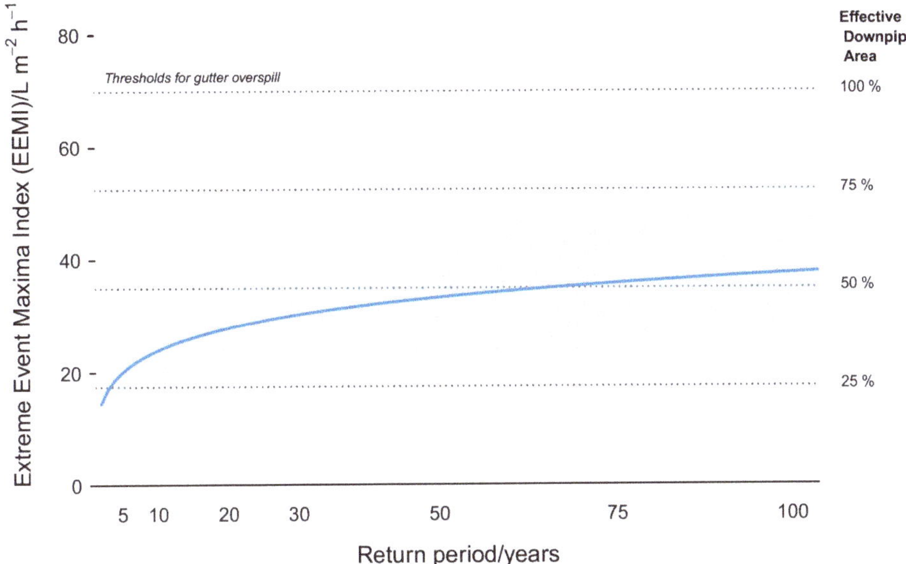

Figure 7. The Extreme Event Maxima Index (EEMI) for the winter calendar season based on semi-empirical wind-driven rain exposure in Plymouth UK between 1986–2015. Shaded areas represent 95% confidence intervals for the return period modelling. The return levels (the EEMI for different return periods) are contextualised within a model of a gutter managing rainshed on a terraced house. The threshold for gutter overspill are determined by several localised factors, including topography, geometry, and context, which would impact the confidence intervals if adapted to reflect other scenarios than the one studied herein.

3. Discussion

3.1. Indices

3.1.1. Empirical Approaches to Wind-Driven Rain

A robust statistical risk index requires a long time series of data to accurately project infrequent events. This often limits the source of input to longitudinal meteorological measurements. Due to this, the wind-driven rain input is derived from empirical relationships, which have associated caveats [16]. However, if sufficient data exists or can be generated, the indices can be applied to measured and modelled wind-driven rain exposure. This would result in more accurate representations of the exposure to extreme weather events.

3.1.2. Urban Complexity

The indices can be further refined by applying factors to account for urban complexities and the influence they have on wind-driven rain exposure, e.g., those included in ISO 15927 [3]. Factors have been developed to account for, among others, topography, adjacent structures, as well as the position within a façade.

3.2. Seasonality

The case study demonstrated the importance of seasonal indicators. This acknowledges the varied seasonal characteristics observed in the northern hemisphere. In all cases, the indices derived for specific seasons were better modelled by the GEV distribution and had narrower confidence interval ranges. Thus, seasonal indices not only identify seasons which represent the greatest exposure to built heritage, but also increase the precision and accuracy of our predictions of their intensity.

3.3. Threshold Assessment

The threshold assessment for the EEMI demonstrates the potential for the EEMI to be used as a risk/impact indicator. This conversion was achieved by contextualising the wind-driven rain exposure within the performance of a gutter on a terraced building. This demonstrated the importance of gutter maintenance.

The threshold assessment undertaken for the case study likely underestimates the frequency of occurrence of gutter overspill. Standard meteorological measurement is reported hourly. However, several weather events (e.g., cloudbursts) often occur on shorter timescales. Since the exposure can only be assumed to occur over an hour (due to the data resolution), it may not represent the sub-hourly risk of gutter overspill. Two approaches could be taken to address this: (1) deriving the index from sub-hourly measurements, or (2) converting the hourly measured exposure into an equivalent exposure assuming a shorter time interval. In the latter, an hourly exposure could be assumed to have primarily occurred in a shorter period. Recent work [17] has shown that sub-hourly (10 min) return levels of extreme WDR events in the UK are frequently an order of magnitude greater than those recorded at 1 h intervals. Thus, the equivalent (hourly) EEMI would be the original multiplied by 10, or greater. If this approach is applied to the original EEMI presented in Figure 7, the threshold for gutter overspill (even with 100% effective downpipe area) is likely to be exceeded in any given 5-year period.

The threshold assessment is not intended to empirically generalise the potential impact for a general building in this location. It can be used for a specific building typology and context, or further adapted to account for other contextual factors (as discussed in Section 3.1.2). If the threshold assessment is applied to a building typology and context that is relevant to several regions and contexts, it can be used as a relative indicator of the potential impact.

3.4. Direct Exposure Assessment

Additional modifications to the EEI could be used to assess the direct exposure of building façades to wind-driven rain. A parameterisation for this would need to consider material properties and environmental conditions. This could be done with hygrothermal modelling [18] or controlled testing regimes.

3.5. Future Work and Outcomes

Maintenance of the urban heritage is highlighted in Historic England policy guidance as an important prerequisite to historic building protection with maintenance plans recommended for older buildings [19]. Despite the predominance in the urban landscape of the English-style terraced house, historic buildings are more varied in their construction. A significant component that could be further explored and developed in the model/index is the roof which can vary widely in materials and construction such as thatched, shingle, wood, concrete, metal or vinyl tiles or membrane with or without accompanying gutting. With the expected increase of extreme wind driven rain, how these materials will perform as an assembly with the external walls needs to modelled so that a typology of indices for different types of historic buildings can be developed.

4. Materials and Methods

4.1. Summary of Index Calculation Procedure

The determination of the proposed indices for extreme wind-driven rain events have several components:

1. Determination of wind-driven rain exposure (herein semi-empirical).

2. Time-binning procedure.
3. Index selection.
4. Extreme value analysis.
5. Impact assessment.

4.2. Determination of Wind-Driven Rain Exposure

The intensity of wind-driven rain exposure was derived according to BS EN ISO 15927-3:2009 [3]. In accordance with the World Meteorological Organisation, statistical evaluations (and therefore indices) should be derived from 30 years or more of hourly data [20]. The semi-empirical relationship is a vector-based approach:

$$I = \frac{2}{9}vr^{8/9}cos(D - \theta) \tag{1}$$

in which the hourly intensity of wind-driven rain, I as a volume per units area and time, is determined from vertical rainfall (i.e., precipitation), r, wind speed, v, the prevailing wind direction, D and the façade orientation θ. The cosine component, despite its shortcomings [21], is used to represent the fractional exposure of a façade that is not directly impacted by the wind-driven rain. The parameters are typically meteorological measurements.

Several factors affect the accuracy of different aspects of the representation. This does not discount its use as a regional exposure index. The climate data may not accurately represent sub-hourly behaviour. For example, a drastic shift in the wind speeds mid-way through the hour may not be represented in the data, depending on how prevailing hourly wind speed has been recorded and calculated. Similarly, CFD modelling has shown that WDR exposure can vary significantly across a façade, primarily due to edge effects and sheltering [22]. For the purposes of this model, which represents WDR for exposure assessment on a fairly exposed roof system, these inaccuracies were deemed acceptable to demonstrate the proposed time-binning procedure. If greater accuracy is needed for a particular scenario or purpose, the index can instead be combined with measured and modelled wind-driven rain input.

4.3. Time-Binning

4.3.1. Existing Rules for Prolonged Exposure

The time binning rules for prolonged exposure are taken from BS EN ISO 15927-3:2009 [3], requiring a break of ≥96 h between WDR events, selected to represent prolonged exposure of masonry façades. Caton experimentally demonstrated that up to 96 consecutive hours with no driving rain are necessary before evaporative losses exceed water ingress due to rain exposure [23]. In contrast to current standards, Caton distinguished between events by periods of 96 h without 'appreciable' driving rain, which was approximated as one tenth (10%) of the intensity for a statistical 'once in three years' event [23].

4.3.2. Proposed Rules for Extreme Events

The proposed time-binning rules are based on previous work by Lacy in which WDR events were determined from three rules based on measured data from the southwest England [24] (p. 101):

1. The long event is deemed to be continuous if the aggregate of any breaks is not more than 10% of the whole.
2. No single break is to be more than 5% of the time.
3. The total duration is defined as the time from the beginning to the end of the event, *including* the dry breaks.

The rationale for these time-binning rules is to identify events (also commonly referred to as spells) with *consistent* exposure. Although it may appear that these rules produce events without regard for durations between events, these two approaches are statistically equivalent [25].

The time-binning was applied to the data series using a recursive function. A preliminary analysis applying these rules to measured data from a more exposed site in southwest England (see Section 2.1) determined that 99.7% of intervals between exposure, the proposed rules were equivalent to defining a WDR event as a period of consistent exposure without any hours of no wind-driven rain exposure. This provides a alternative approximation that is easier to calculate to apply the proposed time-binning rules to timeseries of exposure.

4.4. Index Selection

Before EVA can be applied, one of two indices is selected.

4.4.1. Extreme Event Index (EEI)

The Extreme Event Index (EEI) is derived from the average hourly intensity for each event. It represents a generalised indication of the intensity of these events.

4.4.2. Extreme Event Maximum Index (EEMI)

The Extreme Event Maxima Index (EEMI) is derived from the maximum hourly intensity for each event. The EEMI is useful to evaluate particular impacts on buildings, such as the performance of rainwater goods. In these scenarios, the maxima for each event represent the period of greatest risk. If the data resolution is greater than 1 h, it can be determined for sub-hourly intensities.

4.5. Extreme Value Analysis

Extreme value analysis (EVA) was applied to the time-binned wind-driven rain events. EVA is an established method in environmental applications [26] to evaluate behaviour near the extremes of probability distributions.

An 'Annual Maxima Series (AMS)' approach was taken, as identifying appropriate thresholds for the alternative "Peak Over Threshold" approach remains a subjective and contested challenge [27]. The AMS approach requires that the annual maxima of the data series (i.e., the most intense wind-driven event for each year) to be determined. A benefit of the AMS method is that the results are independent of the "population size" (number of events), i.e., comparisons can be made between sites and periods of time that experience different numbers of events. The AMS is fit to the Generalised Extreme Value (GEV) continuous probability distribution:

$$F(x) = e^{-(1-(\xi\frac{x-\mu}{\sigma})^{1/\xi}}.$$ (2)

Three parameters determine the characteristics of the distribution. The location parameter, μ, describes the horizontal shift, while the scale parameter, σ, and the shape parameter, ξ, describe the spread and behaviour at the tails of the distribution, respectively. The shape parameter is derived from skewness, as it represents where the majority of the data lie, which creates the tails of the distribution [28].

The Gumbel distribution has previously been employed to describe wind-driven events [5,14], which can be considered a special case of the GEV in which $\xi = 0$. Distributions with two parameters (such as the Gumbel) have smaller standard error, but larger bias than three-parameter distributions, especially for small sample sizes [29].

After fitting the distribution parameters, the EVA is combined with the concept of return periods, which estimate the likelihood of an event (or statistically 'expected frequency'). The intensity, Q_n of the most extreme wind-driven rain event expected every n years is determined from:

$$Q_n = \mu + \frac{\sigma}{\varsigma}\left\{1 - \left(-log\left(\frac{n-1}{n}\right)^\varsigma\right)\right\}. \tag{3}$$

The index can be determined for sub-annual periods. Previous work [15] has demonstrated that climate change scenarios are predicted to result in increasingly polarised seasonal behaviour for wind-driven rain exposure. Thus, assessment of wind-driven rain exposure and impact should consider these trends in determining indices. While this is paramount for comparative studies under climate change regimes, presenting historical and contemporary exposure using seasonal indices facilitates comparison, for ease of calculation, it is worth noting that the EEMI can be derived from the annual/season AMS derived from the hourly maximum, without applying the time-binning procedure. These values are equivalent.

4.6. Impact Assessment

A mass balance model was developed to represent the response of rainwater goods (a gutter system) to wind-driven rain events. On the basis of the following equations (Equations (4)–(7)), which refer to the rainwater goods associated with a single building façade, the quantity of overspill within the gutter (the 'system') is determined as a function of the rainfall intensity. The balance equation for mass (Equation (4)) is presented as a volume balance, since the system consists of only one component with constant density. It consists of three terms:

$$\dot{V}_{acc} = \dot{V}_{in} - \dot{V}_{out} \tag{4}$$

relating the rate accumulated volume of water, $\dot{V}_{acc}$, to the net balance of the volume into the gutter system and that which exits it, $\dot{V}_{in}$ and $\dot{V}_{out}$, respectively. The model is integrated over a single time step, as the timescales of dynamic components (e.g., rainwater flowing down the roof surface, travelling through the gutter) are much shorter than the time resolution of the input data (1 h). The same would be true for data at relatively-high time resolutions for meteorological measurement, e.g., 5 min.

The rate of volume entering the system is represented by the volume of wind-driven rain exposure that impacts the roof system and the exposed gutter in a given time period. This is determined by multiplying the specific rate of wind-driven rain, $\dot{r}$, with the surface areas of the roof (A_r) and gutter (A_g):

$$\dot{V}_{in} = \dot{r} \cdot F_R(A_r + A_g) \cdot C, \tag{5}$$

in which C is a runoff coefficient (taken as 1 unless under special circumstances) and F_R is a risk factor. The risk factor for eaves gutters is taken to be 1. However, BS 12056-3 suggests a risk factor of 3.0 should be applied for non-eaves gutters for "buildings where an exceptional degree of protection is necessary, e.g., buildings housing outstanding works of art." [30] (p. 9). We argue this risk factor should be used for built heritage if the edifices are considered to be works of art in their own right, i.e., having artistic/aesthetic/architectural value [31].

The rate of volume exiting the system is primarily dependent on the dimensions of the downpipe. For a round downpipe connected to a rectangular gutter, the flow rate is represented by [30]:

$$\dot{V}_{out} = \frac{k_o D h^{1.5}}{7500}, \tag{6}$$

in which $\dot{V}_{out}$ has units of L s^{-1}, while all other measurements are in units of mm. This represents Wier flow (when $h = D/2$) when there is a gap of at least 5% of the diameter of the outlet. The diameter of the downpipe is twice that of the radius, $r_d = 0.5 \cdot D$. The outlet coefficient, k_o, dimensionless, is taken as 1.0 for unobstructed outlets, and 0.5 for outlets fitted with strainers or gratings. The head at outlet, h, is represented by a determined factor $F_h = 0.47$ [30] multiplied by the maximum flow height (taken here to be the gutter height, h_g). This volumetric flow rate is multiplied by the process duration to determine the total flow rate.

A dimensionless parameter is introduced to represent the effective area of the downpipe, f. This factor, ranging from 0 to 1, represents the effect of blockage which can result from of maintenance. It is applied to the volumetric outflow rate.

The volume of a rectangular gutter is represented by $V_g = l_g w_g d_g$, which represents the maximum volume of water it can hold as a function the length, height, and width of the gutter, respectively. This was further simplified by assuming a square cross-sectional area to $V_g = l_g (d_g)^2$. Thus, a gutter is modelled to overspill when the accumulated volume of water, $V_{acc} = \dot{V}_{acc}\Delta t$, satisfies the condition:

$$V_{acc} > V_g. \tag{7}$$

Parameter selection was derived from a combination of standards and informal observation (Table 3). For the purposes of this paper, the required accuracy for parameters is low, as it attempts to represent a common physical configuration. Each parameter could be adapted to suit a specific scenario, heritage typologies, or built context.

Table 3. Parameters for the overspill model, with affiliated source and units, where appropriate.

Parameter	Source	Value
Width of roof, w_r = length of gutter, l_g	informal observation	5 m
Length of roof (eave to crest), l_r	informal observation	5 m
Area of roof (eave to crest), $A_r = w_g \cdot l_r$	*Calculated*	25 m^2
Depth of gutter, d_g	BS EN 8530:2010 [32]	68.8 mm
Surface area of gutter, $A_g = d_g \cdot l_r$	*Calculated*	0.344 m^2
Radius of downpipe, r_d	BS EN 8530:2010 [32]	0.0602 m
effective gutter area coefficient, f	-	1, unless otherwise stated
downpipe coefficient, k_o	with strainer	0.5
downpipe head, h	BS EN 12056–3:2000 [30]	$0.47 \cdot d_g$ mm
Rate of wind-driven rain, $\dot{r}$	measured climate data, ISO 15927–3:2009 [3]	dependent on return period, L h^{-1}

Author Contributions: Conceptualization, S.A.O. and M.C.; methodology, S.A.O.; software, S.A.O.; validation, S.A.O.; formal analysis, S.A.O.; investigation, S.A.O.; data curation, S.A.O.; writing—original draft preparation, S.A.O. and M.C.; writing—review and editing, S.A.O. and M.C.; visualization, S.A.O. All authors have read and agreed to the published version of the manuscript.

Funding: This research received no external funding.

Acknowledgments: We are grateful to the Editor of the Special Issue for providing insight and guidance during during the conceptualisation of this paper. A note of thanks is due to David Watt and colleagues at Hutton + Rostron who kindly shared views on current practice relating to gutter performance. We also thank Josep Grau-Bové for advising on modelling methods.

Conflicts of Interest: The authors declare no conflict of interest.

References

1. Cassar, M.; Hawkings, C. Engineering Historic Futures Stakeholders Dissemination and Scientific Research Report. 2007. Available online: https://discovery.ucl.ac.uk/id/eprint/2612/1/2612.pdf (accessed on 2 January 2020).
2. BSI. *BS 8104:1992—Code of Practice for Assessing Exposure of Walls To Wind-Driven Rain*; Standard, British Standards Institution: London, UK, 1992.
3. ISO. *BS en ISO 15927-3: 2009: Hygrothermal Performance of Buildings— Calculation and Presentation of Climatic Data—Part 3: Calculation of A Driving Rain Index for Vertical Surfaces From Hourly Wind and Rain Data*; Standard, International Standards Organisation: London, UK, 2009.
4. Kinnison, R.R. *Applied Extreme Value Statistics*; Battelle Press: Columbus, OH, USA, 1985.

5. Pérez-Bella, J.M.; Domínguez-Hernández, J.; Rodríguez-Soria, B.; del Coz-Díaz, J.J.; Cano-Suñén, E. Combined use of wind-driven rain and wind pressure to define water penetration risk into building façades: The Spanish case. *Build. Environ.* **2013**, *64*, 46–56, doi:10.1016/j.buildenv.2013.03.004.

6. Sabbioni, C.; Brimblecombe, P.; Cassar, M. *The Atlas of Climate Change Impact on European Cultural Heritage: Scientific Analysis and Management Strategies*; Anthem Press: London, UK, 19 Number 2010.

7. Winkler, E.M. Important agents of weathering for building and monumental stone. *Eng. Geol.* **1966**, *1*, 381–400.

8. Stephenson, V.; D'Ayala, D. Structural Response of Masonry Infilled Timber Frames to Flood and Wind Driven Rain Exposure. *J. Perform. Constr. Facil.* **2019**, *33*, 04019028.

9. D'Ayala, D.; Aktas, Y.D. Moisture dynamics in the masonry fabric of historic buildings subjected to wind-driven rain and flooding. *Build. Environ.* **2016**, *104*, 208–220.

10. UK Met Office. *MIDAS: UK Hourly Weather Observation Data*; UK Met Office: Exeter, UK, 2006.

11. UK Met Office. *MIDAS: UK Hourly Rainfall Data*; UK Met Office: Exeter, UK, 2006.

12. Gould, J. Architecture and the plan for Plymouth: The legacy of a British city. *Archit. Rev.* **2007**, *221*, 78–83.

13. Strong Winds: Woman Killed and Ferry Travel Disrupted. BBC News. 2 November 2019. Available online: https://www.bbc.co.uk/news/uk-england-50273590 (accessed on 7 November 2019).

14. Orr, S.A.; Viles, H. Characterisation of building exposure to wind-driven rain in the UK and evaluation of current standards. *J. Wind Eng. Ind. Aerodyn.* **2018**, *180*, 88–97.

15. Orr, S.A.; Young, M.; Stelfox, D.; Curran, J.; Viles, H. Wind-driven rain and future risk to built heritage in the United Kingdom: Novel metrics for characterising rain spells. *Sci. Total Environ.* **2018**, *640*, 1098–1111.

16. Blocken, B.; Carmeliet, J. A review of wind-driven rain research in building science. *J. Wind Eng. Ind. Aerodyn.* **2004**, *92*, 1079–1130.

17. Chan, S.; Kendon, E.; Roberts, N.; Fowler, H.; Blenkinsop, S. The characteristics of summer sub-hourly rainfall over the southern UK in a high-resolution convective permitting model. *Environ. Res. Lett.* **2016**, *11*, 094024, doi:10.1088/1748-9326/11/9/094024.

18. Hansen, T.K.; Bjarløv, S.P.; Peuhkuri, R. The effects of wind-driven rain on the hygrothermal conditions behind wooden beam ends and at the interfaces between internal insulation and existing solid masonry. *Energy Build.* **2019**, *196*, 255–268.

19. Historic England. Maintenance Plans for Older Buildings. Available online: https://historicengland.org.uk/advice/technical-advice/buildings/maintenance-plans-for-older-buildings/ (accessed on 12 December 2019).

20. Guttman, N. *Guide to Climatological Practices*; Technical Report 100; World Meteorological Organisation: Geneva, Switzerland, 2011.

21. Blocken, B.; Carmeliet, J. On the validity of the cosine projection in wind-driven rain calculations on buildings. *Build. Environ.* **2006**, *41*, 1182–1189.

22. Blocken, B.; Carmeliet, J. Validation of CFD simulations of wind-driven rain on a low-rise building facade. *Build. Environ.* **2007**, *42*, 2530–2548.

23. Prior, J. *Directional Driving Rain Indices for the United Kingdom : Computation and Mapping (Background to BSI Draft for Development DD93)*; Building Research Establishment: Glasgow, UK, 1985.

24. Lacy, R. *Climate and Building in Britain*; Her Majesty's Station Office (HMSO): Richmond, UK, 1977.

25. Ignaccolo, M.; Michele, C.D. A non arbitrary definition of rain event: The case of stratiform rain. *arXiv* **2009**, arXiv:physics.ao-ph/0911.3941.

26. Finkenstadt, B.; Rootzén, H. *Extreme Values in Finance, Telecommunications, and The Environment*; CRC Press: Boca Raton, FL, USA, 2003.

27. Solari, S.; Losada, M. A unified statistical model for hydrological variables including the selection of threshold for the peak over threshold method. *Water Resour. Res.* **2012**, *48*, doi:10.1029/2011WR011475.

28. Millington, N.; Das, S.; Simonovic, S.P. *The Comparison of GEV, Log-Pearson Type 3 and Gumbel Distributions In the Upper Thames River Watershed Under Global Climate Models*; Technical Report 77; Department of Civil and Environmental Engineering, The University of Western Ontario: London, UK, 2011.

29. Cunnane, C. Statistical Distributions for Flood Frequency Analysis. Technical Report. 1989. Available online: http://agris.fao.org/agris-search/search.do?recordID=XF9090879 (accessed on 2 January 2020).

30. BS. *BS en 15927–2: Gravity Drainage Systems Inside Buildings. Sanitary Pipework, Layout and Calculation*; Standard, British Standards Institute: London, UK, 2000.

31. Martínez, A.H. Conservation and restoration in built heritage: A western European perspective. In *The Ashgate Research Companion to Heritage and Identity*; Ashgate Publishing, Ltd.: Farnham, UK, 2008; pp. 245–263.
32. BSI. *BS 8530:2010—Traditional-Style Half Round, Beaded Half Round, Victorian Ogee and Moulded Ogee Aluminium Rainwater Systems. Specification*; Standard, British Standards Institution: London, UK, 2010.

Review

A Geological Perspective on Climate Change and Building Stone Deterioration in London: Implications for Urban Stone-Built Heritage Research and Management

Sudeshna Basu [1,2,*], Scott Allan Orr [3] and Yasemin D. Aktas [4,5]

1. Department of Earth Sciences, University College London, 5 Gower Place, London WC1E 6BS, UK
2. Department of Chemical Engineering, University College London, Bloomsbury, London WC1E 7JE, UK
3. Institute for Sustainable Heritage, University College London, Central House, 14 Upper Woburn Place, London WC1H 0NN, UK; scott.orr@ucl.ac.uk
4. Department of Civil, Environmental and Geomatic Engineering, Kings Cross, London WC1E 6DE, UK; y.aktas@ucl.ac.uk
5. UK Centre for Moisture in Buildings (UKCMB), London WC1H 0NN, UK
* Correspondence: sudeshna.basu@ucl.ac.uk

Received: 1 June 2020; Accepted: 21 July 2020; Published: 26 July 2020

Abstract: The decay rates of building stones and, the processes leading to their deterioration is governed by intrinsic properties such as texture, mineralogy, porosity and pore size distribution, along with other extrinsic factors related to the climate and anthropogenic activities. For urban cities such as London, the influence of extrinsic factors like temperature and rainfall, as well as the concentrations of air pollutants, such as sulphur and nitrogen oxides, along with the emissions of carbonaceous aerosols, can be particularly significant. While considering the long-term preservation of building stones used in various heritage sites in the city, it is imperative to consider how the stone could be affected by the changing air pollutant concentrations, superimposed on the effects of climate change in the region, including rising average annual temperature and precipitation with a hotter, drier summer and, warmer, wetter winter months. This paper deals with the intrinsic rock properties of the common building stones of London, including limestone, marble, granite, sandstone, slate, flint as well as bricks, building on known characteristics including strength and durability that determine how and where they are placed in a building structure. The study reviews how these stones decay due to different processes such as salt weathering in sandstone, microcracking of quartz with kaolinisation of K-feldspar and biotite in granite and dissolution of calcite and dolomite, followed by precipitation of sulphate minerals in the carbonate rocks of limestone and marble. In the urban environment of London, with progressive build up in the concentration of atmospheric nitrogen oxides leading to an increasingly acidic environment and, with predicted climate change, the diverse stone-built heritage will be affected. For example, there can be enhanced carbonate dissolution in limestone with increased annual precipitation. Due to the prolonged wetter winter, any sandstone building stone will also undergo greater damage with a deeper wetting front. On the other hand, due to predicted wetter and warmer winter months, microcracking of any plagioclase in a granite is unlikely, thereby reducing the access of fluid and air pollutants to the Ca-rich core of the zoned crystals limiting the process of sericitisation. Management of the building stones in London should include routine expert visual inspection for signs of deterioration, along with mineralogical and compositional analyses and assessment of any recession rate.

Keywords: heritage buildings; lithotype; salt weathering; kaolinisation; microcracking; weather events

1. Introduction

The heritage buildings in any city are important as landmarks, creating a sense of identity, integrated with local history and values, often attesting to the geological substrate on which they stand. While globally, modern buildings are predominantly composed of concrete, the building stones of the heritage sites are both compositionally and structurally distinct. Some may be quarried locally and used over a sustained period of time in a given place or region, and referred to as traditional stone with proximity, the main determinant of its use. However, for cities such as London, which stands on underlying sand, gravel and clay which are poorly consolidated and unsuitable as building stones, building stones are mostly procured from other parts of the UK as well as abroad.

The building stones undergo decay affecting their structural integrity, external fabric and the internal environment as a consequence of the natural patterns of rock weathering depending on their intrinsic properties. These are superimposed by additional conditions and factors including the structure of the building and the urban climate. Once placed in a structure, the pattern of their natural decay is altered, influenced by the degree of exposure to sun, wind and the rain affecting their cycles of wetting, drying, heating and cooling. All these in combination, will either result in acceleration or retardation of the decay processes, affecting how well they can be preserved over time. In the current scenario with climate change being one of the greatest challenges facing the world, the management of buildings, both modern and historic, must consider it to ensure their long-term preservation, including planning an appropriate regime of intervention. A notable difference between modern and historic buildings is that, while the former has an expected design life of 20 to 100 years (although used much longer), the latter can be hundreds of years old. The threat posed to cultural heritage by potential climate change effects can be direct, influenced by extreme weather and environmental conditions. It can also be indirect, affecting the social and economic structures in which they are embedded, for example, by affecting the numerous jobs centred around the cultural tourism sector if these heritage sites undergo degradation [1].

Climate change is manifested as an event related to weather or climate persisting for a longer duration than usual and/or, when they differ from average weather and/or climate events, sometimes with significant changes in trends [2,3]. In the past, it was largely due to natural causes, including variations in the Earth's orbit, ocean currents and volcanic eruptions [4], but today is accelerated due to anthropogenic activity. Taking into account different scenarios to estimate future greenhouse gas emissions, climate models have been developed. For the UK, climate change modelling predicts warmer and wetter winters, hotter and drier summers with increased summer temperature maxima, extreme rainfall events and intensification of the urban heat island effect for the future [5]. Some of the predicted trends for temperature and precipitation are summarised in Figure 1. A mean annual increase in temperature of 0.3 °C and 0.9 °C for England between 2008–2017, from 1981–2010 and 1961–1990, respectively, are observed. During the same time period, it has been wetter by 4 to 11% in terms of the annual average rainfall. Based on the report, projected temperature estimates using probabilistic projections from 1981–2000 to 2080–2099 for the UK region, taking into account a high greenhouse gas emission scenario, lie between 5.7 to 6.3 °C [5]. The projected precipitation during the same time period is highly variable, declining by 6% in summer but increasing by up to 48% during the winter [5]. Based on the exponential Arrhenius equation between the chemical reaction rate and the temperature, for a rise of ten degrees in temperature, the power of destruction by such reactions due to decomposition of constituent materials in any building site will essentially double [1]. Of the many consequences of wetter winters, one would be an increased growth of mould fungi. In previous studies, overall impacts of climate change on buildings have been considered in terms of any change in stress conditions and potential impacts on material properties such as strength, durability and permeability [6,7].

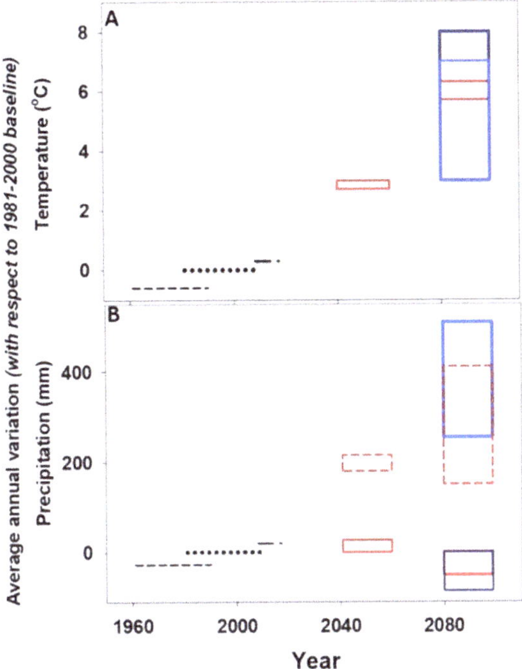

Figure 1. Decadal variation with respect to a baseline of 1981–2000 of annual (**A**) temperature (**B**) precipitation marked by dotted (1981–2000), dashed (1961–1981) and dash-dotted (2008–2017) lines, based on UKCP18 projections (Met Office, 2019) [5]. The boxes denote predictions for England (red) and London (blue) with dark and navy for summer and winter, respectively, for London. Solid and dashed boxes mark seasonal projections for summer and winter, respectively, for England. An increase in temperature is observed throughout for England with a predicted significant rise from 2040. This is even more pronounced for London from 2080. A significant increase and decline in precipitation are predicted for London during winter and summer, respectively, towards the end of this century. See text for further discussion.

Climate change and its impacts are more pronounced in large urban areas compared to the global average. In spite of taking up less than 1% of the Earth's landmass [8], it is well established that cities are the epicentres of long-lived greenhouse gas emissions [9,10]. Some of these phenomena, such as the urban heat island, are specific to cities but depend on population size, built area, density and compactness and, regional location. An interplay of anthropogenic impacts such as the energy usage of a city and reemission of energy absorbed by the built environment, depending on the relative distribution of parks, rivers and buildings [11], along with heat absorption/retention, radiation reflection and evapotranspiration properties of prevalent building materials [12], significantly affect not only the temperature, but also the humidity along with the total emissions of air pollution [13]. Based on regional climate model projections for London [5], the temperature increase of 3–8 °C during the summers in 2080–2099 from 1981–2000, is higher than the predicted national range of 2.3–6.3 °C in the same time period (Figure 1). The precipitation decline during the summer can be up to 10% (~85 mm) in 2080–2099 from 1981–2000, higher than the 0–6% (up to ~50 mm) expected for England, and would increase by 30–60% (~250–500 mm) during the winter, which is significantly higher than the national predicted increase of 18–48% (~150–400 mm). These estimates are strongly dependent on future global greenhouse gas emissions and can change in the future depending on methodological and data choices and implementation of the Paris Climate Agreement [14].

The deterioration of building stones will be impacted by changes in the concentration of air pollutants over time. As a city expands and its population grows, and if the summer temperature becomes particularly high with low wind conditions, there could be increased concentration of air pollutants from transport emissions while, the dispersal of the pollutants will be affected by the temperature lapse rate within the planetary boundary layer. The traditional pollutants, such as sulphur dioxide and smoke from coal, have decreased since the mid-20th century, with rising traffic related pollutants including nitrogen oxides and particulate matter [15]. Some air pollutants, such as nitrogen oxides, undergo chemical transformation and are often reduced in urban centres while they increase in the outer regions.

Initiatives to make the cities more liveable and "green" may have long-lasting implications for their building stones on one hand. Planting deciduous trees to provide shade in summer, permitting solar gain in winter and usage of thermally reflective surfaces can be viable options to reduce greenhouse gas emissions. Building stones have high thermal masses that should help to soak up unwanted urban heat during the day and regulate temperature better. Based on requirement, mechanical ventilation and cooling need to be installed. To cope with any reduced water infiltration into the ground and any increase in surface run off because of their impervious paving stones, the drainage system needs to be improved. On the other hand, considering the direct environmental effects on the building stones that challenge the resilience of the urban heritage can also help to better preserve our heritage sites.

This paper reviews the possible effects of current and projected trends in extreme weather events in urban areas on a range of building stones used in heritage structures of London. Particular emphasis has been placed on the impacts of climate change phenomena such as precipitation and temperature, the intensity and frequency of which have been modified as a consequence of human activity. Detailed attention has been paid to the different forms of decay, both physical and chemical, such as surface recession and erosion by precipitation, biodeterioration, microcracking and decohesion due to salt crystallisation, that together contribute to the overall deterioration of the stone fabric. The decay processes have been considered in the context of the diverse lithotypes of the building stones as they determine their responses to the changing external environment. Both intrinsic properties of the rocks, as well as the extrinsic attributes such as extreme weather events and pollution, have been considered in tandem to understand any decay of the building stones to the alteration-inducing factors.

2. Deterioration of Building Materials

The durability of building stones can be assessed by their ability to resist weathering so that their original size, shape, strength and appearance are retained over an extensive period of time [16]. Rocks weather by a combination of processes including dissolution, salt crystallisation and freeze-thaw activity that are influenced by extrinsic attributes such as meteorological factors and pollution, as well as intrinsic attributes including the petrological and petrophysical properties of the stones. The effect of air pollutants depends strongly on the physical, chemical and mineralogical properties of the lithotypes under consideration as discussed below. The building stones used for important heritage sites in London are listed in Table 1 and shown in Figure 2.

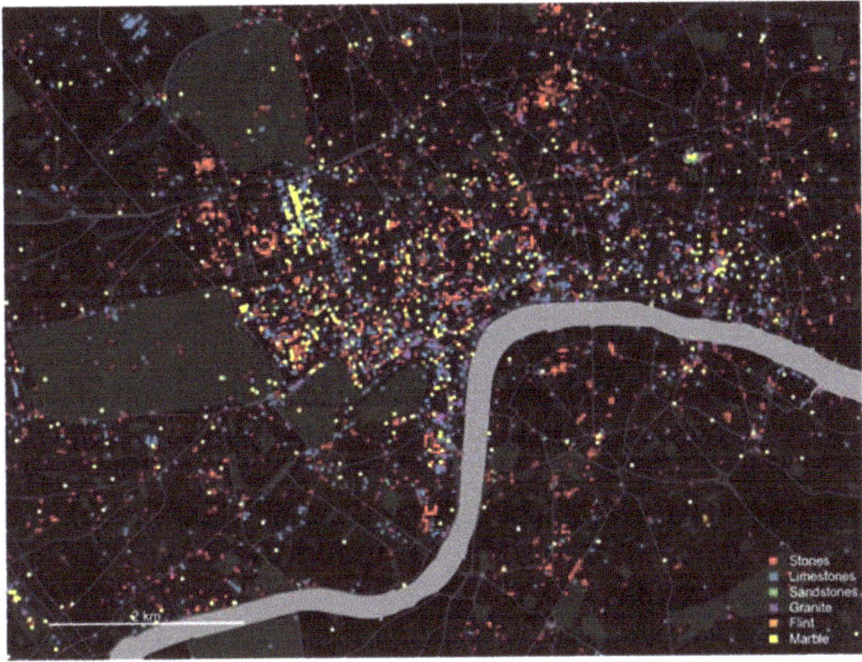

Figure 2. Spatial distribution of common building stones in London including stones (nonspecific), limestone (Portland, Clipsham, Caen and Rag) sandstone (Reigate and others) granite, flint and marble. Portland stone is the most common of the specified varieties.

Table 1. Some of the important heritage sites in London and the building stones used in their construction.

Heritage Site	Building Stones
St Paul's Cathedral	limestone (Portland stone)
Buckingham Palace	limestone (Caen); later refaced (Bath Stone)
Westminster Abbey	sandstone (Reigate stone), limestone (Kentish Rag with Purbeck as decorative columns; refaced with Portland and Yellow Bath stones) and chalk
White Tower of London	limestone (Kentish Rag rubblestone, Caen and Quarr) with sandstone (Reigate stone as upper dressings)
British Library	limestone (Hauteville as paving stone; Portland and Purbeck for flooring; travertine as indoor decorative slabs), granite (Royken granite: facade and steps), sandstone (outdoor, mounted decorative slab), red bricks (Figures 3 and 4)
Marble Arch	marble (Carrara)
Burlington House, Tower Bridge British Museum, Somerset House, Bank of England and Mansion House	limestone (Portland) and granite (Cornish)
New quay and docksides	granite (Cornish and Scottish) and sandstone (Yorkshire)
The Palace of Westminster; New Houses of Parliament	limestone (originally magnesian limestone, Cadeby Formation, from quarries at Bolsover Moor and Mansfield; later substituted from the quarries at Anston for most of the upper fabric; ultimately replaced by Clipsham stone
St. Pancras Grand Hotel	red bricks (made from clay), sandstones (Derbyshire) limestones (Lincolnshire) roofing slates (Leicestershire) and granites (Cumbria)

Table 1. *Cont.*

Heritage Site	Building Stones
Southwark Cathedral	flint cobbles in dark mortar with pale stone quoins
Trafalgar Square	granite (Dartmoor: Nelson's column and base; Aberdeen granite: bollards, walls and statue plinths; Cornish: inlaid strips); limestone (paving); sandstone (red Mansfield: paving)
Albert Memorial	granite (Cornish: walling, steps and lower platform); slates (paving slabs); limestone (fossiliferous: paving slabs; sandstone (Red Mansfield: paving slabs); marble (Campanella: statue)
Big Ben	bricks, limestone (Caen and Anston, in addition to Clipsham for restoration), granite (Cornish)

The most commonly used building stones are limestone, marble, sandstone, granite, flint and slate. It is common for a combination of rock types to be used for any particular building, as seen in British Library [17] (Figures 3 and 4) and Tower of London (Figure 5). The properties of these different stones and any related deterioration over time, are discussed below and summarised in Table 2.

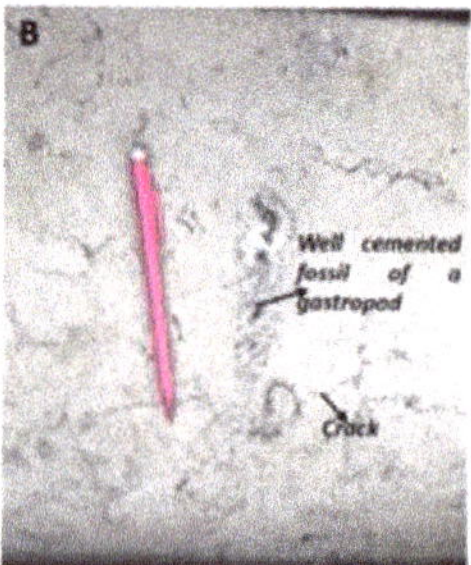

Figure 3. Different building stones used on the British Library exterior, Euston road [17]. (**A**) Red Sandstone: The red colour, related to its deposition under desert conditions and formation of iron oxide as coatings on the grains makes it aesthetically pleasing. While the pore spaces can provide pathways for water transport and salt deposition, iron oxide acts as a barrier for any reaction with the constituent grains. Laminations are visible, restricting transportation across the layers. Closer inspection is required to comment on its maturity, that will have implications for pore size distribution (unimodal versus bimodal) and consequent ability to transport water by capillary flow. (**B**) Hauteville Limestone has a crystalline texture and is thereby hard and compact, making it suitable as a paving stone. Compositionally it is homogenous, dominantly made of lime mud which reduces the possibility of differential weathering, except where pits are formed due to preferential removal of uncemented fossil fragments. However, many of the fossil fragments are well cemented, adding to the overall hardness of the rock. The white colour of the rock is partly discoloured to yellow related to salt formation and deposition. The slabs show microcracks due to differential expansion of the carbonate minerals. (**C**) Royken granite's interlocking mineral grains result in a crystalline texture that gives it compactness and strength, making it suitable for use as a building block for the steps and facade of the library. The reddish colour of the rock can be attributed to the presence of ~60% potassium feldspar (KF), the most abundant mineral present with quartz (Q) and mica (M). The quartz grains are mm sized with no visible fissures under hand lenses or naked eyes. The potash feldspar and mica can be prone to kaolinisation and effects of weathering which is evident from the diffused boundaries of the discoloured feldspar. Its finer grain size makes it prone to chemical weathering because of the large surface area available. Differential weathering between the micas and the quartz can leave pits on the exposed rock surfaces. Currently unpolished, however, polishing later can fold and deform the micas if the polishing plane is oblique to the cleavage planes of the micas.

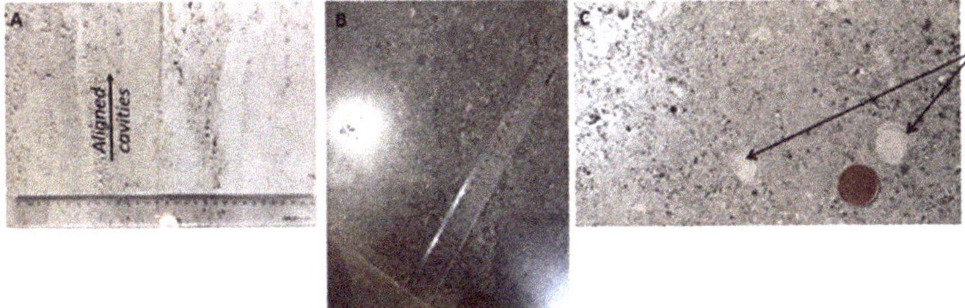

Figure 4. Building stones used in the British Library interior, Euston road [17] including (**A**) Jura travertine which is a spring deposit from volcanic activity. It is compositionally calcareous, and because of its textural homogeneity, it is advantageously used as a building stone. The macroporosity, arising due to the hollow stems of rushes and plants that flourished in the spring water, attributes a pattern to the rocks due to their alignment and contributes to the total rock porosity while reducing the overall strength of the rock. However, they can still be quite compact where they are structurally massive, lacking bandings and laminations. The macroporous bands act as weak points and, when oriented perpendicular to the load direction, attribute minimum mechanical strength to the building stone [18]. Notably, they do not result in an increase in capillary transport if not connected. At many points, the pore spaces have been infilled with synthetic material to provide a more consistent surface that helps with cleaning. Naturally infiltrated clay minerals result in faint banding adding to the beauty of these stones. (**B**) Purbeck limestone is highly fossiliferous with abundant fragments of fossil bivalve shells giving it a well patterned appearance. It is dark in colour as deposition occurred under muddy, shallow, freshwater conditions. Because of the hardness, the limestone can be polished and forms a good substitute for more expensive marbles. Cracks are observed but are often cemented. (**C**) Portland stone is an oolitic limestone, well cemented and compact, and lacks any directional properties making it convenient to cut in all directions. Some calcareous algal pellets are present with some minor oyster shell fragments that have left pits where they were ripped off preferentially by constant foot tread.

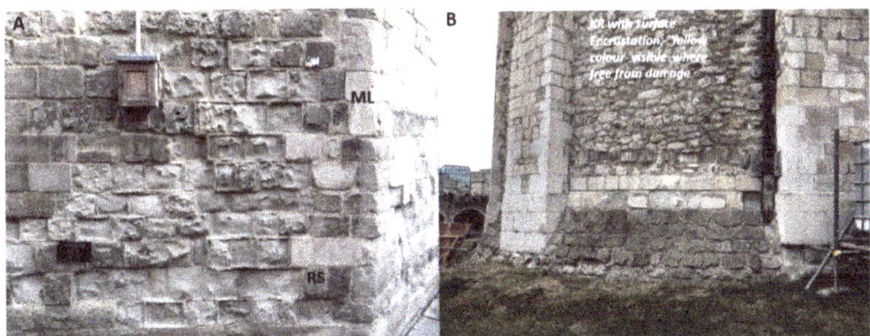

Figure 5. (**A**) The building stones at the Tower of London including Reigate stone (RS) and magnesian limestone (ML). Although a combination of stones including Caen and Quarr have been used [19], because of surficial damage identifying the individual building stone can be sometimes challenging. (**B**) The Kentish ragstone (KR) appears damaged with surficial encrustation. Portland stone has been commonly used as replacement blocks identifiable by their sharp boundaries and pale colour. Image credit: Historic Royal Palaces; used with permission.

2.1. Carbonate Rocks (Limestone and Marble)

These rocks are comprised of more than 50% of carbonate minerals. Although they are compositionally uniform, their physical characteristics such as hardness, fossil content and porosity are variable which contributes towards their differences in terms of durability. For example, the Caen stone has two variants depending on its depositional fabric. The pelleted variety has high microporosity and is susceptible to weathering and the one dominated by bioclasts is less porous and less prone to alteration (Table 2). During formation, their durability is enhanced by cementation and recrystallisation that result in a reduction in total porosity. The degree of interlocking between component minerals is also an important textural attribute contributing towards the strength of a rock, with higher crystallinity resulting in higher strength. In general, geologically older carbonate rocks that have increased overburden pressure tend to have lower porosity and higher Young's modulus which makes them more durable, as evident for example, from the Carboniferous limestone such as Settle in Yorkshire, when compared to the younger Jurassic oolitic Bath Stone [16,20]. This can be understood in the context of the diagenetic alteration and changing character of the calcite cement in the primary pore spaces between original grains in a limestone as it progresses from sub-mature to mature. The cement growth initially starting as small crystals, finally recrystallises to form a denser fabric of fewer larger crystals, cutting across grain boundaries and resulting in a stronger fabric [21]. Any subsequent process of dolomitisation, when magnesium carbonate replaces the calcite or calcium carbonate, can enhance porosity with lowered unconfined compressive strength. Higher porosity also facilitates chemical weathering which leads to further enlargement of pore sizes, providing greater accessibility to oxygen, moisture and water with further chemical breakdown. A highly porous carbonate rock, with a significant fraction of pore diameter less than 2.5 μm, which facilitates capillary absorption of water and saturation of the stone, can be considerably damaged by freeze-thaw action [22]. When the internal surface area of the constituent grains is higher, they are more susceptible to water retention by adsorption and breakdown during weathering, as in the case of any pellets and ooids being present. On the contrary, the presence of crystalline shells can result in a lower internal surface area. For those building stones with lower microporosity, the clay fraction can be significant for its hygroscopicity and, by retarding the appearance of capillary bridges, reduces the rate of mechanical strength loss [23].

Carbonate rocks are strongly affected by dissolution in acidified water with disassociation to Ca^{2+}/Mg^{2+} and HCO_3^- under rainwater pH of 4 to 7 [24,25]. The intensity of the rainfall is an important factor here with atmospheric pollution in urban areas accelerating the weathering rate. This has implications for anticipated increase in rainfall intensity related to climate change and air pollution for London. Even during the summer in which a seasonal decline of rainfall (Figure 1) is predicted, there can be later interaction and mobilisation during subsequent rainfall (or other forms of precipitation as well as humid air) of the accumulated dry deposits of sulphur dioxide, oxides of nitrogen, chlorides, sulphuric and nitric acids and particles such as soot, fly ash, etc., mostly emitted as combustion products of fossil fuels. When stones have a high specific surface area or a high deliquescent salt content, it can promote further dry deposition of NO_x [26]. This results in an enhanced acidity of the rainwater and pronounced carbonate dissolution from the building stones. As calcite dissolution is limited by the rainfall amount and its pH, while sulphur dioxide in the atmosphere plays a major role in the damage of carbonate stone, other factors such as humidity and precipitation intensity are also important [24,25]. For example, the decay rate of Portland limestone shows a decline over time since the 1980s for both Munich and London (Figure 6), but this is not proportional to the reduction in atmospheric SO_2 concentration during the same time period [27–29]. Additionally, the decay rate of Portland stone in London is considerably higher since the initial measurement period in the 1980s, as compared to Dublin or Munich (Figure 6), although the atmospheric SO_2 concentration in 1978 is comparable [28,29]. As traditional air pollutants continue to decline, beside the effects of acid rain and dry deposition of gaseous pollutants, especially SO_2 and NO_x discussed above, the degradation of carbonate stones by carbonate dissolution via the karst effect (with rain at pH ~5.6 in equilibrium with atmospheric CO_2) can be considerable. This is estimated to result in stone weathering of ~18.8 μm/m at 330 ppm CO_2 [30]. It can account for ~96 to 99.5% of the surface recession of carbonate stones from 1990 to 2099, arising not due to changes in rainfall but due to the predicted higher CO_2 concentration of up to 750 ppm [31]. Marbles are less prone to deterioration as compared to the limestones. Owing to their minimal porosity, they are not susceptible to the strong capillary suction that leads to rising damp, thereby limiting any salt damage [32].

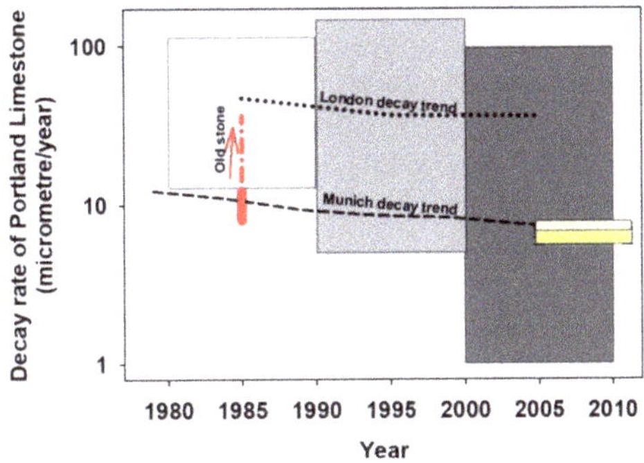

Figure 6. Decay rate of Portland stone in St. Paul's Cathedral, London, over three decades marked by boxes in white (1980–1990), light grey (1990–2000) and dark grey (2000–2010). The dotted line represents the trend indicating decline of the average decay rate. The decay rate for Mainz (dark yellow box) and Munich (light yellow box) and Trinity College Dublin (orange blob) are shown for comparison. The dashed line indicates the decline of the average decay rate for the stone in Munich. The orange arrow indicates an increase in decay rate for an old stone (relatively weathered rock) as compared to its new stone counterpart. Note that the decline in the trend of the decay rate for both London and Munich are not gentle in spite of the drastic reduction in the concentration of atmospheric SO_2 over the same time period in both the cities. This indicates that other parameters are important in controlling Portland limestone deterioration. For further discussion, see text. Data sources: [27–29].

Subsequent to calcite dissolution, salts such as calcium and magnesium sulphates can crystallise at or near the surface, giving rise to efflorescence, or they can remain in solution and be transported by the moisture and rainwater trapped in carbonate rocks. This can result in a breakdown caused by crystallisation pressure, and hydration pressure if the salts recrystallise to different hydrates. Efflorescence often leads to exfoliation with losses of the outer layers of the rocks referred to as contour scaling. Crumpling by powder formation due to the crystallisation of soluble salts such as sodium chloride, sodium sulphate and sodium hydroxide, in addition to black crust formation by a mixture of gypsum and soot are also common deterioration features [16]. Differential thermal expansion can occur as the salts expand and contract at different rates from that of the host rocks during cycles of wetting and drying, and heating and cooling, resulting in the build-up of internal stresses. For example, halite can expand by 0.5% when the temperature increases from 0 to 60 °C. The gypsum formed can stabilise the surface and near surface stonework without any immediate impact on its interior when it maintains a good internal cohesion and adhesion to the underlying stone substrate [33,34]. Such crusts, having different levels of disintegration, can have different surficial manifestation of their morphology from laminar to framboidal [35]. Scaling and flaking of the crust can also be visible with formation of small blisters. When the major stress threshold is breached, there can be more rapid, catastrophic decay with multi layered flaking and large blister formation and, in severe instances crust detachment. Owing to its open crystalline structure, the black crust further accentuates moisture penetration bearing dissolved salts through the building stones.

In addition to crust formation, powder efflorescence can also occur, dominated by magnesium sulphate and only minor gypsum [33]. Due to the high solubility of magnesium sulphate, it only accumulates when water run-off is absent, which occurs in parts of structures that are sheltered. This also has implications for the predicted dry summer months in England where magnesian limestones are in use. Limestones are susceptible to soiling due to the accumulation of carbonaceous matter within the pore spaces, although soiling can be undone with cleaning [36]. Soiling can be locally concentrated, being determined by wind direction and protection from the rain, as well as the application of protective treatments that induce greater particle matter accumulation [37]. The process of limestone deterioration is summarised in Figure 7.

Calcareous stones are prone to biodeterioration as they can be colonised by lithobiontic communities including bacteria, algae, fungi, lichen and mosses, sustained by the moisture retained in the pores on horizontal limestone surfaces as well as the inner zones that are pathways for moisture entrance and/or water accumulation [38,39]. Penetration of lichen hyphae within the substrate leads to physical destruction of the stone microfabric as well as chemical dissolution and corrosion of the underlying substrate affecting calcite grains in particular [39]. The latter occurs as "lichen substances", including organic acids such as oxalic acid, extract metallic cations from the stone to use as nutrients, leading to the formation of pits of up to 2 cm in diameter and depth [40]. This occurs with simultaneous precipitation of clusters of Ca-oxalate crystals, also found in the black crusts. In addition to gypsum, the black crust typically consists of calcite and quartz derived from the substrate, along with other non-carbonate carbon forms. With the relative increase in atmospheric nitrogen and organic compounds, with respect to SO_2 leading to an increase in the biological activities and production of oxalates, formates and acetates in buildings such as the Tower of London, the yellowing may be of more concern than any darkening arising from the presence of elemental carbon [41].

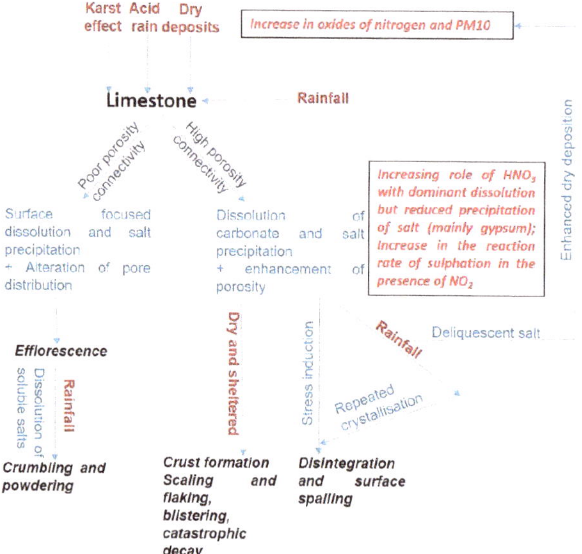

Figure 7. A model of limestone decay in urban settings, modified after [34,42]. Both rainfall and dryness may increase in the future as well as the concentration of atmospheric nitrogen oxides in London and other urban cities (indicated in red font in text boxes) which will affect decay processes (indicated in black font). In marble, microbial communities on stone surfaces are commonly observed. The surfaces can be cleaned by using suitable compounds although they can be subsequently recolonised [43]. Marble is also prone to efflorescence of salt such as sodium sulphate resulting in significant dissolution, with calcite being more susceptible than dolomite, reflected in a loss of gloss [44]. Degradation by microcracking is another major problem for marble, which is more pronounced during cooling than heating. This commences at a higher temperature differential for dolomitic marbles as dolomite has a lower thermal expansion anisotropy than calcite [45]. In addition, microcracking occurs at a larger temperature differential for decreased grain size and increased shape and lattice preferred orientation [45]. However, the effects of grain size and grain boundary were found to be insignificant for some marbles and made no difference even for polygonal grains with interlobate boundaries, but those with strong texture exhibited minimal residual strain following thermal treatment [45]. In the context of London and predicted temperature change, since cooling is expected to decline with time, any microcracking will be relatively limited.

2.2. Granite

Granite is a crystalline rock that is commonly granular with a phaneritic texture. Occasionally, where the quartz, potash feldspar, plagioclase and biotite occur as mm-sized phenocrysts in a finer grained groundmass with interlocking mineral grains, it is referred to as the porphyritic texture. Resistance to any decay is higher when the biotite content is low along with a reduced capillary coefficient and anisotropy [46]. Granite is very often used in the facade of buildings (e.g., the British Library; see Figure 3), very close to a heavy traffic street and exposed to high levels of foot of traffic. Common forms of decay include spalling and granular disaggregation. Granular disaggregation occurs due to kaolinisation of feldspar and biotite, a weathering process that weakens the granitic microfabric due to the chemical and mineralogical transformation of feldspar (hardness = 6–6.5 on the Mohs scale) to kaolinite (hardness = 2–2.5 on the Mohs scale) [47,48]. The process is facilitated by wetting and drying cycles in moist and acid urban environments, accelerated by elevated CO_2 and SO_2 from anthropogenic activity. In a later stage of weathering, biotite can be replaced by chlorite along rims and lamellae under relatively oxidizing conditions, but the process of chloritisation can be of less significance in a generally acidic, urban environment. The spalling can be related to gypsum formation or aerosol deposition, as well as the use of gypsum and other lime-based mortars that can be a source of calcium [49]. This manifests as sulphate efflorescence resulting in decay as the salt crystallises on the surface or within pre-existing cracks or defects [44]. Sometimes, surficial staining is seen due to the salt formation of sulphates and nitrates by the use of cleaning products that were not neutralised [50]. Dark staining in granitic rock surfaces can be correlated with high moisture content, attributed to the formation of waterproof films by the application of water repellents during subsequent intervention [50]. While the decay of the feldspars chemically originated like the biotites and is mainly dependent on kaolinisation and chloritisation, the decay in quartz is dominantly mechanical and characterised as fissurisation under compression [51]. Quartz fissurisation is directly proportional to the mineral size while that of feldspar kaolinisation is inversely related.

In the weathering of granite, SO_2 plays a dual role by promoting sulphate precipitation as well as kaolinisation of feldspar and biotite. Granite's reactivity with SO_2 is low in general requiring hundreds of thousands of years to occur under natural conditions [52]. It has very low porosity (typical values for open porosity are 0.2 to 0.3% by volume, as compared to up to 30% for limestone). In urban areas, the sulphation reaction is due to availability of gaseous SO_2 that can be adsorbed on the surface of the rock, as opposed to SO_2 in aqueous or dissolved states in pore solutions [48]. It subsequently reacts with the calcium provided by the plagioclase sericitisation, accelerated in a CO_2+H_2O rich atmosphere [50]. This explains the occurrence of the weathering product of kaolinite at a depth of up to 4–5 mm from the granitic wall surface in northern Spain, coincident with a time of significant SO_2 release [50]. Formation of an aeriform mixture of SO_2 (gas) + H_2O (vapour) can be more likely in the future in London as a rise in annual rainfall is predicted [5], although the availability of SO_2 is anticipated to be lower. Alternatively, kaolinisation can occur if CO_2 substitutes for SO_2 in the reaction pathway of dissolution and chemical precipitation, but the rate of reaction is going to be slower [53].

Thermal stresses can lead to microcracking that accelerates decay and can be significant over a temperature range of 30 to 80 °C, with progressive coalescence of existing microcracks and generation of new ones [54]. This can lead to intracrystalline microcracking in K-feldspars and biotite along the direction of their cleavage planes or focused at the centre for the zoned plagioclase feldspars, while for the quartz, microcracking is intragranular with irregular patterns [55]. Over time, surface microcracking will lead to crystal disintegration and detachment as well as facade scaling and flaking [55]. The microcracks also accelerate biodeterioration by biological colonisation by lichen and plants. This leads to disaggregation of the stone fabric as any penetrating lichen hyphae enhances pre-existing points of weakness, such as cleavage planes in feldspar and mica, as well as intragranular cracks in quartz grains [50]. The physical disintegration is accompanied by chemical decay. This is manifested as etching pits and dissolution cavities in feldspar and mica, as well as the chemically resistant quartz and apatite, often accompanied by the precipitation of amorphous deposits, suggesting that the biological patina often derives its constituents from the disaggregated granitic surface [50]. As the feldspars, both plagioclase and potash feldspar, undergo microcracking due to physical weathering processes such as freeze-thaw, it results in crystal loss and facilitates chemical alteration processes by providing access pathways to fluid flow and air pollutants [48]. This is in agreement with observations of the higher rate of chemical weathering for plagioclase than K-feldspar, that can be related to the increased rate of physical weathering for plagioclase driven by different mechanisms of microcracking related to their microstructures and crystallographic anisotropies. Zoned, plagioclase crystals with a high number of core microcracks are susceptible to seritisation at the core where there is also more calcium than in the perimeter zone, driven by a combination of deuteric alteration and physical disintegration. However, under predicted climate change for London, due to relatively wetter but warmer winter conditions [5], freeze-thaw processes may be greatly reduced in the future. Sometimes a heterogeneity in relief can be observed for granitic surfaces with darker, microgranular enclaves appearing to be more weathered than the surroundings. This phenomenon is poorly understood but can be related to the lower albedo of the darker enclaves enriched in biotite and plagioclase, as compared to the remaining rock. This promotes differential spalling with only a thin detachment of the stone where the decay is accelerated by the lower conductivity and the heat capacity leading to overheating of the enclaves [52]. The processes of deterioration of a granite is summarised in Figure 8.

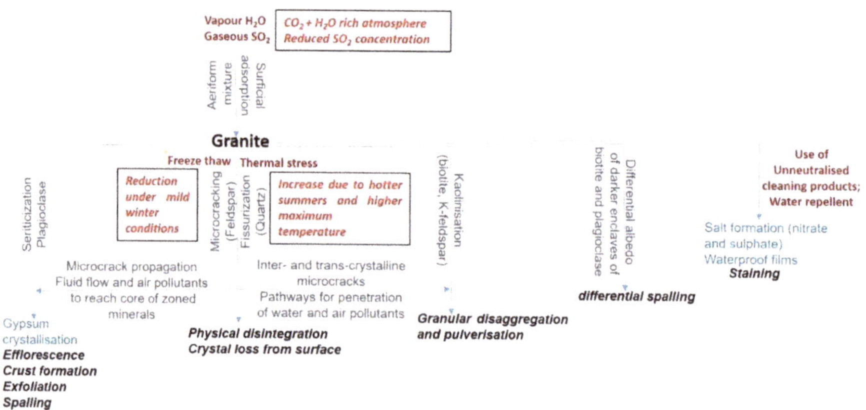

Figure 8. Model of granite decay in an urban setting. Since the concentration of atmospheric sulphur dioxide in London and other urban cities is expected to continue to decrease, the frequency of many decay processes for granite such as kaolinisation and gypsum crystallisation are also likely to be decrease.

2.3. Sandstone

Sandstones are variable in mineralogy depending on their provenance which determines the original deposited clastic material and diagenesis, and in turn controls their subsequent evolution by compaction and cementation. The diagenetic evolution finally determines its effective porosity, along with its pore space distribution, the pore geometries and their interconnections. The major decay process, affecting any sandstone, is salt weathering with the pore spaces being a major determinant as it is where the crystallisation processes take place [56]. Water transport and storage are key factors associated with pore space and sizes. A larger interconnected pore network that can facilitate moisture saturation is often related to a higher rate of salt crystallisation but cannot be solely attributed to it. While crystal growth can occur preferably in the larger pores, the residual solution as well as any absorbed moisture in the smaller pores represents a reservoir for possible subsequent crystal growth in the larger pores [57]. Consequently, the sandstones with a bimodal pore-size distribution or, with a sub-maximum in the smaller pore ranges are more sensitive to salt weathering [58]. For any sandstone with an equal pore radii distribution, a capillary water uptake will occur, though constrained mainly to the driving rain season [59], with some contribution from moisture rising from the foundations or lower wall elevations. The uptake of water and its transportation occurs dominantly across the pores ranging in size from 0.01 to 1000 μm in radii which is most likely to control the effective porosity of the sandstone [59]. Any secondary porosity can be subsequently generated if the water invades the matrix [59]. Because of the porous nature of sandstones, there is likely to be constant accumulation of salts at the point of evaporation from aqueous salt solutions, where the generated crystallisation pressures can lead to degradation. In addition to the repeated dissolution-crystallisation cycles of the salts in the porous network, the pore walls can be subjected to destructive hydration pressures, induced by salts as they undergo a phase transition during wetting. Osmotic swelling of clays, differential thermal expansion and wet/dry cycles by deliquescent salt are other mechanisms of salt weathering [32].

The availability of moisture is regulated by rainfall, capillary rising damp or condensation. Depending on the relative humidity, there might be moisture stains affecting the appearance of stone when there is deliquescence or, salt crystallisation with pressure exertion and material damage if there is evaporation. Beside rainfall and relative humidity, the other important factor is the temperature that controls the evaporation rate and consequently, the degree of supersaturation of the salt and the resulting damage. The type, amount and distribution of the salt present also regulates the intensity of salt weathering. In urban areas, gaseous pollutants may react with the porous building material forming different types of salts including the nitrates, sulphates and the carbonates [60]. The salts present are a complex mixture of ions which have significant implications. For example, when the most common salts, halite and gypsum, coexist in the presence of sodium chloride, the solubility of gypsum increases which aids its migration and recrystallisation [61]. For a mixture of salts, the highest salt content may not imply the greatest damage, as ion mixtures that are strongly hygroscopic will not crystallise under normal conditions [32]. Furthermore, the damage may not be proportional to the component salts, with certain salt mixtures causing more intensified weathering [32].

The porosity and the pore size distribution, important for regulating the moisture content and its accumulation with implications for water absorption, salt loading and hygric dilatation, can be ultimately related to the lithology [62]. For example, a decrease in hygric dilatation and an increase in salt resistance is observed with increasing compositional and textural maturity from litharenites to quartz arenites as the pore distribution changes from bimodal, to unimodal unequal to unimodal equal. For a quartz arenite with a unimodal equal pore-radii distribution, it is characterised by restricted moisture absorption and, in spite of a quick capillary water uptake during driving rain, it has low water retention due to its missing smaller capillaries and a high drying velocity [59]. However, its pore sizes can be significantly reduced during diagenetic cementation. On the other hand, the presence of additional smaller pores (as in a sublitharenite) is characterised by a higher water content as the additional smaller pores allow the absorption of moisture via sorption.

The transportation of the salt is controlled by ion diffusion, which requires continued saturation but is also determined by pore connectivity, mineralogy and sedimentary structures [63]. Ion diffusion leads to the distribution of salt throughout the rock mass as penetrating saline solution that, keeps the sandstone blocks saturated for an extended period of time and facilitates chemical reaction with the mineral constituents present [63]. To maintain continued saturation, there should be consecutive wet days to maintain wetness. However, even if the surface/near-surface zone dries out by evaporation, there can still be a significant reservoir of moisture at depth, driving chemical action in the stone interior [63]. The connectivity of pores allows the ions to move through the stone along the concentration gradient, but can be modified as weathering proceeds, where salt solutions reach their saturation and may crystallise out of solution, blocking pore throats.

The mineralogy also impacts porosity and adsorption capacity. For example, any mica present can disaggregate and accumulate in the pore spaces, thereby reducing the porosity and simultaneously increasing the surface area available for the adsorption of hydrated ions that lowers the overall diffusion [63]. Like the clay minerals, the micas can also "fix" ions in the interlamellar spaces of their structure that decrease the overall diffusion process. Other factors that need to be considered are physical heterogeneities such as the presence of clay layers that influence moisture regimes [59]. The presence of clay layers within a sandstone, induces additional stress due to their high sorption capacity and diffusion resistance and promotes a retarded interaction with the environment by acting as a barrier to water migration. Mixed layers of clays such as chlorite and smectite may undergo intracrystalline swelling in the presence of water, while chlorite–illite may be subjected to osmotic swelling processes facilitated by the presence of sodium chloride in the pore spaces inducing a concentration gradient of Na ions between the clay particles and rock pores [64]. The presence of bedding layers inhibits the rate of diffusion if they are perpendicular to the concentration gradient, although if the diffusion gradient is bedding parallel, it is possible that the rate of diffusion can be accelerated [58]. Hydric swelling of clay minerals is pronounced when their basal planes act as planes of weaknesses with orientation parallel to the original sedimentation beds [64].

Salt weathering in a sandstone, controlled by fluctuations in temperature and moisture, occurs when there is frequent wetting and drying in the near-surface zone. This causes repeated crystallisation of salt with complex three-dimensional distribution in the rock mass, that undergo repeated hydration/dehydration and expansion/contraction [63]. It leads to efflorescence and finally, in surface losses in the form of granular disaggregation, scaling and multiple flaking [63]. Although stone surfaces may appear stable for a considerable period of time, the stress caused by accumulated salts can eventually exceed the threshold of stone strength and trigger material loss [65]. The penetrating saline solution can lead to selective silica dissolution thereby weakening the stone matrix and/or the grain boundary cementing. Biodeterioration of sandstone occurs as a result of the dissolution of the constituent silicates and nitrates and particularly the carbonates that often weaken the stone cement, this is due to the action of carbonic acid formed when the surface water reacts with the produced carbon dioxide from microbial respiration [66]. Under increased moisture conditions, with a relative humidity of 80 to >90%, fungal infestation can be particularly enhanced—especially when the porosity is high, and the pores are interconnected. The fungi further proliferate in the presence of black crust and also activate formation of new biominerals [67]. The prevalent deterioration processes of sandstone are summarised in Figure 9.

In London, considering the effect of climate change (Figure 1) with predicted wetter winter months and higher annual precipitation, ion diffusion may become more pronounced as a mechanism of salt transportation and the salts are likely to stay in solution longer [63]. While this will delay the onset of crystallisation of the salts on one hand, on the other hand it may considerably weaken the rock cement by extending further the depth of the wetting front and increasing further the mobilisation and precipitation of the constituent elements of the rock mass. Consequently, when the salt eventually crystallises out of the solution, the rock will be too damaged to withstand any stress and may damage more easily. The decay initiated at the surface, will vary with depth of the material, with some time delay before the onset of decay at the deeper portions of the building material [60]. Biodeterioration can be more pronounced for the mesophilic fungi that flourish under temperature of 20–45 °C under enhanced relative humidity in the future. The biogeochemical effects of organisms can promote salt attack by increasing the moisture content and pore volume while precipitating sulphates and oxalates [68]. On the other hand, it can decrease salt weathering by reducing the effective permeability, preventing pollutant accumulation and altering thermal characteristics and wetting times of stone. With predicted greater seasonal variability in rainfall and consequent relative humidity, the swelling and shrinkage of different clay phases can have different impacts on the mechanical and fracture behaviour of the building stones. High swelling clay phases such as smectite and glauconite can have great influence on the stone properties, even with low moisture content during the summer, while the less swelling chlorite and interlayers of chlorite–smectite can impact only when the moisture content is higher during the winter months [69].

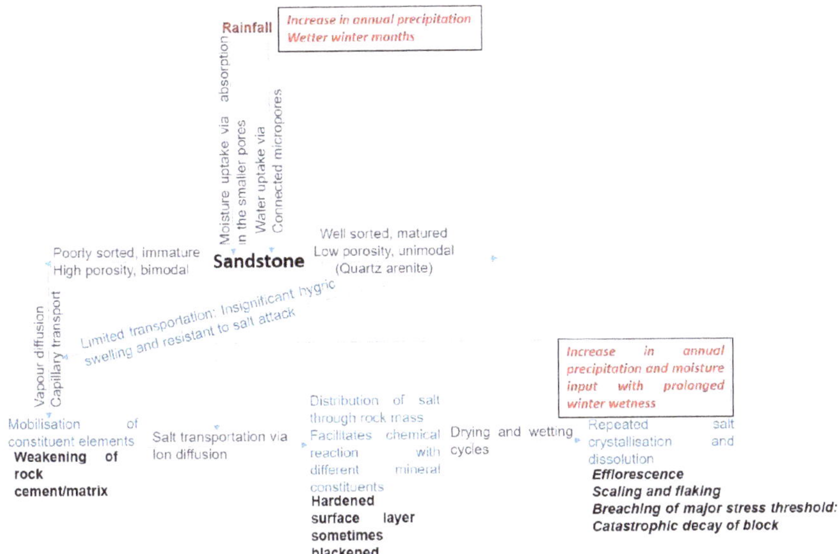

Figure 9. A model of sandstone decay in an urban setting, modified after [34]; predicted overall increase in annual rainfall and prolonged wetter winters for London in the future (indicated in red font in text boxes) can affect corresponding decay processes such as the onset of salt crystallisation and period of wetting.

2.4. Flint

Flint is a compact crystalline silica, black or dark blue-grey in colour with a trace element contamination. The microcrystalline silica can occur as nodules or bands, closely associated with chalk, with calcium carbonate often forming a coating of white cortex around the core in a nodule. They are concretions, formed during the natural growth of precipitated mineral matter around a core. A complex process converts the fine particles to nodules or cobbles of various sizes, as a result of chemical changes in the compressed sedimentary rock formation as it undergoes diagenesis.

The tough, intractable, siliceous nature of flint is the source of its great durability. This durability is restricted to individual units making it difficult to split and shape, and it does not necessarily upscale to masonry walls made with flint [70]. While the intrinsic morphological characteristics of masonry may induce additional decay and failure mechanisms into masonry walls, it is of particular concern for those made of flint units. This is because it is especially difficult to construct a vertical wall using flint in its rough, nodular, field form without resorting to the use of very large quantities of mortar to set the irregular flints, making the behaviour and integrity of the final product, both structurally and under environmental exposure, heavily dependent on mortar (see Section 2.6). For the proper bonding of the flint face to its backing, it is very important that the flint is well anchored in hydraulic lime mortar or supported by stainless steel wire ties or anchors. The backing or core, often built of compacted brick and flint, also need to be well built and consolidated.

While any impact of climate change for the flint itself used in London is likely to be minimum, caution should be executed when selecting the mortar for binding them. Cement mortar if used, can be susceptible to changing thermal conditions and moisture availability.

2.5. Slates

The term slate defines a fine-grained metamorphic rock that underwent low-grade regional metamorphism and possesses slaty cleavage due to the alignment of phyllosilicate minerals, mainly mica and chlorite [71]. Compositionally, the main minerals present are quartz, mica and chlorite with varying quantities of accessories including rutile, zircon, monazite, tourmaline and organic matter, in addition to secondary minerals such as iron sulphides and carbonates. Colour of slates are variable, reflecting variation in mineral content and chemistry, but generally range between light and dark grey. It is commonly used as roof cladding as it can be split into plane, thin and regular tiles. The UK has a long tradition of slate mining with some remaining quarries still providing slate for restoration of architectural heritage and special buildings.

Slate can be degraded by salt attack or freeze–thaw impacts, as well as thermal and hydric variations. However, compared to other building stones, roofing slate is less prone to develop alteration due to the low chemical reactivity of the main constituent minerals, the quartz and the phyllosilicates. Iron sulphides such as pyrite and pyrrhotite, as well as the carbonates can be the most damaging group of minerals in terms of the integrity of any slate [72,73]. For example, a slate containing coarse-grained pyrite as inclusions, is not very stable against temperature changes, due to the difference in thermal expansion among the rock constituents [74]. Iron sulphide oxidation also produces red staining that affects the tile surface and the other tiles below. Particularly, any macro sulphide inclusions, visible with naked eye, are a potential source of oxidation. There can also be some discolouration due to gypsification, when carbonates are converted to gypsum, triggered in the presence of water in an acidic environment. Since gypsum has a higher volume than carbonate, its growth in the slate matrix results in disintegration of the slate [75]. The flexural strength of the slate is anisotropic due to its slaty cleavage, with generally a high-water uptake associated with low flexural strength.

For predicted climate change in London, little impact on slate is predicted but some caution is advocated. For example, with increasing rainfall, in the presence of water along with oxygen, any pyrite present can be oxidised to ferrous and ferric sulphates and sulphuric acid. With the consequent drop in pH value, it can lead to further dissolution of any other metal salts present. The possibility of the phyllosilicates present undergoing alteration and transformation cannot also be entirely ruled out as they are more sensitive to heat than quartz. However, for those slates selected on the basis of their intrinsic properties, providing they have no impurities, they can be very resistant as building stones.

2.6. Lime Mortar

In early times, the mortars were a mixture of lime, ash and brick aggregate and reactive volcanic ash (true pozzolana) [70]. Where true pozzolana was not available it was substituted by ceramic pozzolans. The bricks were fired at low temperatures and crushed to a fine particle size that reacted with some of the lime to form calcium aluminates and silicates, imparting a hydraulic set to the mortar. This mixture of mortar had a permeability that allowed water to evaporate through the pore structures, while also providing sufficient adhesion and flexural strength to enable minor structural movements in the masonry without cracking. Later builders used lime mortars without ceramic powder but still containing significant quantities of kiln residues which could also impart a weak hydraulic set. Alternatively, they used natural hydraulic limes with the addition of tallow or beeswax. The use of traditional compatible mortars in retrofit and conservation maintenance is crucial to ensure physical, chemical and structural compatibility with the original fabric. However, due to a long period of use of cementitious mortars incompatible with traditional materials in many heritage sites, more pronounced shrinkage cracks develop continuously through the mortar to accommodate movement caused by thermal and moisture changes. Cement rich mortar also inhibits the free evaporation of moisture which enters through cracks and contributes to the overall dampness of the wall and can result in mechanical failure due to frost formation. If detachment and major cracking occurs, the area affected has be dismantled and rebuilt. Lime putty for sheltered exposures (sand, porous limestone and brick pozzolan) and hydraulic lime for moderate to severe exposures (sand and porous limestone) can be effective as mortar and, should be placed by ensuring thorough compaction and filling of irregular voids.

Limestone and marble, both calcitic and dolomitic, are commonly used for quick lime production. Like the parent rocks, lime mortar is also affected by higher levels of CO_2 and intense acid rainfall. Sulphation of dolomitic lime mortars can lead to the formation of magnesium sulphate and gypsum due to preferential dissolution during exposure to rain [75]. The less soluble gypsum remains confined in the mortar structure, but the more soluble magnesium sulphate can be washed out by rain, and preferentially efflorescence in the outermost layer of the material accumulates in the rain sheltered part of the facade in the porous building stones and bricks causing major deterioration. In general, for repair mortars, depending on their binder/aggregate ratios, the strength decreases while porosity increases with increasing aggregate content [76]. It is yet to be determined whether such lime mortar can be effectively used in an acidic urban environment such as that predicted for London, which would be exacerbated by increasing rainfall.

2.7. Bricks

Bricks are silicate-based, commonly made of firmly cast clay that is not directly affected by acidic rainwater. The best quality clays, as used in the British Library, are high in pure alumina and without any gypsum [17]. They are fired in a kiln at high temperature with controlled oxygen concentrations creating the conditions to produce the red colour. During the process, when the gas escape, it creates small cavities at the surface. For any lime mortar used, the calcium carbonate binder can be dissolved away when it reacts with the acidic rainwater producing soluble salts such as gypsum which are transported in solution in the pore spaces and cavities of both the bricks and the mortar [77]. When the water evaporates and soluble salts are finally deposited, they accumulate over time within the pores forming a thin outer layer or skin with different moisture and thermal movement compared to the substrate. This induces stresses between the skin and the substrate, leading to superficial detachment and blistering. For bricks, their joints with mortar often occur as points of weaknesses [78]. They are prone to damage by salt weathering but the moisture penetration is limited by depth and height as it is distributed by capillary action, which also implies that while the more soluble salts, such as the chlorides and nitrates, can penetrate to greater depth and height, the sulphates are relatively restricted to the surfaces [78]. Different types of bricks can still continue to be used in London but only with appropriate mortar as a binder, that are not susceptible to reaction with acidic rainwater.

Table 2. Details of some varieties of limestone, sandstone and granite commonly used in London building stones, with their description and distinguishing properties that govern their durability and purpose of use in the site and possible deterioration over time. Depending on their intrinsic properties related to their heterogeneous fabric, their deterioration exacerbated by climate change will be different.

Building Stone	Provenance and Age of Formation	Description	Distinguishing Properties Related to Usage	Deterioration that can be Exacerbated by * Climate Change
			Limestone	
Portland stone [21]	Dorset (UK) Upper Jurassic (152 to 145 Ma)	Formed of "oolites" cemented by calcitic cement. The oolites are formed when tiny sand grains act as nuclei for deposition of concentric layers of carbonate material around it, physically evenly rounded by the action of waves and water currents. Some calcareous algal pellets as well as shell fragments can be present.	The well-cemented nature of Portland Stone contributes to the compactness and strength. The oolites behave like well-distributed ball bearings, and consequently the rocks lack any directional properties and can be cut with equal ease in all directions. The microporous oolites that often bear traces of borings are infilled by diagenetic calcitic cement, thereby reducing the microporosity and adding further strength. Large volumes of primary pore spaces remain uncemented. These interconnected, intergranular macropores offer pathways for drying out after wetting, making the stones more resistant to the impact of cyclic wetting and drying.	Generally resistant, but once already weathered, the added pore spaces can provide increased surface area for dissolution and, water and moisture retention.

Table 2. *Cont.*

Building Stone	Provenance and Age of Formation	Description	Distinguishing Properties Related to Usage	Deterioration that can be Exacerbated by * Climate Change
		Limestone		
Purbeck	Dorset (UK) Upper Jurassic (~145 Ma)	Highly fossiliferous and dark with clay minerals and organic matter and pyrite. Deposited under shallow freshwater conditions. Well cemented cracks are present.	Because of its dark colour and hardness, it can be polished and forms a good substitute for the more expensive marbles.	The expansion and contraction of the clay minerals related to the wetting/drying cycles is facilitated by the condensed moisture on the dense surface of the stone. This results in solubility of part of the surface matrix and the staining of pyrites that leads to the deterioration of the stone. They tend to delaminate along the bedding planes. The role of polishing in preservation is not understood.
Lincolnshire limestone (Clipsham stone) [79]	From Dorset to Yorkshire (Limestone Belt); commonly obtained from the large quarries at Clipsham Middle Jurassic (~165 Ma)	Typically, medium to coarse grained, shelly and/or oolitic. Subordinate silty, sandy or muddy beds with silicate grains of terrigenous origin may be present. The shell fragments and other skeletal remains well cemented by calcitic spar. Post depositional diagenetic alteration and consequent recrystallisation leads to further variability, where it may be shelly or oolitic.	The oolites internally cemented with radial calcite crystals that give them strength and reduce the microporosity. Lack of calcitic cement leaves the primary pore space open facilitating rapid drying. Calcite cement fills up the primary and the secondary pore spaces in the shelly varieties, the latter generated on the dissolution of the aragonitic shells, obscuring any lines of weaknesses that otherwise existed along the boundaries of the bioclasts. Clipsham stone is moderately strong and massively bedded. Performs well in sulphur-polluted atmospheres.	High to moderate porosity as intergranular macropores and micropores associated with the ooliths, sparite cement and micritic intraclasts.
Caen stone [80]	France (Normandy) Mid Jurassic (~167 Ma)	Pelleted and bioclastic fine limesands deposited in the seabed, that pass to shallow deposits of lagoonal sediments and muds fringed with oolites. They are underlain by the deeper water deposits of sponge-rich marls. Large shell fragments present at times with minute pyrite crystals. Overgrowth cement filling up intergranular pore spaces, interlocking the nucleus and the overgrowth for the bioclastic fragments.	Uniform texture that can be attributed to the compact and uniform faecal/psudofaecal pellets, with easy carvability and no obvious sedimentary laminations. Dense structure and low porosity when dominated by the bioclasts.	Severe decay due to gypsum formation facilitated by the presence of micropores in the pellets. Damp conduction when used with impermeable bricks. Oxidation halo around pyrite.
Quarr [81]	Isle of Wight Palaeogene	Bioclastic, freshwater with two contrasting lithologies. Fine-grained with bioclasts of thin-walled bivalve fragments in a micritised matrix present. This contrasts with coarsely bioclastic, porous limestone with fragmented mollusc shells replaced by calcite cement. The layers of these broken and abraded fossil shells result in a characteristic laminar texture.	The framework of fossil fragment and cement where present, gives it strength.	Decay due to gypsum formation facilitated by the presence of the highly porous framework of the coarsely bioclastic limestone where the fragmented mollusc shells are replaced with calcite cement.

Table 2. *Cont.*

Building Stone	Provenance and Age of Formation	Description	Distinguishing Properties Related to Usage	Deterioration that can be Exacerbated by * Climate Change
Limestone				
Bath stone [21]	Bath region, England, UK Upper Middle Jurassic (~195 to 135 Ma)	Oolitic, with diameter of ooliths ranging from 0.2 to 0.8 mm, along with a smaller proportion of larger, mm-sized shell fragments in the cemented, calcitic matrix. The bioclastic fragments are distributed along laminations. The secondary pore spaces along dissolved bioclasts are well cemented with calcite.	Strong, rigid and resistant due to the low porosity crystalline, calcitic matrix and the cemented bioclasts. The overall mechanical strength is not compromised by the weak ooliths.	Individual ooliths, being soft and crumbly are preferentially weathered as compared to the more resistant shell fragments and the calcite cement. Holes formed by preferential dissolution of the microporous ooliths, leaving behind the shell fragments and the encasing calcite cement, contribute to enhanced salt dissolution and water/moisture retention during intense rainfall.
Magnesian limestone [82,83]	Quarries at Bolsover Moor (Derbyshire, Mansfield (Nottinghamshire) and Anston (South Yorkshire)	Diagenetically altered—the original bioclastic fabric remains intact when the alteration is minimal otherwise a coarse crystalline fabric develops with no relict of its primary depositional structure or fabric.	An interlocking, porous framework with a high proportion of fine sand quartz grains. The crystalline texture along with the highly resistant quartz grains contributes to the durability. The common occurrence of large open or carbonate crystal-lined cavities or vugs as a consequence of recrystallisation, can favour drying.	In urban settings, magnesium salts are more soluble than calcium rich varieties. Once affected by acid rain, the derived magnesium sulphate by-products within the pore spaces have a greater volumetric expansion as compared to gypsum. For the Anston stones, the original bioclastic texture has been preserved despite the olomitization, with the framework consisting of accumulations of abraded, dolomitised, bioclastic fragments. This bioclastic framework with the interconnected pore network compromises the durability of the stones. Affected by surface discoloration, efflorescence, blistering and ultimately severe surface exfoliation.
Kentish ragstone (Lower Greensand Bed) [19]	Kent (~115 to 110 Ma)	Grey in colour, consisting of rounded detrital grains of quartz and the green iron silicate mineral glauconite, with associated bioclasts, cemented by diagenetic calcite.	Hard due to cementing that contributes to the overall strength.	Numerous fractures, identified under optical microscope, distributed throughout the rock contribute towards a high secondary porosity that weakens the rock structure. An inhomogeneous texture and high microporosity that can be attributed to diagenetic recrystallisation contributes to the degradation of the stone.
Sandstone				
Reigate stone (Upper Greensand Bed) [84]	Surrey Lower Cretaceous (105 Ma)	Calcareous sandstone/sandy limestone. Sandy and glauconitic, well cemented with silica and calcite cement. Bioturbation features are common. Beside glauconite, bioclastic debris also comprises the framework, with dominant fine to medium grained quartz.	Variation in the proportion of quartz grains, glauconite and carbonate cement/matrix. Though, if well-cemented, it can be highly durable and requires low maintenance over its life [85]. Additionally, it has a high thermal expansion capacity.	Often soft and weakly compacted and porous, and therefore swells when wet. Exhibits pronounced contour scaling and flaking when exposed (Figure 3) and surface powdering in sheltered areas.
Red and White Mansfield	Nottingham Permian (299–251 Ma)	Dolostone transiting to dolomitic sandstone	Durable due to high quartz sand content and as the sand grains are cemented by the dolomite.	Poor weathering quality because of the calcareous constituents that rapidly destroys the structure of the stone.

Table 2. *Cont.*

Building Stone	Provenance and Age of Formation	Description	Distinguishing Properties Related to Usage	Deterioration that can be Exacerbated by * Climate Change
			Granite	
Aberdeen Scottish Cornish: (Dartmoor, Bodmin Moor (Cheesewring); Cumbria Shap Granite	Variable	Many varieties. Coarsely crystalline with quartz, feldspar and biotite mica with various accessory minerals. Large phenocrysts of K-feldspar may be set in the groundmass (porphyritic texture) but in some cases may be relatively finer grained.	The interlocking crystals provide cohesion which adds strength and makes them suitable for polishing without plucking of the grains. The predominance of silica and other relatively stable minerals make it particularly strong and durable.	Incipient kaolinisation causing the feldspars to become cloudy; deferruginisation resulting in the release of iron from the biotite and its dispersal throughout the rock body, thereby discolouring the rock to some rusty shade due to iron oxide staining. Water leakage through joints when of inferior quality comprising smaller block size. Smearing by white lime from the mortar between such blocks.
Royken	(Norway, ~250 Ma)	An interlocking texture of prominent coarse, feldspar laths seen, set against a background groundmass of finer grey-to-dark-grey quartz, with flakes of black biotite mica with some silvery muscovite mica also present Figure 3C. Textures show some contrasts, with variable size of the feldspar laths from very coarse to slender.	The crystalline texture of the rock with the interlocking mineral grains give it compactness and strength. The reddish colour of the rock can be attributed to the presence of potassium feldspar that adds to the aesthetic. The granite surface is roughened underfoot by the tougher crystals of feldspar, a perfect non-slip surface on a wet day.	As above

* In London, a temperature rise of 3–8 °C during the summer with a precipitation drop by up to 10%, and a winter precipitation increase of up to 60% is predicted by 2080–2099 from a 1981–2000 baseline [5].

3. Deterioration of Building Stones in London: Future Directions of Study

From the previous discussion, it is clear that the building stones in large urban areas such as London are likely to be largely affected by increasing atmospheric pollution compounded with change in rainfall intensity and urban temperatures. The SO_2 concentrations have a strong influence on stone deterioration but its concentration in London has decreased by four times from 220 $\mu g/m^3$ since the period 1960 to 1980 [86]. The decay rate of Portland limestone, a common building stone in London, measured in St Paul's Cathedral in London is 220 μm/year based on the runoff analysis of the calcium ions being added, however, it is lower at 130 μm/year based on the direct measurement of erosion [16]. This can be attributed to differences in the relative contribution of solution and particulate losses to the total loss that is strongly controlled by porosity, often enhanced in weathered samples [27]. Higher porosity offers more surface area for chemical dissolution and increased retention of water further facilitating solution. A decay rate of 35–49 μm/year for the same site in London, over 30 years assessed as surface lowering [28], is still higher compared to the same building rock in other cities such as Dublin and Munich (Figure 2). However, in Munich the assessment was based on available dose–response functions for Portland limestone, pre-derived from exposure experiments at different locations with different climate and pollutant concentrations [29]. In Dublin, the assessment was based on runoff analyses of exposed micro-catchment units, for both freshly quarried and relatively weathered rocks [27]. Beside the surficial runoff, it is very important to consider that the dissolved decay products may be re-precipitated within the pore spaces of the building stones. For all practical purposes, a comparison between different studies may not be always straightforward until an equivalent methodology is used, or diverse methods can be universally calibrated.

It is important to consider that a rise in the level of the oxides of nitrogen acts as catalyst for the formation of sulphates, the concentrations of which have increased two-fold in London over the last fifty years [86]. Consequently, current SO_2 concentrations can still affect stone deterioration, especially in the context of increasing particulate matter. NO_2 can function as an oxidant and significantly increase the reaction rates of the sulfation processes. In addition, NO_x gases can be oxidised by lithotrophic bacteria to nitric acid, which contributes towards stone decay [87]. The annual recession rate of Portland limestone in Munich has dropped down from approximately 12 to 7 μm between 1979 and 2009 (Figure 2), accompanied by a decrease in SO_2 but with an increasing influence of acidity from HNO_3 as a consequence of higher NO_2 from emitted combustion gases, along with rising PM10 particulates [37]. This influence of NO_2 and PM_{10} particulates is reflected in the higher recession rates of the limestone in the traffic hot spots of both Munich and Mainz with higher NO_2 and PM_{10}. Such a study is lacking for London but required to draw better conclusions on the long-term deterioration of Portland limestone.

Higher humidity and rainfall in Munich results in a higher recession rate of Portland limestone as compared to Mainz [29]. Relative humidity is another important factor that needs to be considered as it controls the deposition velocity of SO_2 on building surfaces along with atmospheric concentration and must also affect the deposition of nitrogen oxides and other particulate matter, but notably is overlooked in current studies. As discussed in the previous section, the decay of Portland limestone as a consequence of gypsum formation can be enhanced in London due to predicted wetter winters. Gypsum has a comparatively lower solubility and shows little migration, often accumulating in the pore space [82]. However, with higher moisture and rainfall conditions, there will be an increased solubility and migration of the less soluble gypsum [88], a small amount of which will be re-diluted. Following crystallisation in the fissures or interstitial areas at grain contacts, there is an induction of stress in the construction due to high crystallisation pressure [49]. With less severe winters, mechanical decay processes, such as frost weathering, are retarded, as is evident from the decrease in the annual number of the predicted freeze-thaw cycles in the long term up to 2100 [89]. However, moisture dilatation can be more active under the wetter conditions. This can have a feedback effect as moisture dilatation is intensified in the presence of salts and irreversible in contrast to salt-free systems, accelerating further deterioration as scaling, flaking and crumpling [90]. Such processes of salt crystallisation and moisture dilatation will be seen in other rocks, too, including sandstone and granite, but not adequately investigated although the long term trend of salt weathering for London, estimated on the basis that predicted thenardite–mirabilite, exceeding 10 MPa, clearly indicate an increasing effect [89]. Furthermore, if stones are wetter, the possibility of the freeze-thaw cycling being more effective, cannot be ruled out.

With continued cycles of hydration/salt migration/crystallisation, there will be increased salt penetration and crystallisation in the pore spaces (crypto efflorescence) leading to progressive deterioration. The extent of damage is more profound in the case of crypto efflorescence when compared to efflorescence, the latter being only confined to the surface. Only if the rate of rehydration from within exceeds the rate of water evaporation from the surface, efflorescence will be observed as the surface remains hydrated and evaporation from the surface will continue to take place [88]. Relative humidity < 75% can escalate crystallisation–hydration cycles, so drier, hotter summers in London in the future can be a potential threat, especially for carbonate and sandstone but no estimate is yet available related to their correlation.

Increased moisture content will translate to greater depth of wetting front, with penetration known to exceed a depth of 25 cm in a sandstone rock [91]. With longer and consequent deeper wetting, ion diffusion will become a prominent mechanism shifting emphasis to chemical rather than physical damage, related to salt weathering [62]. However, the delay in the onset of crystallisation as salt remains in solution for longer periods during the prolonged winter wetness, versus the exaggerated and accelerated material decay when it finally occurs after the onset of crystallisation has not yet been assessed. In addition, further studies are required on whether increasing surface moisture will encourage colonisation by algae and other biological agents [92], further aiding in moisture retention at and below the surface [62] for various rock types. Focus should also be given to the specification and application of appropriate mortar, which are more resistant to thermal and moisture changes in an acidic urban environment, to be used in conjunction with flint, slate and bricks in London.

Finally, it remains to be seen in future studies if stone units in masonry should be replaced by concrete blocks, timber or cast stone features that replicate the appearance of natural stone. Such modern substitute materials, while well covered by standards and tests, do not always weather in the same way as natural stone. This can lead to detrimental aesthetic effects and building performance failure. These materials should be replaced on a "like-for-like" basis, using the same type of stones as the original construction or a replacement that is petrographically similar and compatible. Any replacement should consider holistic heritage values in addition to the potential decay of the building material over time.

4. Monitoring Building Stones for Deterioration

The key to long-term preservation of building stones is regular monitoring and maintenance to take timely intervention or replacement to reverse, minimise and arrest further deterioration (Figure 10). Some key methods for monitoring are discussed below.

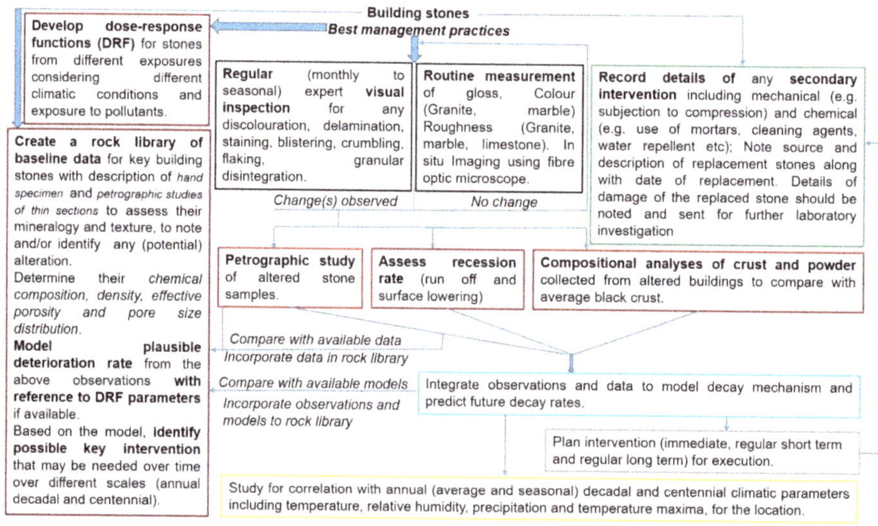

Figure 10. Best management practices for assessment and preservation of building stone deterioration with time.

4.1. Visual Inspection

To assess if a building stone has been affected by weathering, it is important to visually inspect for evidences of deterioration such as blistering, crumbling and flaking, granular disintegration, discolouration (commonly reddening, darkening and general deepening of the original colour), surficial staining (commonly orange or brown related to iron oxide from rusting; blue green streaking from copper carbonate and copper sulphate run off; bright to very dark green related to algal growth) [84]. Deterioration can also be identified on the basis of any change in colour and texture in comparison with the sample from the same quarry, when possible. Any damage observed needs to be categorised to identify the best intervention. For example, a noticeable crack of width 5 to 15 mm or a series of them of ~3 mm can require superficial repair with limited reconstruction [93]. On the other hand, when cracking is more severe with a width of 15–25 mm, considerable replacement and reconstruction may be involved. Visual inspection by the naked eye can be complemented by a magnifying glass inspection. In situ, high resolution (up to 600× magnification) and contrasted images can be obtained using a fibre optic microscope. This requires no surface preparation but can be used to assess any preliminary decay based on the information on the surface texture and morphology, as well as evaluation of cleaning and consolidation interventions [94].

Roughness, gloss and colour are the other key properties to assess surface decay of building stones like marble and granite [33,49,51]. Weathered granite shows noticeable colour variation, a higher surface roughness and lower gloss [51]. A detailed, visual inspection can also be combined with the use of contact surface analysis devices and glossmeters [33]. An increase in surface roughness in a limestone also indicates deterioration, as it is related to the solution of the calcite [49]. There are several ways of conducting these measurements, including using a hand profilometer and 3D laser scanning [95].

4.2. Optical Microscopy and Other Imaging Techniques

Petrographic analyses involving study of thin sections can be very informative. For limestones, the distribution of the bioclasts, matrix and cement can be used to assess crystallinity. For granite, it can be used to identify kaolinization and sericitisation and, therefore, the impacts of chemical weathering, as well as fissurisation in quartz and feldspar to look at any alteration related to stress-related physical weathering. For slate, the identification of inclusions can be a clue to any potential oxidation. Petrographic studies can be complemented with scanning electron microscopy to aid in close visualisation of individual minerals and micro computed X-ray tomography (micro-CT) to give 3D information about the structure and distribution of the metallic minerals [70].

4.3. Mineralogical and Geochemical Methods

Powder X-ray diffraction (XRD) for mineralogical composition and fluorescence (XRF) for chemical composition, coupled with scanning electron microscopy or energy-dispersive X-ray spectroscopy (SEM–EDX), can be used for assessing potential deterioration in building stones. Such elemental–mineralogical composition, correlated with imaging, can indicate alteration related to changes in the fabric, such as enhancement in the porosity of a building rock. Analysing the powder deposit collected from stone surfaces can be informative: the inorganic and carbonaceous fractions of limestone can be used to monitor the soluble salt content. This will not only help to constrain the degradation mechanism but can also correlate to the source of the accumulated particulate matter [44]. Care should be exercised to distinguish any depletion of atmospheric particulate matter, such as nitrate in the analysed powder, to the source with deeper migration of pollutants into the stone structure through ion exchange reaction in the presence of moisture or rainwater Analyses of the powder can employ techniques such as XRD and Fourier transform infrared (FTIR) spectroscopy. XRD and FTIR can also identify if a sulphation process is still active as a result of the interaction between the stones and the atmospheric SO_2 pollutants [95].

In a dolomitic limestone, the dolomite/gypsum ratio can be determined to constrain the decay extent of the stone material as a function of sulphation degree, with value < 2, indicating a higher degree of alteration [75]. For silicate building stones, the depletion associated with silicate phases (SiO_2, Al_2O_3, Na_2O, K_2O), correlated with an increase in sulphates, can be an indicator of the degree of decay [49]. The detection of other chemical ions in the stone samples can indicate presence of secondary pollutants (e.g., oxalates) or application of non-documented treatment to the stones. Analyses of the powder should be routinely compared to that of the reported values for black crust in order to understand if the incoherent powder deposits will eventually evolve into black crusts.

4.4. Other Attributes

The measurement and comparison of physical parameters such as ultrasonic velocity, bulk density, open porosity and water absorption of crystalline rocks such as granite and limestone can be useful [50]. Non-destructive moisture measurement, indicating the distribution of moisture, is often used as a broad indicator of any decay mechanism related to water transfer. The differences found in these parameters between the construction stones and the same rocks in the quarry can indicate any decay of the stones. The density of a rock is directly correlated to its compressibility strength as well its hardness. Although that may not be necessarily related to the durability, the maximum density as in a limestone of 2.74 g/cc, corresponds to a pure, un-porous limestone [79]. In such a variety, with limited pore spaces for circulating fluids and moisture absorption, there will be minimal decay and higher durability. The crystallinity and the compactness of the rock can be assessed on the basis of the ultrasonic velocity while lower porosity can be related to lower accessibility to water and water absorption. In marble too, reduction in the ultrasonic velocity is accompanied by an advancement in the weathering front, with a decrease in the mechanical strength and the Young's modulus [56]. The moisture response of the building stones should be tested both in the presence of hygric moisture (related to relative humidity) and hydric moisture (under water immersion).

Dose–response functions (DRF), expressed as surface recession due to the prevailing attack of SO_2 and its secondary products, can be used for monitoring by comparing with data available for standard material, such as Portland limestone [96] which is a widely used building stone for a number of London heritage sites. DRF can be used for assessment in the context of both former and future pollution loads. Considering the decreased SO_2 concentration, the relative influence of other pollutants has to be considered from a multi-pollutant situation in future studies [29].

5. Conclusions

The predicted changes in the UK climate, especially pronounced for London, indicate warmer and wetter winters, hotter drier summers with increase in summer temperature maxima, extreme rainfall events and higher annual precipitation with a decline in summer but an increase during winter. Such phenomenon will affect the decay of the common building stones used in the heritage sites of London. The deterioration mechanisms and monitoring methods discussed here are not limited to London, but are transferrable to other urban contexts too, although consideration should be given to regional climate variation.

The dissolution of carbonate minerals in limestone will be enhanced due to the predicted increase in precipitation and the rising concentration of air pollutants, such as nitrogen oxides, which may also accumulate as dry deposits on the stone surfaces to be later mobilised by subsequent rainfall. This results in near-surface sulphate crystallisation and efflorescence that leads to exfoliation and crumpling by powder formation, which contributes to the development of internal stresses in the rock mass. Any immature sandstone, such as litharenite, with high porosity and bimodal pore distribution is prone to salt weathering as capillary water uptake and moisture absorption are facilitated by the micropores present. Salt transportation occurs by ion diffusion that is likely to be more pronounced in the future in London due to the wetter, winter months and higher annual precipitation. Due to the prolonged wet winters, the saline solution circulates and interacts extensively with the rock matrix and cement, weakening it considerably, even before the onset of the salt crystallisation. More significantly, for predicted CO_2 concentration of up to 750 ppm by 2099 and decline in traditional air pollutants, 96 to 99.5% of surface recession of carbonate stones can be due to karst effect.

In contrast, marble that deteriorates mostly by microcracking, as cooling is expected to decline with time, will not be as adversely affected (as with the limestone and the sandstone). Any granitic building stone will also remain unaffected due to the expected continued decline of SO_2 concentration with time in London that should reduce the incidence of kaolinisation. Microcracking of feldspars will be greatly reduced too, due to the predicted wetter and warmer winters in London that will lead to a decrease in any freeze-thaw processes. Consequently, access pathways of fluid flow and air pollutants will be limited, thereby further reducing the possibility of kaolinisation and sericitisation. Flint and slate, as well as the red bricks, commonly used as building stones in London will not be significantly affected by predicted climate change. However, care should be taken with the use of mortar that will be required to be more resistant to thermal and moisture changes. Additionally, any mineral impurities that may be present should be noted. As an example, any pyrite in the slate will be more susceptible to oxidation under higher rainfall leading to the deposition of iron sulphate salts.

In the future, studies on London building stones should focus on the assessment of their recession rates and any influence of emerging air pollutants such as NO_2 and PM_{10} particulates. The role of humidity, and consequently, the effect of hotter, drier summers and warmer, wetter winters in London on the crystallisation-hydration cycles for carbonate and sandstone, needs to be better understood. Under the predicted wetter conditions, the role of moisture dilatation may also be enhanced and requires better understanding.

The best practices for the management of building stones in London, also applicable to other cities, is summarised in Figure 10. It involves regular observations for alteration features and routine measurements of gloss, colour and roughness. This can be complemented with further petrographic studies and chemical analyses as necessary. It would be important to create a rock library with baseline data of mineralogy and chemical composition of the common building stones (Figure 10). Finally, dose–response functions (DRF), for different building stones from different exposures that consider different climatic conditions and exposure to pollutants, should be developed.

Author Contributions: The study was conceptualised and designed by S.B., S.A.O. and Y.D.A. The formal analysis and funding acquisition were carried out by S.B. The visualisation was done by S.B. and S.A.O. The original draft of the manuscript was prepared by S.B and reviewed and edited by S.A.O. and Y.D.A. All authors have read and agreed to the published version of the manuscript.

Funding: This research was funded by UCL Global Engagement Funds 2019/20 award number 177785.

Conflicts of Interest: The authors declare no conflict of interest.

References

1. Leissner, J.; Fuhrmann, C. *Cultural Heritage and Climate Change: Are We at the Tipping Point?* Instituto Italiano di Cultura: Brussels, Belgium, 2017; p. 221.
2. Stocker, T.F.; Qin, D.; Plattner, G.K.; Tignor, M.; Allen, S.K.; Boschung, J.; Nauels, A.; Xia, Y.; Bex, V.; Midgley, P.M. (Eds.) *Climate Change 2013: The Physical Science Basis: Working Group I Contribution to the Fifth Assessment Report of the Intergovernmental Panel on Climate Change*; IPCC 2013; Cambridge University Press: Cambridge, UK, 2013.
3. Field, C.B.; Field, C.B.; Barros, V.; Stocker, T.F.; Dahe, Q. (Eds.) *Managing the Risks of Extreme Events and Disasters to Advance Climate Change Adaptation*; IPCC 2012; Cambridge University Press: Cambridge, UK, 2012; p. 582.
4. Fookes, P.G.; Lee, E.M. Climate variation: A simple geological perspective. *Geol. Today* **2007**, *23*, 66–73. [CrossRef]
5. Lowe, J.A.; Bernie, D.; Bett, P.; Bricheno, L.; Brown, S.; Calvert, D.; Clark, R.; Eagle, K.; Edwards, T.; Fosser, G.; et al. *UKCP18 Science Overview Report Version 2.0*; Met Office, Hadley Centre: Exeter, UK, 2018; Updated March 2019.
6. Nathanail, J.; Banks, V.J. Climate change: Implications for engineering geology practice. In *Geological Society London Engineering Geology Special Publications*; Geological Society: London, UK, 2009; Volume 22, pp. 65–82.
7. Viles, H.A. Implications of future climate change for stone deterioration. In *Natural Stone, Weathering Phenomena, Conservation Strategies and Case Studies*; Geological Society: London, UK, 2002; Volume 205, pp. 407–418.
8. Schneider, T.; O'Gorman, P.A.; Levine, X.J. Water vapor and the dynamics of climate changes. *Rev. Geophys.* **2010**, *48*, RG3001. [CrossRef]
9. Satterthwaite, D. Cities' contribution to global warming: Notes on the allocation of greenhouse gas emissions. *Environ. Urban.* **2008**, *20*, 539–549. [CrossRef]
10. UN Climate Neutral Strategy. Greening The Blue. 5 June 2007. Available online: https://www.un.org/en/sections/general/un-and-sustainability/ (accessed on 26 July 2020).
11. Aram, F.; García, E.H.; Solgi, E.; Mansournia, S. Urban green space cooling effect in cities. *Heliyon* **2019**, *5*, e01339. [CrossRef]
12. Aktaş, Y.D.; Stocker, J.; Carruthers, D.; Hunt, J. A Sensitivity Study Relating to Local Urban Climate Modelling within the Built Environment. *Procedia Eng.* **2017**, *198*, 589–599. [CrossRef]

13. Hunt, J.C.; Aktas, Y.D.; Mahalov, A.; Moustaoui, M.; Salamanca, F.; Georgescu, M. Climate change and growing megacities: Hazards and vulnerability. *Proc. Inst. Civ. Eng. Eng. Sustain.* **2017**, *171*, 314–326. [CrossRef]

14. Paris Agreement under the United Nations Framework Convention on Climate Change. In Proceedings of the 21st Conference of the Parties, Paris, France, 12 December 2015.

15. Siegfried, S.; Brimblecombe, P. Editorial to the Special Issue "urban use of rocks". *Environ. Earth Sci.* **2013**, *69*, 1067–1069.

16. Bell, F.G. Durability of carbonate rock as building stone with comments on its preservation. *Environ. Geol.* **1993**, *21*, 187–200. [CrossRef]

17. Robinson, E. A Geology of the British Library. Available online: https://www.webarchive.org.uk/wayback/archive/20100427150116/http://www.bl.uk/reshelp/experthelp/science/science@blevents/futureevents/geology_of_the_bl.pdf (accessed on 14 October 2016).

18. Garcia-del-Cura, M.A.; Benavente, D.; Martinez-Martinez, J.; Cueto, N. Sedimentary structures and physical properties of travertine and carbonate tufa building stone. *Constr. Build. Mater.* **2012**, *28*, 456–467. [CrossRef]

19. Sabbioni, C.; Bonazza, A.; Zamagni, J.; Ghedini, N.; Grossi, C.M.; Brimblecombe, P. The Tower of London: A case study of stone decay in an urban area. In *Air Pollution and Cultural*; Saiz, C., Ed.; AA Balkema: London, UK, 2004; pp. 57–62.

20. Building Stones and the Landscape British Geological Survey. Available online: https://www.bgs.ac.uk/discoveringGeology/geologyOfBritain/limestoneLandscapes/resourcesConflictsSustainability/buildingStones.html (accessed on 7 March 2020).

21. Palmer, T.J. Limestone petrography and durability in English Jurassic freestones. In Proceedings of the England's Heritage Stone: Proceedings of a Conference, Tempest Anderson Hall, York, UK, 15–17 March 2005; pp. 65–78.

22. Ingham, J.P. Predicting the frost resistance of building stone. *Q. J. Eng. Geol. Hydrogeol.* **2005**, *38*, 387–399. [CrossRef]

23. Cherblanc, F.; Berthonneau, J.; Bromblet, P. Role of hydro-mechanical coupling in the damage process of limestones used in historical buildings. In Proceedings of the 13th International Congress on the Deterioration and Conservation of Stone, Glasgow, UK, 6–10 September 2016.

24. Morse, J.W.; Arvidson, R.S. The Dissolution Kinetics of Major Sedimentary Carbonate Minerals. *Earth Sci. Rev.* **2002**, *58*, 51–84. [CrossRef]

25. Reddy, M.M. Acid rain damage to carbonate stones: A quantitative assessment based on the aqueous geochemistry of rainfall runoff from stone. *Earth Surf. Process. Landf.* **1988**, *13*, 335–354. [CrossRef]

26. Grossi, C.M.; Murray, M. Characteristics of carbonate building stones that influence the dry deposition of acidic gases. *Constr. Build. Mater.* **1999**, *13*, 101–108. [CrossRef]

27. Cooper, T.P.; O'Brien, P.F.; Jeffrey, D.W. Rates of Deterioration of Portland Limestone in an Urban Environment. *Stud. Conserv.* **1992**, *37*, 228–238.

28. Inkpen, R.; Viles, H.; Moses, C.; Baily, B. Modelling the impact of changing atmospheric pollution levels on limestone erosion rates in central London, 1980–2010. *Atmos. Environ.* **2012**, *61*, 476–481. [CrossRef]

29. Auras, M.; Beer, S.; Bundschuh, P.; Eichhorn, J.; Mach, M.; Scheuvens, D.; Schorling, M.; Schumann, J.V.; Snethlage, R.; Weinbruch, S. Traffic-related emissions and their impact on historic buildings: Implications from a pilot study at two German cities. *Environ. Earth Sci.* **2013**, *69*, 1135–1147. [CrossRef]

30. Lipfert, F.W. Atmospheric damage to calcareous stones: Comparison and reconciliation of recent experimental findings. *Atmos. Environ.* **1989**, *23*, 415–429. [CrossRef]

31. Bonazza, A.; Palmira, M.; Sabbioni, C.; Grossi, C.; Brimblecombe, P. Mapping the impact of climate change on surface recession of carbonate buildings in Europe. *Sci. Total Environ.* **2009**, *407*, 2039–2050. [CrossRef]

32. Doehne, E. Salt weathering: A selective review. In *Geological Society London Special Publication Natural Stone, Weathering Phenomena, Conservation Strategies and Case Studies*; Siegesmund, S., Weiss, T., Vollbrecht, A., Eds.; Geological Society of London: London, UK, 2002; Volume 205, pp. 51–64.

33. Gulotta, D.; Bertoldi, M.; Bortolotto, S.; Fermo, P.; Piazzalunga, A.; Toniolo, L. The Angera stone: A challenging conservation issue in the polluted environment of Milan (Italy). *Environ. Earth Sci.* **2013**, *69*, 1085–1094. [CrossRef]

34. Smith, B.J.; Viles, H.A. Rapid, catastrophic decay of building limestones: Thoughts on causes, effects and consequences. In *Heritage Weathering and Conservation*; Taylor and Francis: London, UK, 2003; pp. 191–197.

35. Török, A. Morphology and detachment mechanism of weathering crusts of porous limestone in the urban environment of Budapest. *Cent. Eur. Geol.* **2007**, *50*, 225–240. [CrossRef]

36. Haynie, F.H. Theoretical model of soiling of surfaces by airborne particles in aerosols: Research, risk assessment and Control Strategies. In Proceedings of the Second US-Dutch International Symposium, Williamsburg, VA, USA, 19–4 May 1986; Lee, S.D., Ed.; Lewis Publishers: Williamsburg, VA, USA, 1986.

37. Grossi, C.M.; Esbert, R.M.; Alonso, F.J. Soiling of building stones in urban environments. *Build. Environ.* **2003**, *38*, 147–159. [CrossRef]

38. Ascaso, C.; Wierzchos, J.; Souza-Egipsy, V.; de los Ríos, A.; Delgado, J.; Rodrigues, J. In situ evaluation of the biodeteriorating action of microorganisms and the effects of biocides on carbonate rock of the Jeronimos Monastery (Lisbon). *Biodeterior. Biodegrad.* **2002**, *49*, 1–12. [CrossRef]

39. Schiavon, N. Assessment of Building Stone Decay: A Geomorphological Approach. In Proceedings of the International Conference of Stone Cleaning and the Nature, Soiling and Decay Mechanisms of Stone, Edinburgh, UK, 14–16 April 1992; Webster, R.G.M., Ed.; Donhead Publishing: Shaftesbury, UK, 1992.

40. Sterflinger, K.; Piñar, G. Microbial deterioration of cultural heritage and works of art—Tilting at windmills? *Appl. Microbiol. Biotechnol.* **2013**, *97*, 9637–9646. [CrossRef] [PubMed]

41. Bonazza, A.; Brimblecombe, P.; Grossi, C.M.; Sabbioni, C. Carbon in black layers at the Tower of London. *Environ. Sci. Technol.* **2007**, *41*, 4199–4204. [CrossRef]

42. Eyssautier-Chuine, S.; Marin, B.; Thomachot-Schneider, C.; Fronteau, G.; Schneider, A.; Gibeaux, S.; Vazquez, P. Simulation of acid rain weathering effect on natural and artificial carbonate stones. *Environ. Earth Sci.* **2016**, *75*, 748. [CrossRef]

43. Hallmann, C.; Stannek, L.; Fritzlar, D.; Hause-Reitner, D.; Friedl, T.; Hoppert, M. Molecular diversity of phototrophic biofilms on building stone. *FEMS Microbiol. Ecol.* **2013**, *84*, 355–372. [CrossRef]

44. Vázquez, P.; Luque, A.; Alonso, F.J.; Grossi, C.M. Surface changes on crystalline stones due to salt crystallisation. *Environ. Earth Sci.* **2013**, *69*, 1237–1248. [CrossRef]

45. Shushakova, V.; Fuller, E.R., Jr.; Heidelbach, F.; Mainprice, D.; Siegesmund, S. Marble decay induced by thermal strains: Simulations and experiments. *Environ. Earth Sci.* **2013**, *69*, 1281–1297. [CrossRef]

46. Fort, R.; Varas, M.J.; de Buergo, M.A.; Martin-Freire, D. Determination of anisotropy to enhance the durability of natural stone. *J. Geophys. Eng.* **2011**, *8*, S132–S144. [CrossRef]

47. Goudie, A.S. Laboratory simulation of the 'wick effect' in salt weathering of rock. *Earth Surf. Process. Landf.* **1986**, *11*, 275–285. [CrossRef]

48. Schiavon, N. Kaolinisation of granite in an urban environment. *Environ. Geol.* **2007**, *52*, 399–407. [CrossRef]

49. Graue, B.; Siegesmund, S.; Oyhantcabal, P.; Naumann, R.; Licha, T.; Simon, K. The effect of air pollution on stone decay: The decay of the Drachenfels trachyte in industrial, urban, and rural environments—A case study of the Cologne, Altenberg and Xanten cathedrals. *Environ. Earth Sci.* **2013**, *69*, 1095–1124. [CrossRef]

50. Perez-Monserrat, E.M.; de Buergo, M.A.; Gomez-Heras, M.; Muriel, M.J.V.; Gonzalez, R.F. An urban geomonumental route focusing on the petrological and decay features of traditional building stones used in Madrid, Spain. *Environ. Earth Sci.* **2013**, *69*, 1071–1084. [CrossRef]

51. Sousa, L.M.O. The influence of the characteristics of quartz and mineral deterioration on the strength of granitic dimensional stones. *Environ. Earth Sci.* **2013**, *69*, 1333–1346. [CrossRef]

52. Gomez-Heras, M.; Smith, B.J.; Fort, R. Influence of surface heterogeneities of building granite on its thermal response and its potential for the generation of thermoclasty. *Environ. Geol.* **2008**, *56*, 547–560. [CrossRef]

53. Wilson, M.J. Weathering of the primary rock-forming minerals: Processes, products and rates. *Clay Miner.* **2004**, *39*, 233–266. [CrossRef]

54. Gräf, V.; Jamek, M.; Rohatsch, A.; Tschegg, E. Effects of thermal-heating cycle treatment on thermal expansion behavior of different building stones. *Rock Mech. Min. Sci.* **2013**, *64*, 228–235. [CrossRef]

55. Freire-Lista, D.M.; Fort, R.; Varas-Muriel, M.J. Thermal stress-induced microcracking in building granite. *Eng. Geol.* **2016**, *206*, 83–93. [CrossRef]

56. Ruedrich, J.; Kirchner, D.; Seidel, M.; Siegesmund, S. Deterioration of natural building stones induced by salt and ice crystallisation in the pore space as well as hygric expansion processes. In *Geowissenschaften und Denkmalpflege. Zeitschrift Deutsche Geologische Gesellschaft*; Siegesmund, S., Auras, M., Ruedrich, J., Snethlage, R., Eds.; Schweizerbart Science Publishers: Stuttgart, Germany, 2005; Volume 156/1, pp. 59–73.

57. Wellmann, H.W.; Wilson, A.T. Salt Weathering, a Neglected Geological Erosive Agent in Coastal and Arid Environments. *Nature* **1965**, *205*, 1097–1098. [CrossRef]

58. Ruedrich, J.; Knell, C.; Enseleit, J.; Rieffel, Y.; Sigesmund, S. Stability assessment of marble statuaries of the Schlossbrucke (Berlin, Germany) based on rock strength measurements and ultrasonic wave velocities. *Environ. Earth Sci.* **2013**, *69*, 1451–1469. [CrossRef]

59. Stück, H.; Koch, R.; Siegesmund, S. Petrographical and petrophysical properties of sandstones: Statistical analysis as an approach to predict material behaviour and construction suitability. *Environ. Earth Sci.* **2013**, *69*, 1299–1332. [CrossRef]

60. Godts, S.; Hayen, R.; De Clercq, H. Investigating salt decay of stone materials related to the environment, a case study in the St. James church in Liège, Belgium. *Stud. Conserv.* **2017**, *62*, 329–342. [CrossRef]

61. Charola, A.E.; Pühringer, J.; Steiger, M. Gypsum: A review of its role in the deterioration of building materials. *Environ. Geol.* **2007**, *52*, 339–352. [CrossRef]

62. Stück, H.; Plagge, R.; Siegesmund, S. Numerical modeling of moisture transport in sandstone: The influence of pore space, fabric and clay content. *Environ. Earth Sci.* **2013**, *69*, 1161–1187. [CrossRef]

63. McCabe, S.; Smith, B.J.; McAlister, J.J.; Gomez-Heras, M.; McAllister, D.; Warke, P.A.; Curran, J.M.; Basheer, P.A.M. Changing climate, changing process: Implications for salt transportation and weathering within building sandstones in the UK. *Environ. Earth Sci.* **2013**, *69*, 1225–1235. [CrossRef]

64. Sebastián, E.; Cultrone, G.; Benavente, D.; Fernandez, L.L.; Elert, K.; Rodriguez-Navarro, C. Swelling damage in clay-rich sandstones used in the church of San Mateo in Tarifa (Spain). *J. Cult. Herit.* **2008**, *9*, 66–76. [CrossRef]

65. Smith, B.J.; Turkington, A.V.; Warke, P.A.; Basheer, P.A.M.; McAlister, J.J.; Meneely, J.; Curran, J.M. Modelling the rapid retreat of building sandstones: A case study from a polluted maritime environment. In *Natural Stone, Weathering Phenomenon, Conservation Strategies and Case Studies*; Seigesmund, S., Weiss, T., Vollbrecht, A., Eds.; Geological Society: London, UK, 2002; Volume 205, pp. 347–362.

66. Caneva, G.; Nugari, M.P.; Salvadori, O. *Biology in the Conservation of Works of Art*; ICCROM: Roma, Italy, 1991.

67. Jain, A.; Bhadauria, S.; Kumar, V.; Chauhan, R.S. Biodeterioration of sandstone under the influence of different humidity levels in laboratory conditions. *Build. Environ.* **2009**, *44*, 1276–1284. [CrossRef]

68. Bjell, T.; Thorseth, I.H. Comparative studies of the lichen-rock interface of four lichens in Vingen, western Norway. *Chem. Geol.* **2002**, *192*, 81–98. [CrossRef]

69. Tiennot, M.; Mertz, J.-D.; Bourgès, A. Influence of Clay Minerals Nature on the Hydromechanical and Fracture Behaviour of Stones. *Rock Mech. Rock Eng.* **2019**, *52*, 1599–1611. [CrossRef]

70. Ashurst, J.; Williams, G.; Bourgès, A. Flint and the Conservation of Flint Buildings. The Building Conservation Directory. 2005. Available online: https://www.buildingconservation.com/articles/flint/flint.htm (accessed on 7 May 2020).

71. Cardenes, V.; Cnuddle, V.; Cnuddle, J.P. Petrography of roofing slate for quality assessment. In Proceedings of the 15th Euroseminar on Microscopy Applied to Building Materials, Delft, The Netherlands, 17–19 June 2015.

72. Cárdenes, V.; Rubio-Ordonez, A.; Monterroso, C.; Calleja, L. Geology and geochemistry of Iberian roofing slates. *Geochemistry* **2013**, *73*, 373–382. [CrossRef]

73. Gómez-Fernández, F.; Castaño, M.A.; Bauluz, B.; Ward, C.R. Optical microscope and SEM evaluation of roofing slate fissility and durability. *Mater. De Construcción* **2009**, *59*, 91–104.

74. Demarco, M.M.; Oyhantcabal, P.; Stein, K.M.; Siegesmund, S. Dolomitic slates from Uruguay: Petrophysical and petromechanical characterization and deposit evaluation. *Environ. Earth Sci.* **2013**, *69*, 1361–1395. [CrossRef]

75. Siedel, H. Magnesium sulphate salts on monuments in Saxony (Germany): Regional geological and environmental causes. *Environ. Earth Sci.* **2013**, *69*, 1249–1261. [CrossRef]

76. Szemerey-Kiss, B.; Török, A.; Siegesmund, S. The influence of binder/aggregate ratio on the pore properties and strength of repair mortars. *Environ. Earth Sci.* **2013**, *69*, 1439–1449. [CrossRef]

77. Henry, A.; McCaig, I.; Willett, C.; Godfraind, S.; Stewart, J. (Eds.) Historic England Practical Building Conservation. In *Earth, Brick & Terracotta*; Ashgate/Routledge: Farnham, UK, 2015.

78. Bates, S.J. A Critical Evaluation of Salt Weathering Impacts on Building Materials at Jazirat al Hamra, UAE. *Geoverse* **2010**, 1758–3411. Available online: http://geoverse.brookes.ac.uk/article_resources/batesSJ/batesSJ.htm (accessed on 9 July 2020).

79. Bide, T.P.; Barron, A.J.M.; Evans, D.J. *An Assessment of the Aggregate Properties of the Lower Lincolnshire Limestone in South Lincolnshire and Surrounding Areas*; British Geological Survey Commissioned Report: CR/15/083; Keyworth, Nottingham British Geological Survey: Nottingham, UK, 2015; 22p.

80. Palmer, T.J. Understanding the Weathering Behaviour of Caen Stone. *J. Archit. Conserv.* **2008**, *14*, 45–54. [CrossRef]

81. Historic England Strategic Stone Study A Building Stone Atlas of East Sussex. Available online: https://www.southdowns.gov.uk/wp-content/uploads/2016/12/East_Sussex_Building_Stone_Atlas.pdf (accessed on 7 May 2020).

82. Lott, G.K.; Cooper, A.H. Field guide to the building limestones of the upper Permian Cadeby formation (magnesian limestone) of Yorkshire. In Proceedings of the England's Heritage Stone: Proceedings of a Conference, Tempest Anderson Hall, York, UK, 15–17 March 2005; pp. 80–89.

83. Lott, G.K.; Richardson, C. Yorkshire stone for building the Houses of Parliament (1839–1852). *Proc. Yorks. Geol. Soc.* **1997**, *51*, 265–272. [CrossRef]

84. Heritage, E. *English Heritage Practical Building Conservation Stone*; Odgers, D., Henry, A., Eds.; Ashgate/Routledge: Farnham, UK, 2012.

85. Michette, M.; Viles, H.; Vlachou-Mogire, C.; Angus, I. Assessing the Long-term Success of Reigate Stone Conservation at Hampton Court Palace and the Tower of London. *Stud. Conserv.* **2020**, 1–8. [CrossRef]

86. Wiese, U.; Behlen, A.; Steiger, M. The influence of relative humidity on the SO_2 deposition velocity to building stones: A chamber study at very low SO_2 concentration. *Environ. Earth Sci.* **2013**, *69*, 1125–1134. [CrossRef]

87. Fuchs, G. *Allgemeine Mikrobiologie*; Georg Thieme: Stuttgart, Germany, 2006; pp. 321–340.

88. Colston, B.J.; Watt, D.S.; Munro, H.M. Environmentally-induced stone decay: The cumulative effects of crystallization–hydration cycles on a Lincolnshire oopelsparite limestone. *J. Cult. Herit.* **2001**, *4*, 297–307. [CrossRef]

89. Brimblecombe, P.; Grossi, C.M. Millennium-long recession of limestone facades in London. *Environ. Geol.* **2008**, *56*, 463–471.

90. Snethlage, R.; Wendler, E. Moisture cycles and sandstone degradation. In *Saving Our Architectural Heritage: Conservation of Historic Stone Structures*; Baer, N.S., Snethlage, R., Eds.; John Wiley and Sons Ltd.: London, UK, 1997; pp. 7–24.

91. McAllister, D.; McCabe, S.; Srinivasan, S.; Smith, B.J.; Warke, P.A. Moisture dynamics in building sandstone: Monitoring strategies and implications for transport and accumulation of salts. In *Salt Weathering on Buildings and Stone Sculptures 2011*; Ioannou, I., Theodoridou, M., Eds.; University of Cyprus: Nicosia, Cyprus, 2011; pp. 39–46.

92. Cutler, N.; Viles, H. Eukaryotic microorganisms and stone biodeterioration. *Geomicrobiol. J.* **2010**, *27*, 630–646. [CrossRef]

93. Ellison, R.A.; McMillan, A.A.; Lott, G.K. *Ground Characterisation of the Urban Environment: A Guide to Best Practice*; Research Report RR/02/05; British Geological Survey: Keyworth, Notthingham, 2002; p. 37.

94. Menéndez, B. Non-Destructive Techniques Applied to Monumental Stone Conservation. *Non-Destr. Test* **2016** [CrossRef]

95. Tambe, S.; Gauri, K.L.; Li, S.; Cobourn, W.G. Kinetic study of sulfur dioxide reaction with dolomite. *Environ. Sci. Technol.* **1991**, *25*, 2071–2075. [CrossRef]

96. Watt, J.; Jarrett, D.; Hamilton, R. Dose-response functions for the soiling of heritage materials due to air pollution exposure. *Sci. Total Environ.* **2008**, *400*, 415–424. [CrossRef] [PubMed]

Review

Policy Framework for Energy Retrofitting of Built Heritage: A Critical Comparison of UK and Turkey

Negin Jahed [1], **Yasemin D. Aktaş** [2,3,*], **Peter Rickaby** [3,4] **and Ayşe Güliz Bilgin Altınöz** [1]

[1] Department of Architecture, Middle East Technical University, 06800 Ankara, Turkey;
negin.jahed@metu.edu.tr (N.J.); bilging@metu.edu.tr (A.G.B.A.)

[2] Department of Civil, Environmental and Geomatic Engineering (CEGE), University College London (UCL),
London WC1E 6BT, UK

[3] UK Centre for Moisture in Buildings (UKCMB), London WC1H 0NN, UK;
peterrickabyconsultancy@gmail.com

[4] Institute of Environmental Design and Engineering (IEDE), University College London (UCL),
London WC1H 0NN, UK

* Correspondence: y.aktas@ucl.ac.uk

Received: 31 May 2020; Accepted: 22 June 2020; Published: 26 June 2020

Abstract: Energy efficiency is one of the most prominent global challenges of our era. Heritage buildings usually have a poor energy performance, not necessarily because of their intrinsic constructive features but due to their mostly dilapidated condition owed to age and previous damage, exacerbated by other factors such as the limited maintenance allowed by the restrictive legal framework and/or residents not being able to afford retrofit. On both national and international levels, energy efficiency measures are considered the key to answering the global challenge of climate change. This article aims to provide a critical discussion of the policy framework for energy retrofitting targeting built heritage in the UK and in Turkey. To this end, the development of guidance and legislation on cultural heritage, energy efficiency and climate change in both countries were thoroughly reviewed, and the retrofit incentives and constraints were determined in order to identify existing policy gaps and potential problems with implementation in the realm of energy retrofitting and climate resiliency of heritage buildings. As a result of a critical comparative analysis, the paper is concluded with suggestions on policy frames for the retrofitting of heritage buildings for improved energy efficiency.

Keywords: built heritage retrofit; energy-efficient retrofit policy; conservation policy; UK; Turkey

1. Introduction

It is claimed that the current energy use trends could lead to a 2 °C rise in global temperatures by 2030 [1], which is the level considered by scientists as a tipping point for climate catastrophe [2]. Buildings' energy use is estimated to be responsible for more than 40% of all energy use per nation on average [3]. The situation is considered to be more critical for heritage buildings, which despite their environmental credentials in terms of passive heating/cooling, lighting, ventilation, and good orientation, due to age, lack of maintenance, and previous damage, mainly, may demonstrate a poor energy performance that makes them more vulnerable to the consequences of changing climate [4,5]

Although reducing energy usage in buildings is considered key in national and international efforts to minimise carbon dioxide (CO_2) emissions, buildings with heritage values are often excluded from policies and regulatory frameworks on buildings' energy use reduction. A clear example of such exemptions on the international level is stated in the European Commission Directives; 2010/31/EU on the energy performance of buildings [6] and 2012/27/EU on energy efficiency [7]. According to these directives, officially protected buildings and monuments due to their special architectural and historical merits may be excluded from the energy requirements. On the other hand, in the amended directive 2018/844/EU [8], the 'research' for and 'testing' of new solutions for improving energy efficiency of historical buildings are encouraged provided cultural values are preserved. In the case of Turkey, national legislation such as the Turkish Energy Efficiency Law No: 5627 also highlights the exclusion of "protected buildings or monuments" from the scope of the law [9]. Besides, the Turkish National Energy Efficiency Action Plan (NEEAP), released in 2018, highlights the importance of developing a national energy efficiency roadmap to ensure energy demand in various sectors, including buildings, meet global target levels for sustainability. While the action plan calls it a strategic goal to 'reduce building energy demand and carbon emissions; scale up environment-friendly buildings', it does not concern itself with the heritage building stock [10].

This study seeks to assess regulatory frameworks in the UK addressing standards for energy retrofitting of built heritage and then compare these with the existing situation in Turkey. UK was selected as the benchmark case here for a number of reasons. Firstly, the UK's existing housing stock is still one of the oldest and least energy-efficient housing stocks in Europe, 20% of which is composed of pre-1919 homes and another almost 20% were constructed between 1920 and 1939 [11]. Secondly, retrofitting heritage buildings, particularly historic residential buildings, is being given high importance in the UK following the government's commitment on reducing national energy use to achieve 2050 emission reduction target [12]. Thirdly, the UK is one of the first countries in recognizing the economic and security threats of climate change through the Climate Change Act launched in 2008 [13] and has been among the most successful developed countries at growing its economy while reducing emissions [14]. Fourthly, UK policies on energy performance of buildings have been developed towards the strategies to overcome the future overheating problem in building stock [15]. This problem, specifically, makes the UK a comparable case with Turkey, where climate change impacts are mainly characterised by strong warming trends [16].

We believe regulatory policies on energy-efficient retrofitting of historical buildings should be country or even region specific. However, the tensions on the intersection of energy and conservation philosophies are of broader concern. Developing new policy frameworks in this realm, would only be possible in light of an appraisal of both the constraints and incentives in the existing regulatory framework of each country. Several analyses have already been carried out to assess Turkey's current energy efficiency policies [17–19], however, none of them addresses the lack of legal and technical legislative frameworks for improving energy efficiency of historic buildings. The limited number of studies concerning energy-efficient retrofitting of historic buildings in Turkey in comparison to European practices e.g., [20,21] confirm that Turkey needs to be more engaged in research and development activities to be able to increase public awareness and close policy gaps in this field [20]. In this study, we expand on these analyses by discussing the position of energy efficiency within the built heritage conservation legislative and technical guidance frameworks. This paper is organised as follows: Section 2 describes the methodology applied in the study. Section 3 provides a background on national energy efficiency and heritage conservation policies in each country, while analysing the incentives and constraints for retrofit both in the case of the UK and Turkey. In Section 4, a critical comparison of the two countries' policy frameworks is presented and six main suggestions/lessons are pointed out. Following the discussion of other critical factors which play a defining role in the eventual viability/efficiency of policies on the energy efficiency of heritage structures in Section 5, the paper is concluded in Section 6.

2. Methodology

In order to capture the developmental process of policies, the data collection for this study began with mapping the landmark national policies on cultural heritage conservation, energy efficiency, and climate change for both the UK and Turkey in a chronological order. In the international level, only the policies with a specific focus on energy efficiency for the built heritage sector were considered to determine the level of alignment between these two countries with international legislative frameworks. The published policy documents, their end goal, and related institutional structures in both countries were gathered from (a) international and multinational sources, (b) government websites, and (c) non-governmental organisations' websites. In order to provide an in-depth understanding of the retrofit incentives and constraints, various information sources, such as research and review articles, project reports, and conference proceedings focused on energy retrofit policy frameworks for built heritage from both UK and Turkey were identified and accessed through widely used research databases, i.e., ScienceDirect, Web of Science, and SpringerLink. Further, retrofit practitioners or experts from academia, from both Turkey and the UK, provided access to country-specific information sources and in-country contacts to supplement the information collated through web-search. In order to establish an even and reliable basis for the review exercise, we covered sources falling under any one of the categories listed below:

1. In terms of level and type of policy framework: law, regulation, standard, action plan;
2. In terms of institutional structure: governmental and non-profit non-governmental organisations (NGOs);
3. In terms of sector coverage: existing building sector and built heritage subsector.

As a result of this initial search, in order to reveal the position of each country in terms of the overall policy framework for the energy-efficient retrofit of built heritage, we categorised the policy instruments based on where they fall in the following three key streams of developments:

1. Regulatory schemes including legislations, building codes, performance standards, energy labels, energy efficiency obligations, and action plans.
2. Market-based/financial schemes including taxes, finance programmes, loans.
3. Informative approaches including voluntary programmes, awareness raising campaigns, programmes, competitions, online decision-making tools, and published handbooks.

3. Background to the Policy Frameworks: Decoding the Retrofit Incentives and Constraints

3.1. UK's Legal and Administrative Policies on Improving Energy Performance of Buildings: A Chronological Review

UK's energy policy development trends can be classified into four distinct phases of (1) Energy conservation phase (1973–1981), (2) Energy efficiency phase (1981–2000), (3) Energy efficiency and environmental awareness phase (2000–2010), and (4) Near-zero carbon phase (2010–Present). Like many other countries all around the world, under the pressures of oil crises in 1973 and 1979, as well as raising energy prices, the UK's energy conservation phase started in the 1970s. Since 1981, the policies have shifted from 'energy conservation' towards 'energy efficiency' and several regional energy efficiency offices were set up at this time. 'Conservation' meant doing without things, which was not attractive in a consumer-led economy. Instead, 'efficiency' was about doing more with the same amount of resources, which makes it easier to sell it to the public emphasizing the economic and social benefits, warmer houses, lower bills, and higher productivity. The year 1983 is known as the UK's golden age of energy efficiency [22]. Underpinned by the establishment of the Intergovernmental Panel on Climate Change (IPCC) (1988), the UK's first meeting to tackle climate change was held in 1988 and one year later in 1989, energy efficiency was positioned as the central means of delivering emissions reduction. Climate policies gradually gained importance in this period with new programmes and energy prices fell. The 1990s were known as the era of energy efficient appliances and energy rating standards. An Energy Labeling Directive and Standard Assessment Procedure (SAP) were launched in 1992 and 1996, respectively, for the energy rating of dwellings. First building energy efficiency regulation under Part L of Schedule1 of Building Regulations in England and Wales was launched in 1990 [23]. Part L comprised four components referring to both existing and new buildings in residential and commercial sectors and in compliance with the EU Directives the 2002 revised version of the regulation addressed decarbonizing targets. In the late 1990s, the environmental concerns started to emerge through the integration of energy efficiency and climate policies. Government's new climate change programmes were required to comply with the Kyoto Protocol in 1998. At this time, the focus of building performance measures shifted from energy performance towards low carbon dioxide emission. The 2000s saw an explosion of developed regulations in the three main areas of energy efficiency, climate change, and renewable energy. Among the most important ones, are the Royal Commission on Environmental Pollution (2000), the new Climate Change Levy (2001), the 2003 Energy White Paper "Our Energy Future creating a Low Carbon Economy" (the first energy policy statement in 20 years), Energy Efficiency Action Plan [24], and Climate Change Act (2008). Following the requirements of the European Commission Directives (2010/31/EU and 2012/27/EU), UK's energy efficiency policies turned to meet 'near-zero carbon building' standards. The concept of energy-efficient retrofit as a way of improving energy performance of existing buildings appeared in this era. Since 2010, besides the Green Deal there have been several government-sponsored programmes designed to investigate or promote retrofit. There was 'Retrofit for the Future' and 'Scaling Up Retrofit' and now the 'Whole-House Retrofit' competition and a whole series of retrofit supply chain pilots. The Greater London Authority (GLA) has also run domestic retrofit programmes, as have the devolved Scottish and Welsh Governments. None of them focus on historical/listed buildings differently to other buildings, beyond mentioning that they are subject to the requirements of the planning legislation and the Building Regulations Part L.

3.2. UK's Legal and Administrative Policies on Built Heritage Conservation: A Chronological Review

The UK's strength in building conservation has roots in the 19th century, but really took off from the 1930s onwards, when development blight and mass demolition of significant buildings, followed by considerable bomb destruction during WWII, led to social and Governmental response. There was a need to set up preservation schemes starting from empowering local authorities. Empowering local authorities in the UK dates back to early 1930 when the Ancient Monuments Act (1931) was enforced, by which the concept of the "conservation area" was introduced into protective legislation. In 1932 the Town and Country Planning Act introduced Building Preservation Orders to be served by local authorities on historic buildings including occupied dwellings. The system of grading became more specific in the late 1940s. The responsibilities of local authorities were extended towards offering grants for repairing listed or unlisted buildings in 1962 under the Historic Buildings Act. Consequently, in 1968 the Town and Country Planning Act required owners to obtain Listed Building Consent from local authorities. In the 1970s different campaigns for saving endangered historic buildings from redevelopment and demolition activities gained momentum, among which are Local Covent Garden Fruit Campaign (1971) and Save Britain's Heritage (1975). The 1980s was the era that the majority of the UK's important conservation institutions were founded, such as the Association of Conservation Officers in 1982 (later in 1997 reformed as the Institute of Historic Building Conservation), and English Heritage in 1983. General planning legislation was separated from conservation legislation under the Planning (Listed Buildings and Conservation Areas) Act in 1990. The establishment of numerous funding organisations for heritage projects occurred in the 1990s, like the Heritage Lottery Fund (1994), facilitated through the National Lottery Act (1993). The Royal Commission on the Historical Monuments of England (a government advisory body), which was responsible for documenting buildings and monuments of archaeological, architectural and historical importance, was merged with English Heritage in 1999.

The 2000s can be characterised by the explosion of published works and reports in the area of conservation, such as 'Power of Place: The Future of the Historic Environment' [25], 'State of the Historic Environment Report' [26], 'Conservation Principles, Policies, and Guidance for the Sustainable Management of the Historic Environment' [27], and etc. In the 2010s, the UK's conservation community witnessed fundamental transformations of its institutional structure. The first guidance for application of Part L of the Building Regulations to Historic and Traditionally Constructed Buildings was published by English Heritage in 2011. The National Heritage List for England as the first publicly-searchable database with the official records of heritage assets (including listed buildings, scheduled monuments, registered parks and gardens, registered battlefields, and protected wrecks and excluding conservation areas) was launched in 2011. The National Heritage Protection Plan (NHPP), comprising a framework for heritage protection based on a clear set of priorities, published by English Heritage in 2012. In 2015, English Heritage was divided into two separate organisations of Historic England and the English Heritage Trust. English Heritage Trust, as a new independent charity, is responsible for looking after the National Heritage Collection. Currently, Historic England continues as an arms-length body that looks after listing, planning, grants, research, advice, and public information.

Heritage buildings in the UK signify pre-1919 buildings, which are sometimes also referred to as 'historic', 'heritage', and 'conservation' buildings, and sometimes as 'older properties' [28]. The Sustainable Traditional Buildings Alliance (STBA) defines heritage property as, "a property that is generally of solid wall or solid timber frame construction, built before 1919". Although in the UK historic buildings are exempt from full compliance, they must still attempt to "improve energy efficiency as far as is reasonably practicable [29]" as "an informed approach can achieve significant energy efficiency improvements" [30]. The decision as to the level of intervention often remains at the discretion of the building owner and their professional advisors, in liaison with the local conservation officer. Overall, there are several levels of protection of older buildings in the UK:

- Buildings that are 'Listed' as of special architectural or historic interest must be maintained by their owners and cannot be altered without Listed Building Consent, which is obtained from the local authority, advised by the Local Historic Buildings Officer/Advisor and by Historic England.
- Buildings in Conservation Areas, designated by local planning authorities, are subject to a planning consent from the local planning authority, advised by the local Historic Buildings Officer/Advisor and by Historic England.
- Buildings in National Parks and Areas of Outstanding Natural Beauty are protected in a similar way to those in Conservation Areas. Scheduled Ancient Monuments and buildings at World Heritage Sites are treated similarly to listed buildings.

3.3. UK's Regulatory Approaches on Energy-Efficient Retrofitting of Built Heritage

BS 7913:2013 Guide to the Conservation of Historic Buildings is the first British Standard to address the conservation and energy efficiency of heritage buildings with historically appropriate materials and techniques [31]. In 2017, EN 16883:2017 [32] was transposed into UK legislation. The standard is intended for local authorities, building practitioners and building owners and provides a step-by-step guide to the conservation and refurbishment of historic buildings. The heritage-built environment has been identified in government policy as a key component of economic regeneration and urban renewal [33], which helps it receive investment and a legal protection from the UK government. However, UK's built heritage is ferociously defended by a group of powerful NGOs. A number of integrated policy frameworks developed by either the government or NGOs dealing with energy efficiency improvements in heritage buildings are as follows:

- The National Planning Policy Framework (NPPF), contains twelve core principles, two of which relate directly to heritage conservation and energy efficiency. On heritage, it states that "(. . .) *conserve heritage assets in a manner appropriate to their significance, so that they can be enjoyed for their contribution to the quality of life of this and future generations*". On energy efficiency it supports "*. . . the transition to a low carbon future... (and) encourage the reuse of existing resources, including conversion of existing buildings and encourage the use of renewable resources.*" The NPPF does not seek to arbitrate between these two principles but instead provides a framework for assessing heritage significance and weighing the degree of harm to it against the public benefit of reducing energy use [34].
- Historic England (HE) has published a wide range of practical guidance to help owners, managers, and any other relevant stakeholders through the process of energy efficiency improvements to historic buildings [35]. The guidance provides detailed information on energy efficiency

improvements (both in planning and implementation stage) of heritage buildings (built before 1919), buildings listed in a conservation area, or older buildings. Most importantly, these technical advices advocate the Whole House Retrofit approach. Historic England supports the government's efforts on improving the energy efficiency of existing buildings through Part L of the Building Regulations, which makes it clear that *"a reasonable compromise on the energy efficiency targets may be acceptable in order to preserve character and appearance and to avoid technical risks"*. They do this by specifically including some 'exemptions' where 'special considerations' apply for historic buildings and those of traditional construction.

- The Sustainable Development Foundation (SDF) as a non-profit organisation works to deliver a radical step-change in sustainability performance for the UK built environment and is registered under Society's Act 1860 since May 2006. SDF continues to lobby governments and policy makers, conduct research, and influence policy makers. The Sustainable Traditional Buildings Alliance (STBA) as a programme of this foundation has aimed to deliver a sustainable traditional built environment in the UK since 2011. It develops policies, guidance and training to minimise risks and maximise benefits to traditional buildings and their owners with a focusing on human comfort, durability of the building fabric, energy consumption attributed to the building/occupant, and impact on our communities, culture, and natural environment.

- Following the *Each Home Counts review* [36], UK has a 'quality mark' (TrustMark) for domestic retrofit, and the new comprehensive domestic retrofit standard; Publicly Available Specification (PAS) 2035 *Retrofitting Dwellings for Energy Efficiency: Specification and Guidance* [37], which forms part of the British Standards Institution (BSI) Retrofit Standards Framework. PAS 2035 applies to traditionally constructed and protected buildings and makes special provision for them as high-risk categories. It defines traditionally constructed buildings as "constructed with solid brick or stone walls, or timber-framed walls with any infill"; and defines protected buildings as listed buildings, buildings in conservation areas or World Heritage Sites.

- The UK Centre for Moisture in Buildings (UKCMB) is an independent, non-profit organisation run by University College London and the Building Research Establishment as an academia–industry partnership to work on identification and mitigation of moisture-induced damage and moisture risk in UK buildings, including heritage/traditional buildings [38].

3.4. Incentives for Energy Retrofitting of Built Heritage in the UK

Heritage tourism sector is an important part of the UK economy for both domestic, and international visitors, with the purpose of visiting historic towns. In the 2019 report of the Nations' Brand Index Survey of 50 nations, in which nations are ranked upon their universal reputation, UK was ranked fourth for criteria including 'rich in historic buildings and monuments' [39]. To own a historical building or to live in a city rich in these buildings is indicative of a high socio-economic status. Both prestige, cultural and economic incentives to invest in ongoing maintenance and repair are common contexts for retrofit of historic buildings rather than a desire to reduce energy costs alone [40,41]. This is proved in the annual report by the Centre for Economics and Business Research on behalf of Historic England, examining the links between organisations in the heritage sector and the local economies. The study claims that keeping the historic/cultural properties in active use as businesses, homes, tourism attractions or a combination of all three helps to stimulate environmental, economic and community regeneration [42]. This is also addressed by the planning policy guidance by Historic England [43], which enables development of a significant place (e.g., a historic building) to ensure it remains in continued use whilst minimising damage to its heritage value.

One of the outstanding driving forces behind energy retrofitting of heritage buildings is climate change. The frequency, intensity and duration of heatwaves are projected to increase worldwide, including in the UK. All the UK regions are projected to become warmer, especially in summer [44]. While at the time of writing this article the heating demand remains the main energy use driver in buildings, it is estimated that, even today, 20% of the UK housing stock suffer from overheating during summer [45], which will lead to an increased energy use due to rising cooling demands [46]. Consequently, this has led to a substantial interest in policy and research towards reducing indoor overheating risk in UK homes and integrated retrofit designs encompassing both adaptation and mitigation strategies [47].

3.5. Constraints for Energy Retrofitting of Built Heritage in the UK

A result of the high socio-economic value of heritage in the UK is the proliferation of several civil agencies and organisations, involved in conservation activities. Grant funding is provided through these organisations for maintenance, repair, and upgrading of historic buildings, and due to variations in sizes and operation mode of these organisations, the advice and guidance they provide on various matters are not always compatible [48]. This leads to the emergence of fragmented retrofit approaches in local policies. Particularly, the planning approvals needed to be sought for interventions on historic buildings defined by independent local planning authorities can substantially neglect technical innovations on energy efficiency. For example, in the case of listed buildings' applications for double glazing, there is no consistent approach across the country. Some Councils rigidly insist on authentic material regardless of the energy performance of the building, while some allow it [49]. Historic England's recent report [50] suggests in the case of steel windows or window frames capable of carrying double glazed units, their addition to existing windows may be considered acceptable. A similar conflict was reported by [51] with regards to the use of slim profile double glazing (SPDG) for the energy efficiency retrofit of historic buildings.

The potential conflict between energy reduction and conservation of heritage values in local policies is also observed in case of unlisted historic buildings. In a study by [52] carried out in Cambridge, a town which had a boost of energy efficient retrofits following the introduction of the Green Deal Communities Fund, the application of measures incompatible with the historic neighbourhood is criticised. According to this fund, for unlisted historic buildings the use of external wall insulation, which may substantially change the appearance, was indicated as a 'permitted development'. In this way, in the planning application for retrofitting unlisted historical buildings, the decision on prioritising heritage or energy values is completely left to the subjective knowledge of local officers. A similar conflict between different local authorities is also observed in the North of England as shown by the findings of a survey composed of 48 participants (comprising practitioners and local officers), pointing out challenges and conflicts between housing and planning officers over the planning permission for external wall insulation [53].

3.6. Turkey's Legal and Administrative Policies on Improving Energy Performance of Buildings: A Chronological Review

Turkey has set forward comprehensive policy packages over buildings' energy demand for the last two decades. Among the important policies currently in effect are Energy Efficiency Law (2007), Buildings' Energy Performance Regulation (2008), Buildings' Energy Efficiency Regulation (2011), Energy Efficiency Strategy (2012), TS 825 (2008), National Energy Efficiency Action Plan (2018), and the Green Buildings Regulation (2017). The legislative background on improving energy performance of buildings is discussed within two distinct periods in Turkey: during 1970–2000 and during the period from 2000 until present.

Similarly, to the UK, Turkey was not exempted from the results of the energy crisis of the 1970s. The first Turkish Thermal Insulation Standard (TS 825) was developed in 1970 [54]. The first application of thermal insulation in buildings started using imported external insulation materials in 1991, when double glazing units also began to be used in window frames [55]. In 1992, in spite of Turkey's membership in the OECD, Turkey did not sign the UN Framework Convention on Climate Change when it was adopted in 1992, nor has it made emissions reduction commitments under the Kyoto Protocol. The reason was the fact that Turkey's per capita CO_2 emissions were lower than OECD norms at that time [56]. In 1999, a new version of TS 825 was published [57] and after 30 years it was "recommended"; its use became mandatory as of 2000, though only for new buildings, to define the maximum allowable heat losses and calculate heating requirements. The Thermal Insulation in Buildings Regulation (2000) is considered to be the first main regulation dealing with building energy performance in Turkey. In 2007, Turkey adopted The Energy Efficiency Law [9], whose main objectives included increasing energy efficiency, reducing environmental impacts, and reducing the load of energy costs on the economy. There are a number of policy measures outlined in the legislation to be implemented on the built environment related to the sustainable architectural design and green buildings. For the first time, through this legislation, increasing public awareness has come to the forefront of energy efficiency activities. The revised version of TS 825 issued in 2008 extended its scope to also existing buildings with emphasis on dwellings [58]. Although TS 825-2008 is still the mandatory standard, it overlooks cooling energy requirements and heat store capacity [59]. In accordance with European Union's Framework Directive 2002/91/EC and Energy Efficiency Law (No. 5627), Buildings' Energy Performance Regulation (BEP TR) [60] was published in 2008, targeted at both new and existing buildings. The Ministry of Environment and Urbanization defines the objectives of BEP TR as (1) increasing efficient use of energy and applicability of renewable energy systems in buildings, (2) reducing greenhouse gas emissions, and (3) determining performance criteria and application principles in buildings and environmental protection activities. The energy identity certificate for both new and existing buildings is issued based on this regulation.

Important strategy documents for energy efficiency policies have been put forward in recent years in Turkey including Urban Development Strategy Plan 2010–2023 [61], Climate Change Strategy Plan 2010–2020 [62], and Energy Efficiency Strategy Document 2012–2023 [63]. Medium-term targets encouraging applicable energy retrofitting strategies for existing buildings, such as the Energy Identity Certificate by implementation of thermal insulation, are addressed in Climate Change Strategy Plan [62]. In 2011, Buildings' Energy Efficiency Regulation, issued by the Ministry of Energy and Natural Resources [MENR], covering applicable technical measures on improving the efficiency of heating, cooling and lighting systems for both new and existing buildings [64]. Green Building Regulation issued in 2014 by the Ministry of Environment and Urbanization with the aim of evaluating the sustainability of new and existing buildings and settlements in terms of their environmental, social, and economical performances. Very recently, Turkey's government has set up a series of energy efficiency goals and policy frameworks to achieve the 2023 Energy Efficiency targets in the scope of the EU accession negotiations. The National Energy Efficiency Action Plan [10], is one of them that represents a concrete energy efficiency strategy in the building sector in terms of both technology and investment. Decreasing primary energy use in Turkey by 14% in 2023 has been set as its main objective. Although the NEEAP's actions comprise a broad set of domains containing technology, finance and policy, it suggests the use of the existing policy instruments for their implementation, which begs the question of whether these will be sufficient or not.

Turkey's energy efficiency policies are carried out under the responsibility of the MENR and its branches, like the General Directorate of Renewable Energy. The department supports the investments on efficiency improvement in the industrial sector in accordance with the framework of Energy Efficiency Law. The Ministry of Environment and Urbanization is the other governmental body with substantial responsibilities regarding energy efficiency in new and existing buildings, and settlements. National strategies are developed by the Energy Efficiency Coordination Board as mandated by Energy Efficiency Law.

3.7. Turkey's Legal and Administrative Policies on Built Heritage Conservation: Chronological Review

Legal and administrative basis of conservation activities in Turkey dates back to the Ottoman Era, with the enforcement of the Ancient Monuments Regulation (AMR) in 1869. While initially the AMR focused on the archaeological remains and findings, it was amended in 1874, 1884 and 1906 to extend the definition and the scope of 'monuments'. The AMRs were followed by the Conservation of Monuments Regulation (CMR), issued in 1912, in which the permissible interventions on the historic monuments were defined in more detail [65–67]. In accordance with the AMR and CMR, measures and interventions regarding the historic monuments, archaeological excavations, findings and museums were managed by the Ancient Monument Conservation Council (AMCC) established in 1917.

After the foundation of the Turkish Republic in 1923, AMCC, which was later on revised as the Council of Conservation of Ancient Monuments, remained as the main authorised body for decision-making and controlling the interventions to historic buildings until the establishment of the High Council for the Historical Real Estate and Monuments in 1951 [65]. Similarly, AMR and CMR, continued to be the main legal instruments concerning the conservation of historic buildings during the Turkish Republican Era, until the acceptance of the Antiquities Law (No: 1710) in 1973. Different from the previous regulations, this law brought the concept of 'conservation site' setting up the legal basis for area based conservation, not only focusing on the historic monuments but also dealing with the conservation of historic tissues and more modest historic buildings. Accordingly, the High Council's authority was extended to cover decisions not only for individual monuments, but also to conservation areas.

Antiquities Law (No: 1710) remained in act for 10 years, until the acceptance of the Law for the Conservation of Natural and Cultural Properties (No: 2863) in 1983 [68], which is still in act together with some later amendments. This new law; offers a detailed classification of cultural properties and conservation sites in different conservation statuses and degrees, while explaining their registration, documentation, project preparation, decision-making, intervention and control processes. In addition to these, together with this law, decentralisation of the decision-making process was attempted by redefining the authorised bodies in the decision making process as the Regional Councils for Conservation of Cultural and Natural Heritage and High Council for Conservation of Cultural and Natural Heritage. Amendment of the Law for the Conservation of Natural and Cultural Properties (No: 2863) in 2005 with the Law No: 5226, even increased the ongoing decentralisation process by enhancing the roles and responsibilities of the local authorities.

In Turkey, according to the Law No: 2863, historically or culturally important immovable properties built until the end of the 19th century, as well as the ones having special values although built after the 19th century, are in the category of 'cultural properties'. Currently the identification of cultural properties and determining their registration degrees and statuses are the responsibility of the General Directorate of Cultural Properties and Museums, a division of the Ministry of Culture and Tourism. Historic buildings and structures are listed in the national registry either as a cultural property, or as part of larger areas designated as conservation sites, or both. The vast majority of conservation areas and registered historic buildings in the country are first identified as historically or culturally important by the Government and listed in the national registry. Once they are registered, they are made distinct from other immovable property and development rights are restricted. Moreover, whether the property is private or not, a registered building acquires the status of public good, meaning that the owner's freedom to intervene is firmly restricted. All kinds of actions and interventions related with the registered buildings and sites are subject to the approval of the Regional Conservation Council.

3.8. Potential Incentives for Energy Retrofitting of Built Heritage in Turkey

According to the statistics released by the General Directorate of Cultural Properties and Museums, by the end of 2019 [69], there were 113,137 registered cultural properties in Turkey. Moreover, there are 460 registered conservation sites having historic urban/rural tissues composed of a vast number of historic buildings. When combined with unregistered historic buildings, this number rises to a substantial portion of building stock. On the other hand, according to the Ministry of Tourism and the Statistics Institute, visiting historical sites and buildings is ranked second most commonly reported purpose of foreign tourists visiting Turkey. In the latest report of Future Brand Country 2019 Index [70], Turkey is ranked 4th for the 'Heritage and Culture' criteria among 75 countries, surveyed in terms of their potential in Heritage, Culture and Tourism. This shows that following the footsteps of the UK, which has been successful at capitalizing on its historical buildings, Turkey's rich built heritage has also the potential of defining it as a tourism destination.

With regard to the climate change crisis, like the UK, Turkey's building stock is expected to face warmer climatic conditions in the near future, which is another factor increasing the vulnerability of the country's built heritage. Accordingly, IEA [3] urges the government to set a longer-term energy policy agenda for 2030. As a response, The Turkish National Energy Efficiency Action Plan [10] set a strategic goal towards 'low building energy demand and carbon emissions; scale up sustainable, environment-friendly buildings', further highlighting the need for developing energy retrofitting policies covering efficient cooling standards, specifically, for historic houses. Even in the case of existing buildings, Turkey's policy framework suffers from the lack of concrete minimum standards for efficient cooling, which becomes critical in terms of energy use and comfort as the cooling demands increase in parallel with overheating climatic conditions. This need can be transformed to a potential incentive for developing adaptation and mitigation retrofit measures towards a climate-resilient built heritage stock.

3.9. Constraints for Energy Retrofitting of Built Heritage in Turkey

As previously mentioned, the European energy efficiency directives exclude historic buildings from implementation of energy-efficient retrofit measures. This translates into Turkey's attempts to upgrade its legislations in alignment with the EU, making the position of built heritage in the greater scheme of energy efficiency rather uncertain. Moreover, the meaning and scope of the term "retrofit" is vague in Turkish regulations, and it is merely mentioned even in the latest reports and action plans i.e., Green Buildings Regulation (2017), without any definitive explanation.

In Turkey, if a historic building is deemed to attain particular cultural/historical values it is placed completely under the responsibility of the Ministry of Culture and Tourism and the energy efficiency regulations defined for existing buildings by the Ministry of Environment and Urbanization do not apply. Unfortunately, there is no legal and collaborative action between sectors responsible for heritage protection and energy efficiency of buildings in Turkey.

Another challenge in Turkey's built heritage subsector is that a substantial portion of the registered immovable cultural heritage belongs to the historic houses (known as civil architecture) [69]. The residents of these dwellings in Turkey suffer seriously from, in cases of buildings with listed status, the prohibition of development rights on the properties, and in general, the costly burden of the repair, maintenance and restoration of their homes. This is particularly alarming in the case of heritage houses owned by lower-middle income and low-income people, which are mostly left to their fate.

A summary of incentives and constraints for the UK and Turkey with reference to the national and international legislation, as well as the role of NGO's and other financial drivers in the development of guidance directed at the energy-efficient retrofitting of heritage buildings is given in Table 1.

Table 1. Comparative synthesis on energy retrofitting of historic buildings in the UK and Turkey.

Synthesis Realm	Attribute	UK	Turkey
Energy-efficient retrofitting of historic buildings in national legislative frameworks	Incentive	BS 7913:2013 Guide to the Conservation of Historic Buildings is the first British standard on the energy efficiency in historical buildings [31].	
	Constraint		1. No technical reports or guidelines has been published, neither under the responsibility of governmental bodies, nor by NGOs. 2. The scope of the term "retrofit" in Turkish regulations, including the latest reports and action plans is vague and does not refer to a technically established content.

Table 1. *Cont.*

Synthesis Realm	Attribute	UK	Turkey
Alignment with international legislative frameworks	Incentive	British Standard Institute launched BS EN 16883:2017 'Guidelines for improving the energy performance of historic buildings', aligned with EN 16883:2017.	
	Constraint		Aligned with the EU's process in EPBD instructions and exemption of historical buildings from energy-efficient improvements.
Socio-economic value of heritage tourism	Incentive	1. UK ranked 4th, as a nation brand in Nations' Brand Index Survey 2019 of 50 countries for richness of historic buildings and monuments. 2. Majority of the UK's historic houses, are in continued use [43].	1. Turkey is ranked 4th for the 'Heritage and Culture' criteria among 75 countries in Future Brand Country 2019 Index [70]. 2. Users of registered cultural properties are obliged to maintain, repair and restore them in line with the principles of Law no. 2863 [68].
	Constraint		1. The economic value of heritage tourism in Turkey is seen as a reason to rush decisions for poor façade interventions. 2. There is no well-defined official statistical data as to the active use status of, especially, unlisted residential heritage buildings in Turkey.
Climate change mitigation strategies	Incentive	1. The UK switched from its earlier goal on reducing building energy use to reducing CO2 emissions. 2. The problem of overheating in the indoor environments has forced the research and policy actors to develop climate resilient retrofits.	1.The issue of efficient heating standards is already well addressed in Turkish regulations (TS825/2008). 2. The problem of overheating risk in indoor environments is observed also in Turkey, which needs to be considered by policy actors to develop efficient cooling standards [16]
	Constraint		
Cross-sector institutional collaboration (NGOs and governmental bodies)	Incentive	1. There is a bilateral partnership between NGOs and governmental organisations on the subject of developing retrofit schemes and guidelines for historical buildings. 2. Several technical reports and guidance are published by the Historic England (non-departmental body of British Government).	
	Constraint	Incompatibility between local planning authorities in their guidance, including the listed buildings consent applications for double glazing or in the use of external wall insulation in unlisted historic buildings [18,52,53].	There is no legal and collaborative action between bodies responsible for heritage protection and energy efficiency in Turkey.
Built Heritage NGOs' active in the development of energy efficiency guidelines	Incentive	The energy efficiency of the UK's built heritage is ferociously defended and contributed to by a group of powerful NGOs.	
	Constraint		None of the NGOs active in heritage conservation neither proposed nor developed guidelines for improving energy efficiency in historic buildings.

4. Comparative Analysis of the UK's and Turkey's Regulatory Approaches on Energy-Efficient Retrofitting of Built Heritage

In this study, the policy frameworks regarding 'energy efficiency in buildings' and 'built heritage', which are currently in place in the UK and Turkey, have been thoroughly reviewed. A comparative analysis of these frameworks was conducted with the aim of discussing the level of integration between energy and built heritage policies and identifying existing gaps in legislation and organisational structure to make these viable. The conclusions drawn from this comparison are summarised below.

With regard to the evolution of UK's regulatory approaches on energy-efficient retrofitting of built heritage, 2008 can be set as a benchmark date, when the first guide [71] to improve energy efficiency in historic homes was published as the result of a project titled, "Energy Heritage" carried out in Edinburgh. It has become quite clear that in the UK since 2008 until the present, several studies, projects, workshops, technical reports and guidance plans, encompassing all the three levels of policy frameworks were published (Figure 1). Whilst, in Turkey, except for the published one handbook and one workshop under the scheme of 'Energy Efficiency for Historic Buildings' led by the Association for the Protection of Cultural Heritage (APCH) [72], no other technical reports and guidelines have been published, neither under the responsibility of public governmental bodies, nor by NGOs.

Landmark Dates	Target Domain	International	UK	Turkey
2008	Energy Heritage: A Guide to Improving Energy Efficiency in Traditional and Historic Homes (Changeworks, Edinburgh)		■	
2011	Sustainable Traditional Building Alliance (STBA) workshops		■	
2011	A Guide to improving the Energy Efficiency of Traditional Homes in the City of Bath		■	
2012	Fabric Improvements for Energy Efficiency in Traditional Buildings (HE)		■	
2012	The National Planning Policy Framework		■	
2013	BS 7913:2013 Guide to the conservation of historic buildings		■	
2014	UNESCO Guidebook on energy efficiency of cultural heritage	■		
2015	Heritage 2020 Framework: Strategic Priorities for England's Historic Environment 2015-2020		■	
2015	ICCROM Assembly: Climate Change and Cultural Heritage	■		
2015	Bristolian's Guide to Solid Wall Insulation (STBA)		■	
2015	Planning Responsible Retrofit of Traditional Buildings (STBA)		■	
2015	Online Retrofit Guidance Tool (STBA)		■	
2015	Guidance on Energy Certificates for Historic Buildings (HE)		■	
2016	Guideline 34P ASHRAE: Energy Guideline for Historical Buildings		■	
2016	Guidance on Historic Building Fabric Alterations (HE)		■	
2016	The City Plan Part One / Energy Efficiency in Historic Area		■	
2017	EN 16883 Standard for Energy Efficiency in Historic Buildings	■		
2017	BS EN 16883:2017 'Guidelines for Improving the Energy Performance of Historic Buildings'		■	
2017	Application of part L of the Building Regulations to Historic Buildings		■	
2017	Energy Efficiency Handbook for Historic Houses			■
2018	Technical Guidance on Energy Generation Options:Solar Photovoltaics (HE)		■	
2019	Energy Efficiency in Historic Buildings Workshop (APCH)			■
2020	"Retrofitting Traditional Buildings" Course (STBA)		■	

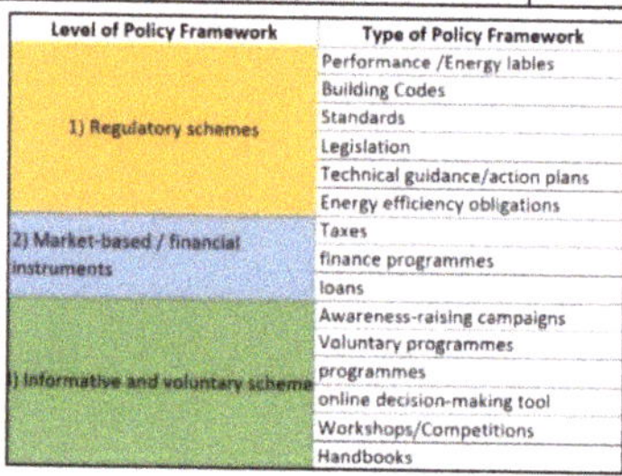

Figure 1. Comparison of the level of policy frameworks on energy efficiency in historic buildings

It can be seen that Turkey uses a predominantly informative and voluntary scheme (third level of policy framework in Figure 1) to approach the energy retrofit of the heritage building stock, which still needs to be improved in other domains of this level, for example through awareness-raising campaigns, competitions, and courses, as there are several examples from the UK, e.g., those run by the STBA, including the "Retrofitting Traditional Buildings" and "Energy Efficiency Measures for Older and Traditional Buildings" courses. As discussed in Section 3.9, since the scope of the term "retrofit" is vague and does not refer to a technically established content in Turkish regulations, firstly, an increased public and industry awareness should be developed through technical training events and courses aimed at communities and practitioners, introducing the general concept of retrofit and viable measures for improving energy performance of historic buildings.

Secondly, in the UK context, the retrofit approaches are substantially streamlined towards climate change mitigation and adaptation schemes, and aimed at making the indoor environments more resilient to heat waves [73,74], while the most common retrofitting measures, such as increased insulation and airtightness may lead to higher levels of indoor pollutants and condensation problems. This becomes more critical in the case of historic/traditionally constructed buildings whose semi-permeable fabrics keep the internal moisture and temperature in balance with the ever varying outdoor hygrothermal conditions [75]. This issue is well-addressed in PAS 2035, now in force in the UK, requiring management of moisture balance and upgrading of ventilation when insulation is installed, to reduce condensation and mould risks. On the other hand, in order to minimise the potential unintended consequences of energy retrofit measures for historic buildings, which are becoming more and more apparent [71], one of the primary concerns of both owners and policy makers should be applying regular maintenance and repair prior to retrofit, which, in the UK case, is a requirement of the PAS 2035 [38]. Therefore, Turkey should pay utmost attention to integrate measures in its policies so as not to disrupt the buildings' moisture balance irreversibly by making it mandatory to (a) repair prior to retrofit, (b) ensure a good management of moisture balance, and (c) upgrade ventilation when insulation is installed.

It is noteworthy that in addition to the projected strong heat waves as one of the results of the climate change crisis, both countries still face the problem of fuel poverty insomuch that of the total energy used by the average UK home, almost 80% was used for heating and hot water demands [76]. The number of households in fuel poverty is estimated to be approximately 10.9% of all English households [77,78]. In the case of Turkey, about one quarter of households are energy-poor and about half of the lowest income households face the problem of fuel-poverty [79]. In this regard, the number of standards and regulations for efficient heating in residential buildings in the UK and Turkey is considerable compared to efficient cooling standards. Although it is observed that the ratio of energy-poor households is decreasing in both countries, an efficient heating problem still is a challenging issue for both cases.

Thirdly, in the UK, cross-sector institutional relationships have been fitted well through the collaborative research and projects carried out between non-profit (governmental/non-governmental) organisations to help policy agencies to understand and enforce the requirements of the energy performance of historic buildings. As an example, in 2011, STBA's first research work on analysis of the gaps in the performance of UK's traditional buildings was funded by Construction Skills and English Heritage, which are two different organisations; the former linked to the Department for Communities and Local Government, and the latter, a charity managing the National Heritage Collection. However, in Turkey, the establishment of NGOs active in the field of conservation of heritage buildings, is much delayed compared to the UK. Two important early examples of these are Vehbi Koç Foundation (1969) and the Sabancı Foundation (VaKSa, 1974), while others were mostly founded in the 1990s [67]. In contrast to the UK, where several technical reports and guidance are published either by non-departmental Governmental bodies (e.g., Historic England) or NGOs (e.g., STBA), none of the NGOs in Turkey proposed, led or developed guidelines towards improving energy efficiency in heritage buildings, and their engagement with the Governmental efforts towards energy efficiency of heritage buildings should be further encouraged. In addition, individual governmental bodies responsible for energy efficiency and heritage conservation should be brought together through cross-sector regulations to address this cross-disciplinary problem jointly.

Fourthly, as discussed previously, UK's planning policy guidance [43] secures the long-term future of historic buildings through "its continued use for a sympathetic purpose", while minimising damage to its heritage values. Turkey's regulations also encourage putting protected cultural properties in continued use, in line with the functions prescribed by the Regional Conservation Councils. Users of such properties are currently obliged to maintain, repair and restore them in line with the principles of Law no. 2863. However, Turkey needs to produce official statistical data as to the active use and maintenance status of, especially, unlisted heritage dwellings, as this is currently nonexistent.

Fifthly, as discussed in Section 3.5, the lack of a clear conservation framework in retrofitting practices leads to a strong inconsistency between independent local planning authorities in their promoted advice and guidance, including the listed buildings consent applications for double glazing or in the use of external wall insulation in unlisted historic buildings. In fact, the UK heritage building sector appears to be over-legalised with many organisations and policy frameworks managing the field. In this regard, a less-is-more approach could have a positive impact by bringing unity and transparency to the practice within the heritage building sector in the UK both when it comes to preserving their heritage values and implementing energy efficiency measures.

Finally, although it is clear that the UK has a better developed policy framework for the energy retrofit of heritage buildings, robust data regarding the actual numbers of the retrofitted heritage building stocks with reference to measurable, concrete effects resulting from, and a critique of the existing policy frameworks, do not exist in either country. All stakeholders of the energy efficiency of heritage buildings should therefore jointly work towards creating this information for a thorough appraisal of the efficiency and viability of their policy framework and practice in the mid- and long-term.

5. Is Policy Enough?

The decision as to whether and how to retrofit heritage buildings for an enhanced energy performance should be informed, among other factors, by original building fabric and construction technology characteristics. A detailed understanding is needed as to how much gain in energy performance is possible by restoring the fabric through repair, and by actually retrofitting it, and therefore life-cycle analysis remains to be one of the most powerful tools for researchers. In any case, strengthening conservation processes of listed and unlisted built heritage through encouraging constant use, and not only allowing but also developing incentives to carry out regular maintenance in accordance with the original fabric characteristics to ensure these buildings are in a better condition is extremely important to close the energy efficiency gap between actual and targeted performances; this should be achieved through simple elemental interventions which have been shown to demonstrate a higher cost-to-benefit advantage [80], and hence to make some retrofitting measures redundant. In this respect, enhanced public awareness is, again, a critical factor to ensure that the policies fulfil their goals.

Historic buildings are commonly labelled as problematic for being 'draughty', 'leaky', and 'inefficient'. However, these buildings often offer a lot to be learned regarding contextual design, and use of architectural and structural detailing in effective ways to tackle the microclimatic conditions they are exposed to, and hence their poor performance may be owed to age, lack of maintenance, change of lifestyle and resident profile, rather than their intrinsic constructive features. Therefore, energy retrofitting of these structures, as a very case/context-specific issue, requires a deep understanding of their original fabric characteristics, as well as the society and community it is valuable for. People's perception of the 'history' and 'heritage', in fact, shapes their perception of 'acceptable changes' and the value of the heritage buildings [5]. That is why despite all the arguments on the alignment of legislation and guidance related to energy retrofitting of heritage buildings, decision-making in this arena still requires case/context-specific assessments and a case-by-case approach as in all interventions on heritage buildings.

6. Conclusions

More efficient energy use in buildings continues to be one of the most valuable untapped potential resources. The constraints to accessing this untapped resource, especially in historic buildings, are numerous and complex, but can be overcome through raising awareness and developing appropriate policies. To achieve this, it is important to draw lessons from convincing examples that demonstrate various possibilities. In a world struggling to confront climate change, a holistic and synergistic approach for improving the energy efficiency of all buildings needs to be a priority. Even though the regulations for buildings exempt most listed heritage buildings from energy performance improvements, many of these buildings can and should be able to accommodate some improvements through options available for long-term sustainability that are compatible with their heritage values.

The UK's current policy framework supports improving energy efficiency of heritage buildings, aiming at combining energy efficiency goals with heritage values. The establishment of such a decision-making system requires knowledge of both building physics and heritage values, and should have financial support and expertise from governmental bodies or NGOs. This begs the need for a strong cross-sector and multi-stakeholder collaboration involving both arenas of built heritage management and energy efficiency in buildings. Last but not least, we believe that the energy retrofitting of historic buildings should be country, even region specific. Accordingly, policy frameworks should be shaped accounting for the socio-economic and cultural backdrops in any given context. Planned policies in this respect need to be promoted through a range of economic incentive programmes and public awareness campaigns as auxiliary means, without which they will fail despite their technical and organisational merits.

Author Contributions: N.J., A.G.B.A.; conceptualisation and methodology. N.J.; investigation. N.J., Y.D.A.; formal analysis. A.G.B.A., Y.D.A., P.R.; provided resources. N.J.; writing-original draft. N.J., Y.D.A., A.G.B.A., P.R.; writing-review and editing. A.G.B.A., Y.D.A., P.R.; funding acquisition. All authors have read and agreed to the published version of the manuscript.

Funding: This study has been carried out as part of the "PROcesses for sustainable retrofit of Traditional dwellings in Turkey for Climate-resilience, Conservation and ComforT (PROT3CT)" project jointly funded by British Council (no. 527666821, funding Y.D.A and P.R.) and TUBITAK (no. 119N514, funding N.J. and A.G.B.A.) through Newton Funds Institutional Links and of "Tales of cold and draft: Establishing retrofit needs of Turkish vernacular architecture for energy-efficiency, comfort and conservation" funded by 2019–2020 UCL Small Grants Grand Challenges scheme. The APC was funded by the UCL Open Access Office. The authors are grateful to all funders.

Conflicts of Interest: The authors declare no conflict of interest. The funders had no role in the design of the study; in the collection, analyses, or interpretation of data; in the writing of the manuscript, or in the decision to publish the results.

References

1. Wagner, L.; Ross, I.; Foster, J.; Hankamer, B. Trading Off Global Fuel Supply, CO_2 Emissions and Sustainable Development. *PLoS ONE* **2016**, *11*, e0149406. [CrossRef]

2. IPCC. *Climate Change 2014: Synthesis Report*; Pachuari, R.K., Meyer, L.A., Eds.; Contribution of Working Groups I, II and III to the Fifth Assessment Report of the Intergovernmental Panel on Climate Change; The Core Writing Team; IPCC: Geneva, Switzerland, 2014; p. 151

3. International Energy Agency (IEA). *The Future of Cooling, Opportunities for Energy-Efficient Air Conditioning*; IEA: Paris, France, 2018.

4. Khalil, A.M.R.; Hammouda, N.Y.; El-Deeb, K.F. Implementing Sustainability in Retrofitting Heritage Buildings. Case Study: Villa Antoniadis, Alexandria, Egypt. *Heritage* **2018**, *1*, 57–87. [CrossRef]

5. Yarrow, T. Negotiating Heritage and Energy Conservation: An Ethnography of Domestic Renovation. *Hist. Environ. Policy Pr.* **2016**, *7*, 340–351. [CrossRef]

6. Directive 2010/31/EU of the European Parliament and of the Council of 19 May 2010 on the Energy Performance of Buildings (Recast). Available online: http://www.eceee.org/buildings/EPBD_Recast (accessed on 31 May 2020).

7. Directive 2012/27/EU of the European Parliament and of the Council of 2012 on the Energy Efficiency in Buildings (2012/27/EU). Available online: http://www.eceee.org/buildings/EPBD_Recast (accessed on 31 May 2020).

8. Directive 2018/844/EU of the European Parliament and of the Council of 2018 on the Energy Efficiency of Buildings. Available online: https://eur-lex.europa.eu/legal-content/EN/TXT/PDF/?uri=CELEX:32018L0844&from=EN (accessed on 25 April 2020).

9. Energy Efficiency Law, no. 5627 (Enerji Verimliliği Kanunu). T.C. Resmi Gazete, 02/05/2007 Date and No. 26510. 2007. Available online: https://www.mevzuat.gov.tr/MevzuatMetin/1.5.5627-20070418.pdf (accessed on 7 January 2020).

10. Turkish National Energy Efficiency Action Plan 2017-2023 (NEEAP). Ministry of Energy and Natural Sources. 2018. Available online: http://www.yegm.gov.tr/document/20180102M1_2018_eng.pdf (accessed on 8 March 2020).

11. Department for Communities and Local Government (DCLG). *English Housing Survey: Headline Report 2013-14 Department for Communities and Local Government*; DCLG: London, UK, 2015.

12. European Commission. A Roadmap for Moving to a Competitive Low Carbon Economy by 2050. 2011. Available online: http://eur-lex.europa.eu/resource.html?uri=cellar:5db26ecc-ba4e-4de2-ae08-dba649109d18.0002.03/DOC_1&format=PDF (accessed on 31 May 2020).

13. The UK Government. *Climate Change Act 2008*; The Stationery Office Limited: London, UK, 2008. Available online: http://www.legislation.gov.uk/ukpga/2008/27/contents (accessed on 31 May 2020).

14. Department for Business, Energy and Industrial Strategy (BEIS). Policy Paper Clean Growth Strategy: Executive Summary. 2018. Available online: https://www.gov.uk/government/publications/clean-growth-strategy/clean-growth-strategy-executive-summary (accessed on 31 May 2020).

15. DEFRA. *Climate Change Bill Final Impact Assessment, UK Government*; DEFRA: London, UK, 2008.

16. Tayanc, M.; Im, U.; Dogruel, M.; Karaca, M. Climate change in Turkey for the last half century. *Climate Change* **2009**, *94*, 483–502. [CrossRef]

17. MWH. *Politika Boşluk Analizi ve Enerji Verimlilik Programı Değerlendirmesine İlişkin Danışmanlık Hizmetleri*; MWH Mühendislik ve Müşavirlik Ltd.: Istanbul, Turkey, 2015.

18. NIRAS. *Binalarda Enerji Verimliliği AB ve Türk Mevzuatı*; Türkiye Cumhuriyeti Çevre ve Şehircilik Bakanlığı: Ankara, Turkey, 2015.

19. Saygin, D.; Ercumen, Y.; De Groote, M.; Bean, F. Enhancing Turkey's Policy Framework for Energy Efficiency of Buildings, and Recommendations for the Way Forward Based on International Experiences. SHURA Energy Transition Center and the Buildings Performance Institute Europe (BPIE). 2019. Available online: \protect\unhbox\voidb@x\hbox{https://www.shura.org.tr/wp-content/uploads/2019/06/Buildings-Energy-Efficiency-Policy-Working-Paper.pdf} (accessed on 31 May 2020).

20. Ulu, M.; Durmuş Arsan, Z. State of the art survey for energy-efficient retrofit of historic residential buildings in both the EU and turkey. In Proceedings of the REHAB 2017—3rd International Conference on Preservation, Maintenance and Rehabilitation of Historical Buildings and Structures, Braga, Portugal, 14–16 June 2017.

21. Sahin, C.D.; Arsan, Z.D.; Tunçoku, S.S.; Broström, T.; Akkurt, G.G. A transdisciplinary approach on the energy efficient retrofitting of a historic building in the Aegean Region of Turkey. *Energy Build.* **2015**, *96*, 128–139. [CrossRef]

22. Owen, G. Energy Efficiency and Energy Conservation: Policies, Programmes and Their Effectiveness. *Energy Environ.* **2000**, *11*, 553–564. [CrossRef]

23. Department for Communities and Local Government (DCLG). Conservation of Fuel and Power (Part L). 2010. Available online: http://www.planningportal.gov.uk/buildingregulations/approveddocuments/partl/approved (accessed on 31 May 2020).

24. DEFRA. The Government's Strategy for Combined heat and Power to 2010, UK Government. 2004. Available online: www.%20defra.gov.uk (accessed on 31 May 2020).

25. English Heritage. *Power of Place: The Future of the Historic Environment*; Department of Culture, Media and Sport; Department of the Environment, Transport and the Regions: London, UK, 2000.

26. English Heritage. *State of the Historic Environment Report*; English Heritage: London, UK, 2002.

27. English Heritage. *Conservation Principles: Policies, and Guidance for the Sustainable Management of the Historic Environment*; English Heritage: London, UK, 2008.

28. May, N.; Rye, C. Responsible Retrofit of Traditional Buildings, STBA. 2012, p. 11. Available online: http://stbauk.org/stba-guidance-research-papers (accessed on 31 May 2020).

29. HM Government. Energy Efficiency: Building Towards Net Zero, Twenty-First Report. 2016. Available online: https://publications.parliament.uk/pa/cm201719/cmselect/cmbeis/1730/1730.pdf (accessed on 31 May 2020).

30. English Heritage. English Heritage Annual Report and Accounts 2012 to 2013. 2012. Available online: https://www.gov.uk/government/publications/english-heritage-annual-report-and-accounts-2012-to-2013 (accessed on 31 May 2020).

31. British Standard Institute (BSI). *BS 7913:2013 Guide to the Conservation of Historic Buildings*; BSI: London, UK, 2013.

32. European Committee for Standardizations (CEN). *EN 16883: Conservation of Cultural Heritage. Guidelines for Improving the Energy Performance of Historic Buildings*; CEN: Brussels, Belgium, 2017.

33. Mansfield, J.R. Heritage protection in England: The New Labour legacy. *Struct. Surv.* **2013**, *31*, 6–20. [CrossRef]

34. The National Planning Policy Framework (NPPF); Ministry of Housing, Communities and Local Government. Para 129. 2012. Available online: https://assets.publishing.service.gov.uk/government/uploads/system/uploads/attachment_data/file/810197/NPPF_Feb_2019_revised.pdf (accessed on 31 May 2020).

35. Pickles, D.; McCaig, I.; Pender, R. *Energy Efficiency and Historic Buildings: How to Improve Energy Efficiency*; Historic England: London, UK, 2018.

36. Bonfield, P. *Each Home Counts: Review of Consumer Advice, Protection, Standards and Enforcement for Energy Efficiency and Renewable Energy*; BEIS; DCLG: London, UK, 2016.

37. British Standard Institute (BSI). *PAS 2035/2030:2019, Retrofitting Dwellings for Energy Efficiency: Specification and Guidance. Department for Business, Energy & Industrial Strategy, UK Government*; BSI: London, UK, 2019.

38. UK Centre for Moisture in Buildings (UKCMB). Ukcmb. Available online: https://ukcmb.org/ (accessed on 29 May 2020).

39. Ipsos. Report of the Nations' Brand Index Survey of 50 Nations. France. 2019. Available online: https://www.ipsos.com/sites/default/files/ct/news/documents/2019-11/nbi_release_final_november_2019_0.pdf (accessed on 31 May 2020).

40. Lloyd-Jones, T. Retrofitting sustainability to historic city core areas. *Proc. Inst. Civ. Eng. Munic. Eng.* **2010**, *163*, 179–188. [CrossRef]

41. Lubeck, A.; Conlin, F. Efficiency and Comfort Through Deep Energy Retrofits: Balancing Energy and Moisture Management. *J. Green Build.* **2010**, *5*, 3–15. [CrossRef]

42. Centre for Economics and Business Research (CEBR). The Heritage Sector in England and Its Impact on the Economy, A Report for Historic England. 2018. Available online: https://historicengland.org.uk/content/docs/research/heritage-sector-england-impact-on-economy-2018/ (accessed on 31 May 2020).

43. Historic England. *Enabling Development and the Conservation of Significant Places*; Historic England: London, UK, 2015.

44. UK Climate Projections (UKCP). Met Office Hadley Center. UKCP18 Science Overview Report. 2018. Available online: https://www.metoffice.gov.uk/pub/data/weather/uk/ukcp18/science-reports/UKCP18-Overview-report.pdf (accessed on 31 May 2020).

45. Pathan, A.; Mavrogianni, A.; Summerfield, A.; Oreszczyn, T.; Davies, M. Monitoring summer indoor overheating in the London housing stock. *Energy Build.* **2017**, *141*, 361–378. [CrossRef]

46. Frontier Economics Ltd.; Irbaris LLP; Ecofys. Economics of Climate Resilience Buildings and Infrastructure Theme, Overheating in Residential Housing, London, UK. 2013. Available online: http://randd.%20defra.gov.uk/Default.aspx?Module=More&Location=None&ProjectID=18016 (accessed on 31 May 2020).

47. AECOM. Investigation into Overheating in Homes: Literature Review. Department for Communities and Local Government: London, UK. 2012. Available online: https://assets.publishing.service.gov.uk/government/uploads/system/uploads/attachment_data/file/7604/2185850.pdf (accessed on 31 May 2020).

48. Funding for Historic Buildings (FFHB). Complete List of Funding Sources. Online. 2013. Available online: http://www.ffhb.org.uk/results.php?action=full (accessed on 31 May 2020).

49. Changeworks. Double-glazing in Listed Buildings: Project Report. 2008. Available online: http://www.changeworks.org.uk/publications.php (accessed on 31 May 2020).

50. Historic England. Traditional Windows: Their Care, Repair and Upgrading. Guidance. London. 2017. Available online: https://historicengland.org.uk/images-books/publications/traditional-windows-care-repair-upgrading/ (accessed on 31 May 2020).

51. Ginks, N.; Painter, B. Energy retrofit interventions in historic buildings: Exploring guidance and attitudes of conservation professionals to slim double glazing in the UK. *Energy Build.* **2017**, *149*, 391–399. [CrossRef]

52. Sunikka-Blank, M.; Galvin, R. Irrational homeowners? How aesthetics and heritage values influence thermal retrofit decisions in the United Kingdom. *Energy Res. Soc. Sci.* **2016**, *11*, 97–108. [CrossRef]

53. Fylan, F.; Glew, D.; Smith, M.; Johnston, D.; Brooke-Peat, M.; Miles-Shenton, D.; Fletcher, M.; Aloise-Young, P.; Gorse, C. Reflections on retrofits: Overcoming barriers to energy efficiency among the fuel poor in the United Kingdom. *Energy Res. Soc. Sci.* **2016**, *21*, 190–198. [CrossRef]

54. Altun, M.; Meral Akgul, C.; Akcamete, A. Effect of Envelope Insulation on Building Heating Energy Requirement, Cost and Carbon. *J. Fac. Eng. Archit. Gazi Univ.* **2018**. Available online: https://dergipark.org.tr/tr/download/article-file/839664 (accessed on 31 May 2020).
55. Bayraktar, D.; Bayraktar, E.A. Mevcut Binalarda Isı Yalıtımı Uygulamalarının Değerlendirilmesi. *Mehmet Akif Ersoy Üniversitesi Fen Bilimleri Enstitüsü Dergisi* **2016**, *7*, 59–66. [CrossRef]
56. Kuygunsuz, K. Energy policy and climate change in Turkey. *Energy Explor. Exploit.* **2003**, *22*, 145–160.
57. Türk Standardlari Enstitüsü (TSE). Thermal Insulation Standards. 1998. Available online: https://intweb.tse.org.tr (accessed on 31 May 2020).
58. Türk Standardlari Enstitüsü (TSE). Thermal Insulation Standards. 2008. Available online: https://intweb.tse.org.tr/ (accessed on 31 May 2020).
59. Aktacir, M.A.; Büyükalaca, O.; Yılmaz, T. A case study for influence of building thermal insulation on cooling load and air-conditioning system in the hot and humid regions. *Appl. Energy* **2010**, *87*, 599–607. [CrossRef]
60. BEP, TR. *Binalarda Enerji Performansı Yönetmeliği*; T.C. Çevre ve Şehircilik Bakanlığı: Ankara, Turkey, 2008.
61. Yüksek Planlama Kurulu. *Kentsel Gelişme Strateji Belgesi (KENTGES). T.C. Resmi Gazete, 04/11/2010 Tarih ve 27749 Sayı*; Yüksek Planlama Kurulu: Ankara, Turkey, 2008.
62. Yüksek Planlama Kurulu. *Ulusal İklim Değişikliği Strateji Belgesi (İDES). Yüksek Planlama Kurulu, 3/5/2010 Tarih ve 2010/8 Sayılı Karar*; Yüksek Planlama Kurulu: Ankara, Turkey, 2010.
63. T.C. Enerji ve Tabii Kaynaklar Bakanlığı. *National Energy Efficiency Action Plan 2017-2023*; T.C. Enerji ve Tabii Kaynaklar Bakanlığı: Ankara, Turkey, 2012.
64. T.C. Enerji ve Tabii Kaynaklar Bakanlığı. Enerji Verimliliği Yönetmeliği. Available online: https://resmigazete.gov.tr/eskiler/2011/10/20111027-5.htm (accessed on 31 May 2020).
65. Madran, E. Cumhuriyetin İlk Otuz Yılında (1920–1950) Koruma Alanının Örgütlenmesi-1. *J. METU Fac. Archit. ODTÜ MFD* **1996**, *16*, 59–97.
66. Madran, E. *Tanzimattan Cumhuriyete Kültür Varlıklarının Korunmasına İlişkin Tutumlar ve Düzenlemeler: 1800–1950*; ODTÜ Mimarlık Fakültesi: Ankara, Turkey, 2002; ISBN 975-429-185-3.
67. Şahin Güçhan, N.; Kurul, E. A History of The Development of Conservation Measures In Turkey: From The Mid-19th Century Until 2004. *METU J. Fac. Arch.* **2009**, *26*, 19–44. [CrossRef]
68. Kültür Varlıkları ve Müzeler Genel Müdürlüğü. Law no. 2863 on the Conservation of Cultural and Natural Property. 1983. Available online: https://kvmgm.ktb.gov.tr/TR-43249/law-on-the-conservation-of-cultural-and-natural-propert-.html (accessed on 31 May 2020).
69. Kültür Varlıkları ve Müzeler Genel Müdürlüğü. 2019. Available online: https://kvmgm.ktb.gov.tr/TR-44798/turkiye-geneli-korunmasi-gerekli-tasinmaz-kultur-varlig-.html (accessed on 31 May 2020).
70. Future Brand Country Index. 2019. Available online: https://www.futurebrand.com/uploads/FCI/FutureBrand-Country-Index-2019.pdf (accessed on 20 April 2020).
71. Changeworks. Energy heritage: A Guide to Improving Energy Efficiency in Traditional and Historic Homes, Edinburgh. 2008. Available online: https://www.changeworks.org.uk/sites/default/files/Energy_Heritage.pdf (accessed on 31 May 2020).
72. Association for the Protection of Cultural Heritage (APCH) (Kültürel Mirası Koruma Dernegi). 2014. Available online: http://www.kmkm.org.tr/ (accessed on 31 May 2020).
73. Wheeler, B. Green Deal Could Lead to Deadly Summer Overheating. BBC News Online. 2013. Available online: http://www.bbc.co.uk/news/%20uk-politics-23180965 (accessed on 31 May 2020).
74. Collins, M.; Dempsey, S. Residential energy efficiency retrofits: Potential unintended consequences. *J. Environ. Plan. Manag.* **2018**, *62*, 2010–2025. [CrossRef]
75. May, N.; Griffiths, N. Planning Responsible Retrofit of Traditional Buildings. Sustainable Traditional Building Alliance. 2015. Available online: https://historicengland.org.uk/images-books/publications/planning-responsible-retrofit-of-traditional-buildings/responsible-retrofit-trad-bldgs/ (accessed on 31 May 2020).
76. Historic England. *Energy Efficiency and Historic Buildings: Application of Part L of the Building Regulations to Historic and Traditionally Constructed Buildings*; Historic England: London, UK, 2017.
77. Hinson, S.; Bolton, P. *Fuel Poverty*; The House of Commons Library: London, UK, 2020.
78. BEIS. *Annual Fuel Poverty Statistics in England, 2020 (2018 Data)*; Department for Business, Energy, and Industrial Strategy; BEIS: London, UK, 2020.
79. Selçuk, I.Ş.; Gölçek, A.G.; Koktas, A.M. Energy Poverty in Turkey. *Sosyoekonomi* **2019**, *27*, 283–299. [CrossRef]

80. Jones, P.; Lannon, S.; Patterson, J. Retrofitting existing housing: How far, how much? *Build. Res. Inf.* **2013**, *41*, 532–550. [CrossRef]

 *atmosphere* 

Communication

The Road Not Taken: Building Physics, and Returning to First Principles in Sustainable Design

Robyn Pender [1,*] and Daniel J. Lemieux [2,3,*]

1   Historic England, 4th Floor Cannon Bridge House, 25 Dowgate Hill, London EC4R 2YA, UK
2   Wiss, Janney, Elstner Associates, Inc., Northbrook, IL 60062, USA
3   Wiss, Janney, Elstner Limited, Bentinck House, 3–8 Bolsover Street, London W1W 6AB, VA 22042, USA
*   Correspondence: robyn.pender@historicengland.org.uk (R.P.); dlemieux@wje.com (D.J.L.)

Received: 22 April 2020; Accepted: 9 June 2020; Published: 11 June 2020

Abstract: The path we are currently following towards 'sustainable design' is a result of the accidents of the past 300 years of history. If we look further back, to before the exploitation of fossil fuels, we find a very different approach to building envelopes, and to building use and comfort. This was necessarily very low carbon, and demonstrably effective, but, unfortunately, we have forgotten many of the fundamental principles on which it rested. This paper argues that our current choice of retrofit pathway is leading us away from, rather than towards, a sustainable built environment. Current efforts to reduce carbon and energy based on modern 'layered' envelopes and misunderstandings of thermal comfort are proving much less effective than predicted. We would further argue that they are too often delivering unintended consequences: contributing to the overuse of carbon and energy, and derailing the development of a sustainable built environment. We draw on research and case studies, as well as on the lessons from history, to show how the problem derives from a neglect of first-principles thinking and fundamental building physics. Equally, though, we show how combining good building physics with a re-evaluation of older approaches to construction and building use delivers some powerful and effective tools for tackling the climate emergency.

Keywords: thermal comfort; durability; performance; life cycle analysis; historic buildings

1. Introduction

For many, in his poem 'The Road Not Taken', American poet Robert Frost celebrates 20th-century America as a culture rooted in risk-taking and 'can-do' individualism. However, this may well be a misinterpretation: critics point out that Frost rather seems pointing out that the road taken may not be a deliberate choice at all, but random, and only later justified in the traveller's mind as being the 'right and proper' path [1].

Frost's insight can help us to understand why our current efforts to drive down carbon and energy in the built environment have been rewarded with remarkably little success. As this paper tries to show, designers and retrofitters are currently embracing a picture of 'sustainable' architecture that is not so much a result of choosing between options as the end result of a series of accidents of history. The road we are following is not the only pathway towards reducing carbon, nor (more importantly) is it by any means the best. As we continue to follow it, it is forcing us into actions that risk being counterproductive as we battle to reduce carbon emissions. It is already clear that our efforts to create more sustainable buildings have been delivering unpleasant unintended consequences, with little demonstrable benefit in long-term carbon and energy reduction [2–4].

Essentially, the accepted road is problematic because it neglects both fundamental building science and some important lessons from the past, and instead rests on a series of unquestioned assumptions and misunderstandings. In this paper, we draw on history, research and field studies to try to explain how and when these unhelpful dogmas arose, and the implications they have had for the built environment. This, in turn, suggests where the opportunities might exist for swapping to a much more effective and productive pathway.

This is not uncharted territory, for the simple reason that for the many centuries prior to the industrial use of fossil fuels buildings had, from necessity, to be both durable and functional with very little input of energy (and certainly none of fossil fuels). Unfortunately, most of the methods people used to make and operate durable and flexible buildings have been almost entirely forgotten in a world that, for more than two centuries, has relied increasingly on simply adding in more fossil-fuel energy whenever it has met with difficulties in design or management.

2. Thermal Comfort as Air Temperature: A Flawed Paradigm That Has Led to Sealing Envelopes

The fundamental issue around our current approach is that it is based on a definition of thermal comfort as a function of air temperature. Despite this oversimplification being questioned regularly from the very beginning of the era of heating and cooling [5], currently it is almost universally accepted without question that for any indoor space to be comfortable and useable, the air temperature must be controlled. This idea has been central to the commodification of comfort, where the 'perfect' temperature is meant to be provided by a space-heating or air-conditioning system [6]. In fact, as many years of excellent research across the globe has unequivocally shown, thermal comfort is a very much more nuanced concept (perhaps better framed as thermal discomfort), and one in which air temperature plays, at most, a minor role [7–9]. Moreover, the gains in human comfort, health and productivity promised by this approach have not been delivered. At the same time, space heating and cooling are recognised as the principle contributors to the built environment's intensive use of energy and carbon; see, for example, [10–12].

It is also often forgotten that trying to control the air temperature in a space can easily have counter-productive effects on comfort. Put simply, heated air rises and creates draughts (Figure 1). Similarly, air conditioning requires the air to be drawn through ducting, to the great discomfort of any occupant unfortunate enough to be seated beneath an intake or outlet.

The dominance of space conditioning is particularly important for retrofit, because to give it a chance of success without huge wastage of energy, the building envelope must be sealed; and it must be sealed not merely by closing the windows, but by separating the interior from the exterior as completely as possible. Building scientists have been at the forefront of developing ways of achieving this, but it remains extremely challenging, and if poorly handled can lead to serious failures of the envelope such as condensation and the consequent deterioration of materials such as wood, lead and iron; see, amongst many examples, [13]. Sealing envelopes can also lead to problems with indoor air quality, for example high humidity, mould growth, and the trapping of indoor pollutants; see, for example, [14]. High humidity is also closely associated with thermal discomfort [15].

The technical issues resulting from sealing modern multi-layered construction are common currency for building performance assessors. It has also been well established that a particularly worrying consequence of the current approaches to retrofit has been the maladaptation of older buildings by, for example, sealing and insulating solid walls. This can cause the failure of materials and envelopes that had hitherto given excellent service over perhaps hundreds of years [16]. The French word for 'sustainability' is 'durability', and it is easy to see that the most durable building will be the one with the longest usable lifespan, requiring the least energy input for ongoing maintenance and operation.

There is extensive literature on air quality and building failure, but some other critical aspects of retrofit's effect on carbon expenditure and building usability do not seem to be widely discussed in the published literature [17]. One is the through-life carbon cost of retrofitting. The energy, carbon and other resources needed to install, operate and maintain retrofit measures will have a critical bearing on the building's long-term sustainability, and therefore to be able to assess the true impact of retrofit choices on carbon outcomes, we need to know how, why and when the retrofit materials and systems will fail in the field, and what would be needed to maintain them and to replace them when they reach the end of their life. For traditional building materials and systems, this knowledge is highly developed and freely available, but it is much less accessible for modern construction, and virtually non-existent for many retrofit materials (not least because these tend to be proprietary products). To take the example of air sealing and insulation: What do we know about the in-use durability and failure modes of materials such as housewrap? What would need to be done to repair building wraps should they begin to fail, or to replace them when they reach the end of their life? Would the interior and exterior wall surfaces of the building need to be stripped? If so, what would such a major intervention mean for building use and for overall carbon consumption?

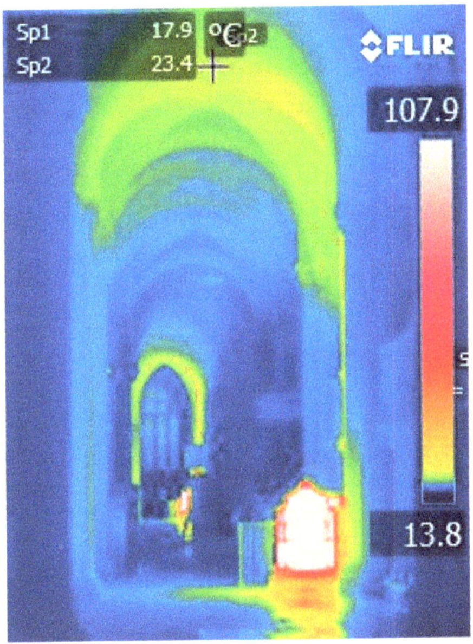

Figure 1. Infrared thermography shows heat generated by large gas-fired radiating units in an English Cathedral rising to the ceiling, causing cracking of the painted wood, strong draughts through the building, and very high energy consumption. Other common issues associated with heating in churches include underside corrosion of lead roof coverings. ©Tobit Curteis Associates.

Even more importantly, we need to find a way of balancing the carbon budget of retrofit measures that might reduce the lifespan of the building, or at least introduce an extra demand for maintenance and repair (both of which may use energy or generate carbon).

Perhaps because occupants do have some awareness of these problems (particularly for older buildings), the uptake of retrofitting in many countries remains low, even in the face of the climate emergency. Indeed, it appears that energy and carbon in the built environment may be continuing to rise [18]. It is clearly imperative that we urgently reconsider our direction of travel. However, is there an alternative route that could deliver better results?

It may help us to recall that our ancestors did not all die of cold or heat (any more than do those living in countries with limited access to high-carbon space-conditioning systems, but a strong tradition of vernacular architecture). They learnt from experience to understand the true causes of thermal discomfort, and knew exactly how those should be combatted.

3. Changes in the Concept of Thermal Comfort

The idea that building usability derives from air temperature is remarkably recent, and would probably have surprised even our grandparents: central heating did not become ubiquitous in Europe until well after the Second World War, and air conditioning is a still newer technology. Even fireplaces did not appear in ordinary houses in the UK until the 17th century, and for the next two hundred years they remained chiefly a means of cooking and providing light and comfort, rather than delivering warm air [6].

Until fossil fuels began to be used extensively, the key to thermal comfort was dealing directly with the causes of discomfort. The primary literature on how the body keeps itself safe as external temperatures change, or as it needs to lose heat in response to exercise, is to be found in the medical journals. There is broad agreement on the causes of heat loss, and on the approximate amounts of body heat given up to conduction, convection, and radiation:

- Some heat is lost by direct conduction into surfaces being touched by some part of the body: how it does this depends on the nature of the surface, and the area of the body that is touching it;
- Some heat is lost by convection into the air: if the air is still, this is no more than 2% of total heat loss. If the air is moving, and the skin is wet (for example, from perspiration), evaporation can raise this to as much as 22%;
- The primary cause of heat loss—60–65%—is the radiation of body heat into the surrounding surfaces [19,20].

These ways of losing heat were arguably much better understood at a time when people were obliged to listen to their own senses, rather than consult a thermometer. The current ubiquity of the thermometer blinds us to the fact it was not invented until the 18th century and not common until the 19th. Air temperature is now easy to measure, but it remains a poor proxy for comfort. Already in 1916 Sir Leonard Hill and his colleagues of the UK's Medical Research Committee noted [5]: "For purposes of controlling the heating and ventilation of rooms the thermometer has been used and has acquired an authority it does not deserve … it affords no measure of the cooling of the human body and is, therefore, a very indifferent instrument for indicating atmospheric conditions which are comfortable and healthy to man."

Interior clothing was more substantial than we are now used to, and buildings were partitioned into smaller spaces. 'Thermal delight' was provided by hearths; and simple tools such as hot bricks, charcoal foot-warmers, and lapdogs helped to heat people directly [21].

Most importantly, the principal comfort issue—heat loss by radiation—was dealt with simply and passively by imposing radiant heat breaks between the occupants and the heat-absorbing surfaces around them. These included some that are still familiar to us, such as mats on the floor, and others, such as wall draperies, that are largely forgotten. We have certainly forgotten their important role in comfort. Cloths could be hung across the entire wall surface or just behind where the occupant was seated, and draped into canopies to cut heat loss upwards. To cut draughts, they could be hung across doors and fireplaces (although in a time of small windows, solid construction, and few fireplaces, draughtiness was not yet the serious concern it would later become; indeed, it was generally considered desirable and healthy [22]).

Wall cloths are almost ubiquitous in contemporary depictions of interiors from the earliest times until the end of the 17th century. Studying these pictures can reveal many interesting details. For example, in chapels (almost the only spaces to have large areas of glazing prior to the industrial production of glass), drapery covered the base of the windows, presumably to capture the air chilled by the glass as it fell. We do not have a name for this type of intervention: it is not 'insulation', because the heat is not passing through the walls and ceiling. However, there is some insulation effect as well, even for light fabric: studies of the impact of net curtains on windows have shown that the temperature of the curtain can be 2.5 to 3.8 °C higher than that of the glass [23].

In England, the preferred means of covering walls was with cloth stretched onto battens and painted to imitate tapestry, and these were found everywhere from the most humble homes and taverns to stately houses. The more expensive option is the best known today: the tapestry, which was the exclusive province of extremely wealthy individuals or institutions (Figure 2). Interestingly, in medieval castles it is not uncommon to find hooks for hanging woollen tapestries directly above decorative wall paintings: we know from the house records that the tapestries were hung up in winter, but taken down again in spring (when they might be at risk from condensation on the wall). Bare walls would also be beneficial in summer, when it became desirable to lose body heat.

Figure 2. Bolsover Little Castle in Derbyshire, England, which dates from the beginning of 17th century, uses every one of the Stuart period options for providing comfort if one was extremely wealthy. In the Star Chamber, rush mats cover the floor, the walls are hung with cloths and tapestries that also block drafts through doorways, and the windows have both secondary glazing and shutters. Other rooms are similar, or paneled in timber. ©Historic England.

Timber was used in a similar fashion to cloth. Wooden paneling provided an excellent thermal break, and depictions of scholars and artists at their desks suggest the desks often had backs and hoods of timber. The elaborate timber canopies constructed in medieval choirs were not just highly decorative, but served a very practical purpose. The cloth hangings on beds are another example of the overlap between the treatment of rooms and furniture.

Occupants and builders alike were in a good position to learn from their experiments into improving comfort, and then to pass that learning on through the guild system. In London, one of the biggest and most powerful guilds was the 'Steyners', who made the painted cloths, but others are likely to have been involved as well: the name of the guild of upholsterers, 'The Worshipful Company of Upholders', suggests it was they who undertook the hanging of cloths.

4. A Paradigm Shift from Radiant Loss to Air Temperature

In England, the use of cloth draperies and matting to make buildings comfortable continued without break until the end of the 17th century. At this point, a number of significant events occur at once.

The first appears to be the Great Plague, which struck London in 1665, and lasted until the destruction of the medieval city by fire the following year. The Rebuilding Act of 1667 required houses to be built in brick or stone. This is usually seen as a response to prevent fire, but fear of the plague is likely to have played a part: the Lord Mayor's Orders of 1665 "Concerning the Infection of the Plague" specify that 'the goods and stuff of the infection, their bedding and apparel, and hangings of chambers, must be well aired with fire ... within the infected house, before they be taken again to use.' [24]. Certainly, depictions of 18th-century interiors show panelling, but no rugs or hanging cloths.

Secondly, in 1709, the first practical thermometer was invented in Germany, setting off a vogue for heat and temperature studies with Enlightenment scientists such as American emigré Benjamin Thompson (Count Rumford). The pivotal point, however—as with so much change in buildings—appears to have been the exploitation of coal as a fuel, and the subsequent dramatic drop in the price and availability of energy. With little by way of radiant breaks, Georgian houses must have been uncomfortable, so people began to turn increasingly to fireplaces to give relief from cold. Coal burning was dirty, though, and carried a high risk of carbon monoxide poisoning.

It was into this landscape that, in 1796, Rumford began promoting a new design for fireplaces that restricted the chimney opening, greatly increasing the updraught so that it carried away soot and fumes. The side walls of the fireplace were angled to reflect heat back into the room, and with this concept, the dominant role of fireplaces began to change from cooking to heating. Rumsford was an astute businessman, and his fireplaces quickly became very fashionable indeed [25].

Unfortunately, the occupants soon discovered the negative consequences of heating the air, coupled as it was with a strong draw though the chimney: draughts became a very serious problem. It is therefore no great surprise to see that within a few decades the fashions had changed back to rooms festooned with heavy curtains and rugs. These would be swept away once again more a century later, when the Modern Movement encouraged a renewed fashion for hard surfaces, this time made palatable by newly introduced building services such as central heating and air conditioning.

5. Changes in Building Envelopes

The emergence of building services is not the only important crossroad along this path: the exploitation of coal and other fossil fuels also led to dramatic changes in the materials and construction of building envelopes.

Traditional building systems operate on principles that are simple, but since they have become unfamiliar to modern architects, engineers and builders, they are perhaps worth explaining in a little detail. They are based on solid walls made of materials that are permeable, such as brick, stone, earth, timber, and lime-based mortars. Water vapour can travel between the voids in modern cavity construction, but it does not travel through permeable walls: if a vapour molecule enters the surface pores, its collisions with the pore walls rapidly cause it to condense. Of course, it may condense onto liquid water in the capillaries and then be drawn elsewhere in the wall by capillary action—perhaps even evaporating out the other side—but actual vapour movement in pores is extremely slow, and independent of the conditions outside the wall [26].

Liquid water could theoretically pass right through the interconnected pores, but in practice it does not, so long as the building is kept in reasonable condition. Sir Frederick Lea (1900–1984, head of the UK's Building Research Station) suggested that traditional construction might be compared to a greatcoat in the way it handles water. A raindrop hitting a 'greatcoat' wall will be held in the pores it hits on the surface, prevented from penetrating further by the pressure of the air it is trapping in the adjacent pores and capillaries. From the surface, it quickly evaporates again, often during the same rainstorm [27].

It is only if a raindrop should happen to hit a surface pore that connects to a capillary that is already filled with water that the rainwater will be drawn into the wall. Thick permeable walls will also resist heat transfer unless they are wet [28]. Traditional architecture is therefore characterised by features such as wide eaves and cornices, or hood mouldings and sills, that are intended to protect the bulk of the wall from rainwater entry at weak points, such as the wall heads (where gutter overflows could inject water into the bulk of the wall) or the window surrounds (where run-off from the glass could be drawn into the fabric). These protective features are often very decorative, but their primary purpose is practical (Figure 3) [29]. Builders learnt quickly from failure: Romanesque buildings, constructed before the invention of window glass, lack the protective window features that are so characteristic of Gothic architecture, with its large areas of stained glass. As windows became larger and larger, run-off from glass was emerging as a new problem [30,31].

Figure 3. The characteristic features of traditional 'greatcoat' architecture, such as wide eaves and hood mouldings, may appear decorative, but their primary purpose is entirely practical: to stop rainwater penetrating weak points of the wall. ©Clive Murgatroyd.

Making sheet glass requires huge amounts of energy, so glazing was rare in domestic architecture until coal began to be used for glassmaking at the beginning of the 17th century [32]. With clear glass windows, a new problem appeared in the form of solar gain. Even in winter, this could cause thermal discomfort [33]. Again, builders responded quickly, developing the vertically sliding sash window, which allowed the finest possible control over ventilation, and could be combined with shutters to allow night flushing without compromising security (Figure 4). Another important invention was the awning, which soon developed in sophistication to allow occupants control over not just solar gain, but ventilation too.

Figure 4. The vertically sliding sash window is one of the cleverest elements of building technology ever invented, allowing ventilation at different heights, and not slamming shut or open in high winds (the perennial problem of hinged casement windows). The flat profile allowed it to be used with awnings and shutters, to give the highest possible degree of control over sunlight, ventilation, and security. Sash windows arguably made the terrace house possible, which characterises townscapes of the 18th and 19th centuries. ©Robyn Pender.

Glass was just the beginning: when fossil fuels began to be used to make ferrous metals, architecture changed even more dramatically. The technologies that allowed steel to be made using coal unleashed a storm of innovation in architecture. Steel structural elements appeared first, alongside machine-made sheet glass, and later high-energy materials such as aluminium. Building development began to center around industrial production and research in enterprise and higher education, rather than learning 'on the job'. As a consequence, vernacular architecture—which developed in response to local materials and local climates—began to disappear.

The technology to make glass facades began with glasshouses, and architects spoke of the potential of this new type of construction for providing 'healthy' buildings in cities. Alas, the Crystal Palace (constructed in London in 1851 for the Great Exhibition) revealed inherent problems in overheating and condensation [34]. Although glass and steel continued to be used for train sheds and factories (which could not yet be lit artificially), this technology did not immediately transfer to other types of building. Early experiments by Liverpool engineer Peter Ellis in building steel-framed offices with glass-heavy facades did not prove popular, and glass and metal architecture might have remained a purely industrial phenomenon were it not for American architect John Wellborn Root, who was visiting Liverpool in the late 1860s when Ellis's building at 16 Cook Street was being constructed. In 1882, Root modified some of Ellis's initial innovations for his Montauk building in Chicago.

America provided a fertile ground in which to develop an entirely new type of construction. Two years after the Montauk building, another Chicago-based architect, William LeBaron Jenney, designed a ten-story building with a complete metal-frame. Then, in 1893, the Chicago World's Columbian Exposition brought together Francis John Plym and Edward Drummond Libbey, who would go on to be regarded as the founders of the modern metal and glass curtain wall. Plym founded the Kawneer Company in the aftermath of the great San Francisco earthquake and fire of 1906, whilst Libbey partnered with Michael Owens in 1912 to patent the world's first 'sheet glass drawing machine', making large sheets of glass commercially viable. Glass curtain walls began to appear in cities across the US.

Despite continuing problems with solar gain, these new building systems gained cachet when, in 1933, the Bauhaus school of architecture in Germany was closed by the Nazis. Many of its teachers found their way to America, where their Modern Movement aesthetic of hard surfaces of metal, glass, and concrete (another material made possible by the exploitation of fossil fuels) proved extremely popular. Eventually, it became fashionable across the world.

The First World War mobilised manufacturing industries in pursuit of a common goal, but it was the Second World War and its aftermath that led to some of the most significant and rapid changes in building materials and technologies. The aircraft industry initiated an explosive growth in aluminum manufacturing, and improvements in material quality and uniformity. New materials appeared, such as silicones, acrylics, and epoxies. These materials and related inventions such as the 'sandwich' panel allowed the curtain-wall industry to develop. As WW2 began to turn in the Allies' favour, the US found itself with a vast supply of surplus aluminum. Architects such as Pietro Belluschi recognised the potential of the embodied energy in this surplus, and embraced the opportunity to design buildings clad almost exclusively in aluminum. His 'Equitable Savings and Loan' building in Portland, Oregon, became " … the first to be sheathed in aluminum, the first to employ double-glazed window panels, and the first to be completely sealed and air conditioned" [35].

6. The Paradigm Shift from Greatcoats to Raincoats

The new envelopes were thin and light, and intended to be waterproof: Frederick Lea noted that they behaved like raincoats. Their surfaces do not have pores to hold the rain: instead, it beads and collects into flows that run down the facade under gravity. The weak points are the joints, which must be very well sealed to avoid run-off being wicked in through the skin, and it was soon discovered that the inevitable leaks were best dealt with by having more than one raincoat layer. These mass-produced, layered 'raincoat' wall systems marked a paradigm shift in the design of building envelopes.

Another problem proved even more challenging than rainwater penetration: raincoat facades trap water in both directions. Water inside the cladding cannot easily pass out through the walls to evaporate, whether that water is the result of rain penetration, plumbing leaks, or condensation. The bread and butter of modern building performance assessment is identifying these types of failures, and finding ways of remediating them.

Despite increasingly obvious issues, after the Second World War raincoat technology completely eclipsed the traditional 'tried-and-true' greatcoat systems that had been the province of practical builders. So many masters and apprentices died during the wars of the first half of the 20th century that the loss of hands-on knowledge was all but complete. By 1946, Swedish scientist C. H. Johansson would feel able to assert [36]: "It is clearly unwise to allow walls, whether of brick or porous cement, to be exposed to heavy rain. They absorb water like a blotting paper and it would be a great step forward if an outer, water-repelling screen could be fitted to brick walls." Similar sentiments remain common today, and indeed Johansson is still much cited by architects and engineers. However, it is very easy to demonstrate that neither blotting paper nor that other familiar analogy for brick, a sponge, will absorb water if they are dry: they resist water uptake until there are some water-filled capillaries to draw the moisture in.

Unfortunately, ignorance of solid-wall construction has led to what amounts to an industry in maladaptation, not least adding coatings that attempt to provide waterproofing to greatcoats. Coatings cannot keep all the water out of the wall—indeed, because they often lead to beading and run-off, they can actively encourage rain penetration – but they greatly slow evaporation, so the moisture content of the wall builds up over time. This has two very well-known consequences: firstly, wetting of the wall (with all the resulting problems, including increased rain penetration); and secondly, powdering and spalling of the surface as salts are deposited below the coating or at the interface between treated and untreated material. Air pressure in the pores may be an additional failure mechanism [16,37].

With body heat loss through radiation as a source of discomfort now poorly appreciated, the thermal behaviour of solid-wall construction is also being misinterpreted. Sealing the envelope, however important it may be for lightweight construction relying on space heating and cooling, is highly detrimental to greatcoat buildings, but maladaptation is common, and indeed is being encouraged by the application of building models that are completely unsuited to either this type of architecture, or to its proper modes of operation [38]. To give the most obvious example, adding insulation to thick solid walls is unnecessary at best, and will slow or prevent evaporation. The end result can be counter-productive: if water gets into the wall (whether from condensation or, more likely, from leaks in plumbing or rainwater goods), moisture levels will build, and the wall will start to transfer heat.

7. The Commodification of Comfort

As the old ways of dealing with discomfort were forgotten over the course of the 20th century, air heating and cooling began to be a common feature of buildings. Raincoat architecture—particularly the curtain-walled skyscrapers—relied ever more heavily on the fledgling building-services industry, which provided (amongst many other requirements) lifts and electric lighting. In 1921, Willis Haviland Carrier patented his 'centrifugal chiller', the first practical approach to controlling the humidity and temperature of the air. Eventually, Carrier's technology gave birth to the idea of "comfort cooling", and a Heating, Ventilation and Air Conditioning (HVAC) industry that would position itself as a very timely means for making the ever-larger spaces behind flat-glass curtain walls livable [38]. Solar gain was no longer tackled with awnings, but with air-conditioning, and then (when this proved insufficient) with air conditioning plus internal blinds. It is ironic that, in many glass-walled buildings, the blinds are almost always drawn to make internal conditions bearable.

With energy cheap, the wastefulness inherent in air conditioning was not yet considered to be a problem, and neither were its impacts on neighbouring areas, including the contribution it was making to urban heat islands. Most worryingly, perhaps, the centralised control of air temperature (whether by heating or cooling) meant that comfort became a commodity to be purchased, rather than something to be achieved by occupants reacting to the quirks of their own building, and to how they were using it [39,40].

A homeowner or a facilities manager can now buy a system that promises to make occupants perfectly comfortable by keeping the air temperature in a narrow band. However, setting aside the technical challenges of this (especially in spaces with partitions and furniture), the ideal temperature will inevitably be different for different occupants, and even for the same occupants at different times. Radiation of body heat and solar gain both play a large part, but so does the level of activity. If the atmosphere is damp, it will feel much colder in cold weather and much hotter in hot weather, regardless of air temperature [5].

With so many factors contributing to thermal comfort, it is not to be expected that any space heating or cooling system could possibly deliver perfect comfort to everyone using the building. However, having been sold a dream of comfort, the common response of occupants to discomfort is to assume they are not running the system hard enough, and so they will over-ride the controls and adjust the thermostat. One can speculate whether this is the underlying cause of the well-known 'rebound effect'; see, for example, [41].

Despite the many limitations of space-conditioning, it has been marketed extremely successfully, even to occupants who are apt to complain about the results [33]. This is underlined by the responses made by readers to a recent New York Times article questioning the wisdom of near universal air conditioning in the US [42]. Many respondents claimed the hottest parts of America would be uninhabitable without it, even though the article had pointed out that the take-up of air-conditioning in still hotter climates is currently very low. With the market in the US now nearing saturation (well over 90% of buildings have air conditioning), manufacturers will be looking to expand into new countries [43]. Of the yearly increase in world energy demand, 21% is currently attributed to the increasing use of air conditioning [44,45]. With the climate now warming rapidly, this prospect becomes even more alarming.

In Europe (where air conditioning is becoming more and more 'standard'), a similar narrative surrounds central heating. 'Fuel poverty' supposes that the health of people unable to heat their houses to certain air temperatures will suffer, although the evidence linking deaths with indoor air temperature extremes show more deaths from heat stress than from cold; see, for example, [46]. The narratives of fuel poverty, energy efficiency, and climate change have become unhelpfully entangled, leaving occupants confused about the best response to reducing energy consumption and carbon outputs.

8. Where Has Our Current Road Taken Us

We have now travelled so far down our post-industrial road that we have arrived in the somewhat bizarre situation of assessing traditional construction not on its own merits, but on how much it differs from contemporary norms. Older buildings are stigmatised as 'hard-to-treat', or energy-hungry, despite the evidence of several thousand years of proven effectiveness in a low-carbon, low-energy environment, and the well-attested problems of modern construction failing to deliver promised energy efficiencies.

Nonetheless, traditional greatcoat construction remains an excellent and robust envelope system. It is difficult to imagine a more sustainable domestic building than a thatched cob cottage in wet, cold Devon in England. The thick solid walls—earth mixed with straw on a stone plinth—are excellent insulators and extremely durable, and were made with local earth and human labour. The thatch (made of locally grown straw) is also superbly insulating, and can be maintained simply by regular re-ridging and "spar coating" (replacing only the deteriorated outermost layer of thatch, which means that building has never to be roofless even for a short period). Maintenance of the walls consists of little more than regular whitewashing or mud rendering, and glazing is minimal [47]. The building itself is flexible, and can be changed and extended with relative ease to suit changing uses. Inside, thermal comfort can be provided as it was originally by cloth hangings or wooden panelling, and mats on the floor, supplemented as necessary with elements to heat the people rather than the air. The introduction of plumbing has given new challenges, but these are relatively easily dealt with by approaching installation and maintenance with sufficient care.

Many such buildings are more than 600 years old, and indeed they have no inbuilt obsolescence. Kept maintained, they could survive indefinitely. Even climate change should have little impact: very similar buildings are constructed in Sub-Saharan Africa, where the thick earth walls provide superb insulation against the heat.

9. The Situation Today: Collecting Points for Sustainability

As building practitioners in a world faced with a climate crisis, we are often drawn to new products and materials not just for the aesthetic options they offer, but also for the performance cited by industry: "thinner, lighter, more cost-effective and energy-efficient". This approach to design and retrofit is strongly reinforced by the promotion of points-based building rating systems such the Leadership in Energy and Environmental Design (LEED®) of the US Green Building Council (USGBC), or the UK's Building Research Establishment Environmental Assessment Method (BREEAM).

Where is collecting points leading us? Current rating systems threaten to replace first-principles thinking with checklists, and, until recently, most have largely ignored truly quantifiable outcome-based design.

In the US, early implementation of the LEED® green building rating system created new–perhaps unanticipated—challenges around the durability and performance of materials. A good example of unintended outcomes is given by a commercial property that was among the first to be awarded a LEED® Platinum rating (the highest possible). The building was designed and constructed almost entirely of rapidly renewable and recycled materials and remains a fitting reflection of the mission of the owner, a not-for-profit, environmental advocacy group. A particular feature was the engineered structural members being expressed on the exterior of the envelope (Figure 5). Unfortunately, the selection of materials, detailing and their exposure in a coastal region created unforeseen challenges for long-term maintenance and care (Figure 6) [48].

Figure 5. Engineered wood structural members expressed on the exterior envelope of the building. Photo: Wiss Janney Elstner, Inc.

The intentions had been good, to be sure, but the outcome was a building uniquely vulnerable to water penetration, and one that proved extraordinarily difficult and costly to repair. By the time substantial completion was reached, uncontrolled rainwater penetration was already widespread. After only ten years of service, we found significant decay in the structural members. To allow the compromised members to be removed and replaced with the least impact on day-to-day operation, it was necessary to construct an externally applied temporary structural steel framing. The carbon and energy costs of these problems, and of their solutions, will have had a significant impact on the sustainability of the building, as well as its maintenance demands and lifespan.

Similar issues are evident when existing buildings are retrofitted to meet new targets, in one infamous case with truly tragic results. The 2013 BREEAM Pre-Assessment report for the refurbishment of Grenfell Tower in London suggested that the proposed refurbishment project could potentially achieve a BREEAM rating of 'Good', with the highest score (14.83%) achievable under 'energy' [49]; the project was given an overall score of 69% for the materials chosen. The renovation was completed in 2016, but in 14 June 2017 the tower was completely destroyed by fire, with the loss of 72 lives (Figure 7).

At time of writing, the enquiry into the complex causes of the spread of the fire was ongoing. It is interesting to note, however, that it appears that the material selection, placement and detailing which were intended to optimize climate-specific heat, air, and moisture transport performance across the building envelope may not have taken full account of the system's combustibility and potential reaction to fire.

The clear lesson is that materials and design must always be considered holistically. The results of the investigation will reveal more, but the tragedy has already taught us that, as stewards of our built environment, accountability during sustainable design and construction is both warranted and necessary [50]. Failing to return to the first principles of building science and to take a holistic view of the building had devastating consequences. We sympathise with the determination of the Grenfell survivors to ensure that the lessons of the fire lead to changes both in retrofitting and the design of new buildings.

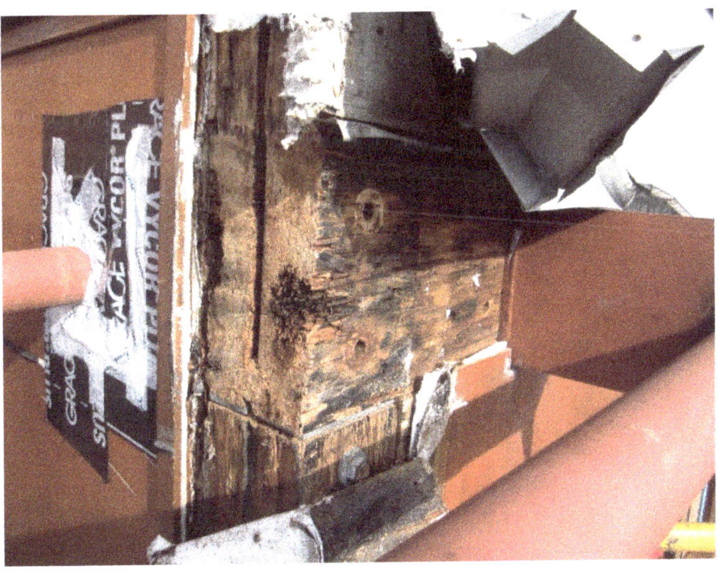

Figure 6. Moisture ingress and decay in a typical engineered wood structural member. Photo: Wiss Janney Elstner, Inc.

Figure 7. The devastating fire in Grenfell Tower in London, which cost 72 lives, was fueled, in part, by the same components we seek to improve thermal performance and reduce energy use in buildings. What can we learn from this? Photo: D.J. Lemieux.

10. Where Do We Need to Be? Developing a New Roadmap

While we cannot unring the bell of history, we need to be more acutely aware that large-scale fossil fuel exploitation has shaped our built environment in curious ways. There is no doubt, either, that the architecture of the future will need to be more environmentally conscious, low-carbon, and sustainable in the broadest sense. If we are to design safe buildings for a zero-carbon future, we must seek to question our current ways of thinking about building envelopes, building services and building use.

In the light of the climate emergency, it is clear we are well overdue for another paradigm shift: one that will finally allow us to reconcile our commoditised present with the many lessons afforded by pre-industrial and vernacular construction.

Can we move away from the road we have been following so blindly, and which has proved so singularly wasteful and damaging? Faced with the challenge of rethinking what we do, right down to its first principles, how do we make the best use of the lessons of history, good and bad? For example, returning to some form of solid-wall construction should allow us to build much longer-lived buildings that do not fail when sealants come to the end of their lives, and need little if any additional insulation. We should not seek to mimic or recreate what has been lost, but rather to find new ways of integrating the tried-and-tested building principles developed before the Industrial Revolution with the best things we have learnt since.

A major stumbling block to achieving a better built environment has been the loss of effective feedback loops; there has been nothing to replace the guilds and apprenticeship systems that served communities so well for so long. Despite being recognised as highly desirable, Post-Occupancy Evaluation (one of the few structured ways of learning from real buildings in real operation) remains very rare. We have shifted towards a reliance on standards, and no longer have a robust way to learn quickly from our mistakes, or to use that learning to refine future practice. It is tempting to simply recommend that better practice be led by professional and vocational organisations, but care would be needed not only to deal with overlapping areas of interest, but perhaps more importantly with the many gaps that are currently not dealt with by any recognised profession. Building performance evaluation, which does cover all the ground from building to occupation, does not yet have a clear structure for training or passing on knowledge.

Currently, we are relying on codification and measurement for passing on practice, and it is here that we would argue that our immediate efforts as building scientists must concentrate. One of the reasons for the unnatural dominance of air temperature is that it is so easy to measure, and the same argument could be advanced for the obsession with R- and U-values. Can we find ways of assessment that take us closer to what it is we really want to know: that is, whether the occupants feel comfortable?

Assessment would need to incorporate not only physical factors (such as loss of body heat by radiation), but also the sense of comfort derived from having immediate control. It will need to take account of the interconnectedness of many sources of discomfort, and also of the measures that could be put in to deal with that discomfort. There must invariably be strong inputs from medicine and from sociology. To take the simple example of an awning: if we restrict our analysis to the impact the awning has on the air temperature of the room by reducing the heating of the glass (and the radiation of that heat), we will fail to appreciate its many other direct and indirect benefits, including preventing the direct solar heating of occupants and the surfaces they are working on, beneficial changes in air circulation, and (not least) the subtle benefits accrued by handing control to the occupant. Perhaps a better marker of success, at least in the early stages of our new journey, would simply be a reduction in the building's demand for space heating and cooling.

Lifecycle analysis (LCA) is another critical component of sustainability, and this must include not only the carbon and energy used to install and run (say) an air-cooling system, but also that used to maintain and repair it, and to decommission and replace it when it reaches the end of its life. These days, many building services are so deeply integrated into the building fabric that efficient maintenance is very difficult; this further reduces what is already a short lifespan for the equipment. Ease of maintainability, durability, and maximum lifespan must be key criteria not just for selecting materials, but when designing complete mechanical services, retrofits, or buildings.

When calculating LCA, it is vital to clearly distinguish between those mechanical services that are integral to the basic use of the building and should therefore be considered as part of the building fabric (for example, lifts and pumps for high-rise structures), and those that are ancillary, such as space heating.

LCA presents a serious challenge if you are attempting to extract very accurate carbon costings: especially in light of the variations in installation and future care that can greatly affect longevity, but also because so many products are proprietary. On the other hand, because consistency is unlikely, trying to be very precise may in fact produce unreliable results. Fortunately, to make choices that do reduce carbon we need to know only enough about the carbon costs of options to be able to triage retrofit options into broad categories: green (little or no carbon involved in creating or using, and little or no risk to the longevity of the building), red (significant carbon inputs or significant impacts on the building); or amber (difficult to label as 'red' or 'green' without further investigation). Interventions using materials obtained locally, with long lifespans, can usually be quickly labelled as green, whereas materials involving high-carbon materials or manufacturing processes, or significant transport, will probably be 'red'. We are then free to concentrate on the more ambiguous 'amber' options. There, too, the aim should be to undertake just enough research to be able to determine whether the option is green or red.

An important aspect of this is that any LCA assessment must include lifespan and whole-life costs over the long term, and all building professionals must get used to thinking of 'the long term' as centuries rather than decades (as building conservators have always been obliged to do when planning conservation and repair).

We will also need to find ways of incorporating the carbon and energy benefits of avoided costs. For example, if using radiant heat loss breaks and local heating in a building allows us to reduce—or, even better, eliminate—space heating, then we will also be able to avoid thermally sealing the envelope to prevent the loss of conditioned air. We will thus be able to avoid not simply the carbon costs of the air heating itself, but also those of the associated retrofitting.

11. Conclusions

Any road directed to a truly sustainable future would lead unambiguously to buildings that are long-lived, good for their occupants, and frugal in their energy and carbon demands. The route we have been travelling along so long and so hopefully may have been paved with good intentions, but its direction is debatable, and it is raises some serious questions, including:

- How is whole-life costing being calculated in new construction for exterior envelopes, and for retrofitting? What are the modes of failure for these interventions? What impact do they have on material durability and performance? Are they always safe, or even desirable?
- Do some retrofit measures decrease the durability of the original construction? If so, what are the consequences for total energy and carbon use over the longer term?
- Are older, vernacular buildings being painted as the villain of climate change simply because we have been comparing apples with oranges? Many important functional elements which originally allowed them to operate have been stripped away. If these were returned, and occupants taught their purpose and how they should be used, would it be possible to reduce or even eliminate space heating and cooling once again? Without space conditioning, there is no need to try to make these buildings airtight or more thermally massive than they already are, avoiding the consequent problems for the fabric and the indoor air quality. How would their energy use and carbon output then compare to those in contemporary buildings, especially over the longer term?

As this paper has tried to demonstrate, history provides many lessons and examples for achieving a low-carbon built environment. What steps do we need to take to build these in to modern low-carbon design?

In terms of research, we should look more attentively through the historic records and at vernacular construction, seeking to understand the functional reasons behind peculiar design features. This should not stop at envelope design: researchers need to be on the lookout for any original fixtures, fittings, furnishings, and patterns of use that enabled these buildings to be effective in a low-carbon environment. From this, we should gain many more useful but forgotten tools that could easily be adopted and adapted, as well as shedding light on our own assumptions.

Assessment is another key issue. We will need to develop reliable methodologies for quantifying thermal comfort (or perhaps discomfort) that take proper account of the occupants and the way they are using the building. Psychology can be a powerful tool. If occupants do feel that radiators are more effective if painted red, or have less need for heating is shown a film of a roaring fireplace, then we would be foolish not to take full advantage of this as part of a low-carbon strategy.

We also need to find robust processes for assessing the through-life carbon costs of both construction and retrofitting, which incorporate lifespans and maintenance needs. Accuracy is probably impossible and undesirable: the aim should be a process that allows us to quickly and transparently triage competing options [51]. In the past, knowledge was developed by trial and error, but we no longer have the luxury of the time that requires, nor do we have the building-skills training systems such as the Guilds that once supported it. Building science can and must step into this breach. Building scientists need to find ways of understanding and assessing impacts that are much more pertinent—albeit possibly more difficult to measure – than air temperature or U-values. Equally, we need to continue developing LCA, and striving to agree simple assessment methods that can support rapid common-sense decision-making.

Climate change is urgent, and the primary remit of building scientists must be to work together to develop the tools the sector needs as quickly as possible. The related responsibility will be to disseminate these tools. We have struggled to pass on to design and construction professionals even the most basic knowledge of building performance, and this must change.

Communication is no less important than research, and it is perhaps even more challenging. How do we influence not only building professionals, but owners, occupants, and policy makers, in a world that no longer has many effective systems for acquiring and passing on best practice? Part of the trick surely lies in making potentially challenging new messages compelling and coherent, and presenting them in ways that make clear sense to our audiences.

Finally, we must disentangle current building practice, which is fraught with problems that we have only alluded to in this paper, but which have a strong bearing on the current situation. For example, the appetite for proprietary building products (often new and untested), is encouraged by a desire for guarantees and warranties, in the hope of transferring risk and providing assured outcomes. Clearly, this is wishful thinking, but it has been a very real barrier to the uptake of traditional materials and systems in modern construction, despite their in situ behaviour being well understood. Siloing of expertise is another problem with very similar underlying causes.

Breaking down these and similar roadblocks will take time and determination. To facilitate that process, we suggest that a few essential aims be embraced by the sector:

1. The training and education of building professionals at all levels should be cross-disciplinary, and begin with the fundamentals of building science. Teaching must cover greatcoat as well as raincoat architectural systems: not only to ensure better outcomes when we are repairing or refurbishing older buildings, but also to give professionals the confidence to draw on a much wider range of building materials and systems when designing sustainable new buildings;

2. We must seek to identify and question all our perceived wisdoms, especially those concerning thermal comfort and sustainability, seeking deep answers to the question of what is really needed to make buildings useable, agreeable, and low-carbon;

3. Climate-driven vernacular design must be re-evaluated as a source of lessons to be drawn on not just for the design and construction of new buildings, but also for their operation;

4. We need to be sure the buildings we design or refurbish do not have embedded requirements for carbon-intensive services. Sustainable buildings are those that can be readily maintained and run for very long periods at minimal carbon cost;

5. Rather than expecting that building services overcome fundamental shortcomings in the envelope or the conception of building use, we must seek to design buildings that are inherently low-carbon in both construction and maintenance, and would able to function in a low-carbon manner in both current and future climates for a wide range of occupants;

6. We must also reconsider the familiar ways we approach services: not just heating and cooling, but lighting, water supplies and sewage. As well as critiquing current practice, this means being much more clear-sighted about the true needs of occupants. We need to work out the best ways of addressing those needs in a low-carbon manner, and we need to be able to communicate this best-practice advice to policy makers as well as building professionals;

7. The unhelpful 'old and new' dichotomy should be abandoned, along with perjorative but ill-informed labels such as 'hard to treat'. Instead, buildings should be assessed on their success in fulfilling the real demands placed on them by real occupants, and their actual outputs of carbon. This means turning away from an excessive reliance on theory and modelling, and towards field

assessment. It is also likely to mean a more broadly applied and creative consideration of what is meant by such terms as 'comfort' and 'passive control'.

Taking these first steps will not be easy, but they will set us firmly back on the road towards a sustainable future.

Authors: A Principal and Director with Wiss, Janney, Elstner Associates, Inc. and Wiss, Janney, Elstner Limited in London, Dan Lemieux is a registered architect. He a leader in the development of international standards for Building Enclosure Consulting and Commissioning, and serves as a Chair on ASTM Committee E06, Performance of Buildings. Dan has more than 25 years of professional practice in commercial building design, including hands-on building envelope assessment, repair design, performance testing and global supply-chain technical support for products and materials sourced from the Americas, UK, EU, UAE, Canada and China.

A technical specialist at Historic England, the arm's-length body advising the UK Government on all aspects of the historic environment, Robyn Pender is also a Commissioner on the planning body for English Cathedrals, and serves on the Environmental Working Group of the Church of England and the editorial board of the Journal of Architectural Conservation. A physicist with a postgraduate degree in wallpainting conservation, her doctoral research examined moisture transport in building materials. She specialises in building performance and the impacts of climate change, two strands which are becoming ever more closely entwined.

Author Contributions: This article was written as a collaboration between the two authors, and reflects their experiences working as building performance specialists in the US and UK. All authors have read and agreed to the published version of the manuscript.

Funding: This research received no external funding.

Acknowledgments: The authors would like to thank their colleagues at Historic England and Wiss, Janney, Elstner Associated for their support both in preparing this article, and when undertaking the research on which it draws. Thanks also to Tobit Curteis for his input, and for the infrared image of heating.

Conflicts of Interest: The authors declare no conflicts of interest.

References

1. Orr, D. The most misread poem in America. In *On Poetry*; The Paris Review: New York, NY, USA, 2015.
2. Preston, J. The Each Home Counts report and traditional buildings. *Context J. Inst. Hist. Build. Conserv.* **2017**.
3. de Selincourt, K. Passiv. House Plus 2018, No. 24. Available online: web.archive.org/web/20200419135721/https://passivehouseplus.ie/news/health/disastrous-preston-retrofit-scheme-remains-unresolved (accessed on 19 April 2020).
4. Jessel, E. Grenfell inquiry: Tower refurbishment failed to comply with building regulations. *Arch. J.* **2019**. Available online: web.archive.org/web/20200419140027/https://www.architectsjournal.co.uk/news/grenfell-inquiry-tower-refurbishment-failed-to-comply-with-building-regulations/10045036.article (accessed on 19 April 2020).
5. Hill, L.; Griffith, O.W.; Flack, M. The measurement of the rate of heat-loss at body temperature by convection, radiation, and evaporation. *Philos. Trans. R. Soc. B Biol. Sci.* **1916**, *207*, 183–220.
6. Crowley, J.E. *The Invention of Comfort: Sensibilities and Design in Early Modern Britain and Early America*; John Hopkins Univesity Press: Baltimore, MD, USA, 2001.
7. Baker, N.; Standeven, M. Thermal comfort for free-running buildings. *Energy Build.* **1996**, *23*, 175–182. [CrossRef]
8. Baker, N.; Standeven, M. A behavioural approach to thermal comfort assessment. *Int. J. Sol. Energy* **1997**, *19*, 21–35. [CrossRef]
9. Schweiker, M.; Huebner, G.M.; Kingma, B.R.; Kramer, R.; Pallubinsky, H. Drivers of diversity in human thermal perception—A review for holistic comfort models. *Temp. Med Physiol. Beyond* **2018**, *5*, 308–342. [CrossRef] [PubMed]
10. UK Green Build. Counc. Website 2020. Available online: web.archive.org/web/20200419182024/https://www.ukgbc.org/climate-change/ (accessed on 19 April 2020).

11. Henley, J. World Set to Use More Energy for Cooling Than Heating. *Guard* **2015**. Available online: web.archive.org/web/20200419181731/https://www.theguardian.com/environment/2015/oct/26/cold-economy-cop21-global-warming-carbon-emissions (accessed on 19 April 2020).

12. Goetzler, W.; Guernsey, M.; Young, J.; Fujrman, J.; Abdelaziz, A. The Future of Air Conditioning for Buildings. In *Report Prepared for the US Department of Energy July 2016*; Department of Energy: Washington, DC, USA, 2016. Available online: web.archive.org/web/20200419182448/https://www.energy.gov/eere/buildings/downloads/future-air-conditioning-buildings-report (accessed on 19 April 2020).

13. Beasley, K.J. Building facade failures. *Proc. Inst. Civ. Eng. Forensic Eng.* **2012**, *165*, 13–19. [CrossRef]

14. Berkeley Lab. Indoor Air Quality Scientific Findings Resource Bank: Building Energy Efficiency. Available online: web.archive.org/web/20200419182639/https://iaqscience.lbl.gov/cc-building (accessed on 19 April 2020).

15. Vellei, M.; Herrera, M.; Fosas, D.; Natarajan, S. The influence of relative humidity on adaptive thermal comfort. *Build. Env.* **2017**, *124*, 171–185. [CrossRef]

16. May, N.; Griffiths, N. *Responsible Retrofit of Traditional Buildings*; Responsible Retrofit Series; Sustainable Traditional Buildings Alliance: London, UK, 2015. Available online: web.archive.org/web/20200419182822/http://stbauk.org/stba-guidance-research-papers (accessed on 19 April 2020).

17. Heritage Counts. *There's No Place Like Old Homes*; Report prepared for Heritage Counts by Historic England; Historic England: London, UK, 2019. Available online: web.archive.org/web/20200525152902/https://historicengland.org.uk/research/heritage-counts/2019-carbon-in-built-environment/carbon-in-built-historic-environment/ (accessed on 15 May 2020).

18. UN Environment Programme. Emissions Gap Report—Report prepared for the UN Environment Programme. 2019. Available online: web.archive.org/web/20200419182957/https://www.unenvironment.org/resources/emissions-gap-report-2019 (accessed on 19 April 2020).

19. Baumgart, B.; Chandra, S. Temperature regulation of the premature neonate. In *Avery's Diseases of the Newborn*, 9th ed.; Gleason, C.A., Devaskar, S.U., Eds.; Elsevier: Philadelphia, PA, USA, 2012; pp. 357–366.

20. McCutcheon, L.J.; Geor, R.J. Thermoregulation and exercise-associated heat stress. In *Equine Exercise Physiology*; Hinchcliff, K., Kaneps, A., Geor, R., Eds.; Saunders Elsevier: Philadelphia, PA, USA, 2008; pp. 382–396.

21. Manwell, C.; Baker, C.M.A. Domestication of the dog: Hunter, food, bed warmer, or emotional object? *J. Anim. Breed. Genet.* **1984**, *101*, 241–256. [CrossRef]

22. Thompson, F. *Lark Rise to Candleford*; "But Lark Rise must not be thought of as a slum set down in the country. The inhabitants lived an open-air life; the cottages were kept clean by much scrubbing with soap and water; and doors and windows stood wide open when the weather permitted. When the wind cut across the flat land to the east, or came roaring down from the north, doors and windows had to be closed; but then, as the hamlet people said, they got more than enough fresh air through the keyhole..." There were two epidemics of measles during the decade, and two men had accidents in the harvest field and were taken to hospital; but, for some years together, the doctor was only seen there when one of the ancients was dying of old age, or some difficult first confinement baffled the skill of the old woman who, as she sadi, saw the beginning and end of everybody; Oxford University Press: Oxford, UK, 1945.

23. Anderson, J. *Net Curtains: British Paranoia or Environmental Friend*. Unpublished case study. 1997.

24. Defoe, D. *Journal of a Plague Year*; E. Nutt: London, UK, 1722.

25. Brown, G.I. *Count Rumford: The Extraordinary Life of a Scientific Genius*; Sutton Publishing: Stround, UK, 1999.

26. Hall, C.; Hoff, W.D. Water Transport. In *Brick, Stone and Concrete*; CRC Press: Boca Raton, FL, USA, 2011.

27. Pender, R.J. The behaviour of water in porous building materials and structures; Reviews in Conservation 5. *Stud. Conserv.* **2004**, *49* (Suppl. 1), 49–62.

28. Historic England. Research into the Thermal Performance of Traditional Brick Walls; English Heritage Research Report 70/2013. 2013. Available online: web.archive.org/web/20200419183319/https://research.historicengland.org.uk/Report.aspx?i=15741 (accessed on 19 April 2020).

29. Historic England. *Practical Building Conservation: Building Environment*; Routledge: London, UK, 2014.

30. Historic England. *Practical Building Conservation: Glass & Glazing*; Routledge: London, UK, 2012.

31. Godfrey, E.S. *Development of English Glassmaking 1560–1640*; University of North Carolina Press: Chapel Hill, NC, USA, 1976.

32. Khan, S. *Learning from History: Traditional Low-Energy Approaches to Comfort*; Research Report for Historic England; Routledge: London, UK, 2020; in press.

33. Schoenefeldt, H. The Crystal Palace, environmentally considered. *Arch. Res. Q.* **2008**, *12*, 283–294. [CrossRef]

34. Wright, S.H. *Sourcebook of Contemporary North. American Architecture: From Postwar to Postmodern*; Van Nostrand Reinhold: New York, NY, USA, 1989; p. 11.

35. Johansson, C.H. The influence of moisture on the heat conductance for brick. *Byggmastaren J. Swed. Build.* **1946**, *7*, 117–124.

36. Ioannou, I.; Hall, C.; Wilson, M.A.; Hoff, W.D.; Carter, M.A. Direct measurement of the wetting front capillary pressure in a clay brick ceramic. *J. Phys. D Appl. Phys.* **2003**, *36*, 3176–3182. [CrossRef]

37. Rose, W.B. *Water in Buildings: An Architect's Guide to Moisture and Mold*; Wiley: New York, NY, USA, 2005.

38. Lemieux, D.J.; Driscoll, M. History of the 20th Century metal and glass curtainwall. In Proceedings of the Restoring Postwar Heritage, Selections from the 2004 DOCOMOMO US Technology Seminar, New York, NY, USA, 26 September–2 October 2004; 2008 DOCOMOMO Preservation Technology Dossier 8.

39. Heschong, L. *Thermal Delight in Architecture*; MIT Press: Cambridge, MA, USA, 1979.

40. de Dear, R. Thermal counterpoint in the phenomenology of architecture—A psychophysiological explanation of Heschon's Thermal Delight. In Proceedings of the Keynote speech, PLEA 2014, Ahmedabad, India, 16–18 December 2014.

41. Economic and Social Research Council. The Rebound Effect. n.d. Available online: web.archive.org/web/20200420075856/https://esrc.ukri.org/about-us/50-years-of-esrc/50-achievements/the-rebound-effect/ (accessed on 20 April 2020).

42. Green, P. Do Americans need air conditioning? *New York Times*, 3 July 2019. Available online: web.archive.org/web/20200514102615/https://www.nytimes.com/2019/07/03/style/air-conditioning-obsession.html (accessed on 14 May 2020).

43. Buranyi, S. The Air Conditioning Trap: How Cold Air Is Heating the World. *The Guard*, 29 August 2019. Available online: web.archive.org/web/20200419183859/https://www.theguardian.com/environment/2019/aug/29/the-air-conditioning-trap-how-cold-air-is-heating-the-world (accessed on 19 April 2020).

44. International Energy Agency. *The Future of Cooling*; Report prepared for the IEA; IEA: Paris, France, 2018; Available online: web.archive.org/web/20200524163428/https://www.iea.org/reports/the-future-of-cooling (accessed on 24 May 2020).

45. Sachar, S.; Campbell, I.; Kalanki, A. *Solving the Global Cooling Challenge: How to Counter the Climate Threat from Room Air Conditioners*; Report for the Rocky Mountain Institute; Rocky Mountain Institute: Basalt, CO, USA, 2018. Available online: web.archive.org/web/20200521085339/http://rmi.org/wp-content/uploads/2018/11/Global_Cooling_Challenge_Report_2018.pdf (accessed on 24 May 2020).

46. Keatinge, W.R. Death in heat waves: Simple preventive measures may help reduce mortality. *Br. Med J.* **2003**, *327*, 512–513. [CrossRef] [PubMed]

47. Historic England. *Practical Building Conservation: Earth, Brick & Terracotta*; Routledge: London, UK, 2015.

48. Lemieux, D.J. Trust But Verify: Building Enclosure Commissioning in Sustainable Design. *Real Estate Issues (Couns. Real Estate)* **2008**, *33*, 29–37.

49. Syntegra Consulting Intelligent and Green Building Solutions. *BREEAM Domestic Refurbishment Pre-Assessment, Rev. C—Grenfell Tower, London*. Unpublished report. 2013.

50. Lemieux, D.J.; Ozog, N.E. Exterior Cladding and Fire Protection: More Than Skin-Deep. *Fire Prot. Eng.* **2018**, Q2.

51. Pender, R. Rethinking retrofit. *Context J. Inst. Hist. Build. Conserv.* **2020**, *164*, 15–17.

MDPI

St. Alban-Anlage 66

4052 Basel

Switzerland

Tel. +41 61 683 77 34

Fax +41 61 302 89 18

www.mdpi.com

Atmosphere Editorial Office

E-mail: atmosphere@mdpi.com

www.mdpi.com/journal/atmosphere